# Space Ethics

# Space Ethics

Brian Patrick Green

ROWMAN & LITTLEFIELD
Lanham • Boulder • New York • London

Published by Rowman & Littlefield
An imprint of The Rowman & Littlefield Publishing Group, Inc.
4501 Forbes Boulevard, Suite 200, Lanham, Maryland 20706
https://rowman.com

6 Tinworth Street, London SE11 5AL, United Kingdom



British Library Cataloguing in Publication Information Available

**Library of Congress Cataloging-in-Publication Data**
Names: Green, Brian Patrick, author.
Title: Space ethics / Brian Patrick Green.
Description: Lanham : Rowman & Littlefield, 2021. | Includes bibliographical references and index.
Identifiers: LCCN 2021025441 (print) | LCCN 2021025442 (ebook) | ISBN 9781786600264 (cloth) | ISBN 9781786600271 (paperback) | ISBN 9781786600288 (epub)
Subjects: LCSH: Astronautics—Moral and ethical aspects. | Astronautics—Social aspects. | Outer space—Exploration—Moral and ethical aspects.
Classification: LCC BJ60 .G74 2021 (print) | LCC BJ60 (ebook) | DDC 174/.96294—dc23
LC record available at https://lccn.loc.gov/2021025441
LC ebook record available at https://lccn.loc.gov/2021025442

∞ ™ The paper used in this publication meets the minimum requirements of American National Standard for Information Sciences Permanence of Paper for Printed Library Materials, ANSI/NISO Z39.48-1992.

# Contents

*Chapter One*

# Why Space Ethics?

**Figure 1.1. Like humans might someday do in space with planets, Polynesian navigators traveled enormous distances across the Pacific Ocean finding and settling innumerable islands.**

Space is everything that is around and beyond the Earth; it is "the final frontier"[1] of human exploration. The universe is vast, spanning billions of light years. In comparison, the Earth is a tiny place. Yet every human ever born has lived on this speck. And all human ethical systems have been developed here, with this planet and human beings in mind. Our perspective has been limited.

However, humankind is leaving the Earth, slowly heading for the stars. In the Apollo missions between 1969 and 1972, humans landed on the Moon and returned to Earth, but no one has returned since. In the interim we have filled orbits around the Earth with satellites and sent probes beyond the edges of the solar system. Space was once about political status and "big science," but now it has also become big business, and the nature of space exploration is changing. Each one of these various endeavors in space raise many questions, such as: Is it technically possible? Is it politically possible? Is it economically worth it? And, ultimately, the question of **ethics**: "Is it good?" And if so, then: "Should we?"

The purpose of this book is to look at the exploration and use of space through an ethical lens, and in so doing not only to provide insight into those specific questions but also to investigate the various human ethical resources available for such examinations and hopefully to find new insights into our present ethical problems here on Earth.

## EXPLORATION

Space exploration is a particular kind of exploration, and as such it is helpful to situate it within the context of the history of exploration of the Earth. The earliest human ancestors originated in Africa and then spread across the world, even reaching Tierra del Fuego and the tiny islands of the Pacific. Later waves of humans also spread forth from various locales on Earth, propelled by technological innovations; political, religious, and economic interests; and population pressure. At this point in time, there are few places on the land surface of the Earth that humans have not explored (though the ice caps of Greenland and Antarctica are not well explored and the deep oceans even less—and that is most of the planet). Instead, in recent decades, with little land to explore, humans have explored more inwardly and deeply, through the arts, sciences, and technology, expanding our knowledge of the universe without going very far spatially. One fruit of this scientific and technological exploration has been incredible knowledge of space. This knowledge of space—that it contains elements usable for human life and new places humans could settle—along with the technology to now reach these places and even survive and flourish there has made people interested in doing so. After exploring the Earth, touching the Moon, and doing a few decades of research before our next step, it seems that it might now be the time to take that next step and go forth into space. It seems the *natural* next step in the progression of history. But *should we*?

While many people today might see exploration as an essentially human activity, it is in fact not.[2] Humans can choose to explore or not to explore, and cultures throughout history have predominantly chosen one or the other, sometimes with abrupt switches between the two. Ming China is a prime example of a society that boldly ventured forth in exploration, in massive expeditions led by Zheng He in the early 1400s, only to later change its mind and let the world come to it. Throughout most of history people simply remained near where they were born or within the territory of their cultural group, and there was no reason to go beyond that. The "frontier" (if such a notion even existed—and the notion can certainly be a point of critique today) was not a place of curiosity and opportunity but of ignorance, disdain, and peril.

However, not all cultures were or are the same. The open perspective of the earlier Ming has appeared in other places in history too. In Western Europe this open perspective on exploration took root and grew in such a way that has, for better and for worse, strongly contributed toward transforming the world into what it is today. The optimistic ideal of exploration that has captured the Western mind for the past few centuries can be condensed into a phrase from English philosopher and mathematician Alfred North Whitehead, who said in 1933, "Without adventure civilisation is in full decay."[3] Is Whitehead right? Are static civilizations actually in decay, or are they in fact "progressive" when they are content with what they have and who they are? How can we know? In this context, Whitehead was referring to the physical exploration of new places and the mental exploration of new intellectual terrain, but we should not underestimate the connection between the two. Indeed, this volume hopes to make that connection clearer.

Beginning in the 1400s, the Age of Discovery began in Europe, as Portuguese, Spanish, Italian, Dutch, English, French, and others ventured forth in search of profit, knowledge, settlement, and the religious conversion—and sometimes destruction—of Indigenous peoples. As Ming China receded from exploration, Europe surged forth, starting colonies around the globe, including in the "New World" of the Western Hemisphere, and beginning new scales of international trade. Two "worlds" became one world. Crops, diseases, trade items, technologies, and peoples—free and enslaved—were transported in a massive process known as the **Columbian Exchange** that has continued to shape events ever since.[4]

Exploration is not without risk, both for the explorer and the explored. While Christopher Columbus is credited with "discovering" the New World in 1492, though he did not recognize what he had found (and it was already "discovered" by those who lived there, as well as by previous Old World explorers, such as Leif Erikson), what he really did was connect multiple civilizations at very different levels of technology and vulnerability. Within a few decades, the New World's Indigenous empires lay in ruins, new empires arose, fortunes were lost and gained and lost again, and millions were dead. This historical case is paradigmatic of careless and exploitative exploration. Despite the great goods that came from the discovery of the New World, such as new foods and resources, massive economic growth, scientific discoveries, and new forms of governance, there were extremely negative effects too. Genocide, extractive colonialism, and the introductions of diseases and invasive species provide negative examples, and in the history of exploration, there are many more cases where, in retrospect, contemporary humans might wish better decisions had been made. While the "new worlds" of space may seem very different from the "New World" of Earth's Western Hemisphere and other colonized lands, humans ought to heed the lessons of the past and prepare to do better than we have done before. The questions explorers face are often ethical questions, and how we decide to act in the face of new opportunities and challenges will set a tone for our endeavors in space. Respecting this past, this book does not refer to space "colonization" but rather to space "settlement," "colonization" not only being offensive but also likely an inaccurate description of the status of future human settlements in space.[5]

## SHOULD WE GO?

While human individuals and cultures can choose not to explore, when they do choose to explore, it can become second nature. For good and for ill, we can be a very curious species. The Western experiment in exploration has now become a global experiment probing science and technology, and the full outcome of the experiment is not yet known. So far the benefits—longer and healthier lives, improved communication and transportation, alleviation of suffering, and so on—seem to outweigh the costs: environmental degradation, gross social inequality, dangerous weapons and emerging technologies, and so on. Perhaps someday a technologically induced disaster may befall us and we shall wish we had never left our past "home" of poverty, weakness, and ignorance. Our descendants may yet curse our explorations. But that day has not yet come and hopefully never will. Frontiers are full of opportunities and dangers, both of which can be tempting. Should we be tempted by them? Why or why not? Humans explore because we want to learn new things, experience adventure, find new resources and new places to live, and do something significant with our lives. But, of course, the ethical question that is the key to this book is "Should we explore?" We have power over our actions, and we have a choice. What shall we do?

Specifically, we ought to ask: "Should humans go forth into space?" When we think of going into space, what is attracting us, "pulling" us toward space? Or is there something on Earth that is motivating us to leave, something "pushing" us away? When we actually do go into space, what are we looking for? When we are thinking or doing things with respect to space, what other important things are we neglecting to think and do instead? What are we ignoring that we ought not?

## In Favor of Going into Space

In favor of going into space are such basics as gaining scientific knowledge and developing beneficial new technologies, both of which space exploration and use have already begun to accomplish with dramatic and sometimes unexpected effects for humankind. Scientific advancements include astronomical and cosmological knowledge from various orbiting experiments and telescopes that have let us gain unprecedented understanding about our universe. But space activities have also contributed to a great deal of scientific knowledge about our Earth, including measurements of environmental status, habitat conversion and destruction, detailed knowledge of anthropogenic **climate change**, and much about Earth's chemistry and geology. We have also learned a great deal about our local planets, for example, that a runaway "greenhouse effect" in the atmosphere of Venus makes the surface scorchingly hot, while too little greenhouse effect on Mars leaves the surface quite cold. There have also been significant contributions made to medical science, especially concerning the behavior of the human body when subjected to **radiation**, microgravity, nutritional restrictions, and so on.

On the technological side, everything with American global positioning system (GPS), Russian Glonass, or other global navigation systems—from smartphones to military vehicles—relies on a network of satellites above us, placed there by rocketry and painstakingly tracked with instruments developed for the task. So many technologies have been pioneered by space exploration and use that it is hard to list them all, but some of the more important ones include weather satellites (which are not only convenient but also allow preparation for and evacuation from severe weather), communication satellites, **solar photovoltaic (PV)** cells, advances in electronics and computers, advances in materials science, and so on.

Space is also an important location for the contention of national interests in a geopolitical and military sense. As the ultimate "high ground" in battle, space allows certain asset classes such as spy satellites to exist in a position unassailable by many or most opponents. While permanent weapons stations and weapons of mass destruction are banned from space by the United Nations **Outer Space Treaty (OST)**,[6] that has not stopped the development of weapons that are impermanent (such as missiles, missile interceptors, and antisatellite weapons) or the research and development of possible space-based weapons platforms, such as were envisioned by U.S. president Ronald Reagan's **Strategic Defense Initiative**, nicknamed "Star Wars." While military and political interests may ultimately seem to be a less noble reason to explore and use space, relative power, safety, and security certainly are very human interests and are valuable to those who feel they are being protected by them.

Space activities are also a key way of promoting international **cooperation** and global awareness. While the international competition of the "space race" fueled one nation all the way to the Moon, shortly afterward, the Apollo-Soyuz program announced a thawing of this competition and commenced a period of cooperation between the United States of America and the Union of Soviet Socialist Republics. Currently the International Space Station continues this cross-national cooperation in space, with five space agencies (representing Canada, the European Space Agency nations, Japan, Russia, and the United States) participating. In addition to cooperation in space exploration itself, the perspective given from space has itself helped to produce some feelings of unity on Earth, with the famous "Blue Marble" and "Earthrise" pictures showing Earth's oneness and scientific discoveries supported by space science, such as those related to climate change, helping to promote international cooperation to address these problems.

Gaining access to new critical resources may be another reason to go into space. Earth is a finite planet, and certain elements on Earth are very rare in the planetary crust, particularly platinum group metals that are very dense and **siderophilic** (iron-loving) and so have tended to sink toward the core over the natural history of the planet. However, asteroids and other objects in space (for example, planets, comets, and moons) can sometimes have these elements in abundance and in more available locations, making them potentially excellent sources for these valuable materials. Now-defunct asteroid-mining startup Planetary Resources once estimated that one "platinum-rich 500 meter wide asteroid contains . . . 1.5 times the known world-reserves of platinum group metals (ruthenium, rhodium, palladium, osmium, iridium, and platinum)."[7] In addition to returning elements to a resource-hungry Earth, further exploration and development of space will require access to resources that are not purely sourced from Earth. In particular, it will be necessary to gain access to water, which is relatively rare in the inner solar system and which would be far too costly to transport in any significant amounts from the Earth's surface.

Another reason that humans may want to explore space would be to create a "**backup Earth**" to hedge against **global catastrophic** and **existential risks** (risks that may cause widespread disaster or human extinction, respectively) on our home planet.[8] Earth has always been a dangerous place for humans, with asteroid impacts, supervolcanic eruptions, pandemic disease, and other natural hazards threatening civilization. Now, in addition to these natural threats, human-made hazards such as nuclear weapons, climate change, biotechnology, **nanotechnology**, and artificial intelligence may threaten not only the viability of technological civilization but perhaps the survival of human life itself. A serious global-scale catastrophe could set back civilization many decades or centuries, and the worst disasters could cause human extinction. In one scenario, in which 100 percent of humanity dies, all of human effort for all of history would be for nothing. However, were the same global catastrophe to happen to Earth, yet humans were a multiplanetary species with just one self-sustaining settlement off-Earth, it would not result in the end of human civilization or human extinction. Instead while the same unimaginable fate would befall the Earth (certainly no mere triviality, with perhaps the deaths of 99.999 percent of all humans and possibly the destruction of the ecosphere and everything in it), at least all of human and planetory history would not be for nothing. Human life and culture would go on elsewhere, as well as other Earth species. This is a dire fate, but less terrible than the first.

Another consideration is that we may not always have access to space, as further chapters of the book discuss. Whether through a setback blocking our access or other disasters that degrade humanity's ability to engage with space, now may be a time within a limited window of opportunity. Before technology advanced sufficiently, space was inaccessible, and there is no necessary reason to assume that this inaccessibility may not return in time, either due to natural or human events.

As a last reason to go into space, many people find inspiring the personal experiences and the sharing of personal experiences and stories of exploration. Only twelve humans have ever walked on the Moon, and those few men, while greatly inspired themselves, also inspired countless other people around the world. I have talked to individuals in less developed nations who, despite their material poverty, felt deeply connected with the idea of space exploration.[9] We should not underestimate the importance of wonder and amazement in human life. With the **existential** turn in some philosophies and cultures, questioning the human *telos*—our goal, purpose, and meaning in life—has become a genuine task, not only for individuals but for all of humankind. In a world where traditional **teleologies** such as those described by religion, nation, ideology, tribe, family, worldview, and other identities are becoming less strong, purposelessness is becoming an increasingly common problem, and evil purposes such as crime, terrorism, ethnonationalism, and other antisocial behaviors seem to be always in the news. Within this shift in our understanding of ourselves and our role in the universe, space exploration may provide a new and more positive outlet for human ambition, identity, and purpose. It can provide a purpose that is creative as well as potentially being compatible with many of our older teleological traditions.[10] Providing wonder and purpose are all worthy goals; they are not insignificant. Many humans constantly seek a group identity with a purpose so as to become parts to a whole, and to some, even what may seem a trivial purpose to others can be perceived as being more important than life itself, for example, maintaining a flag aloft in battle. With a lack of good outlets for this teleological and tribal impulse, pathological forms will erupt forth. Determining the purpose of humanity and of ourselves is an age-old task, but one that is only heightened by some cultures' loss of faith in their traditional **meta-narratives** (the stories and ideas that bind a society together). Turning our eyes toward space may be a good goal to sustain humankind into the future.

## Against Going into Space

There are, of course, also strong arguments against going into space. First, space exploration is extremely costly. For example, the Apollo program cost the United States between 20 and 28 billion U.S. dollars at the time it was operating (the difference depending on what, exactly, is included) or, scaled for inflation (a notoriously difficult task), the equivalent of between approximately 200 and 700 billion 2019 U.S. dollars.[11] Currently, the cost of getting into space is several thousand U.S. dollars per kilogram. Surely, anything in space ought to be worth it, somehow returning more than the likely tens of millions of dollars it costs to get into orbit. With such high costs, spy satellites and interplanetary probes may be acceptable, but activities like settling the Moon or Mars might not.

The cost problem manifests not only in the direct expense of spending the money on projects with specific and limited benefits but also in what is not funded. This ethical issue is particularly acute with government-funded space exploration (private funding being another, though related, issue). When money is spent on space-related projects, it is not spent on other good projects. There are countless pressing problems here on Earth that would seem to be a better use of our limited resources. Surely the requirements of justice and equality would indicate that these resources would be better spent helping the poor gain access to the basic necessities of life like healthy food, clean water, sanitation, education, health care, decent jobs, safety, and security rather than the rich accessing space. And yet, in practice, poverty-alleviation and space exploration programs do not really compete in a budget. Monies subtracted from space do not directly go to feed the poor (though an argument could be made that they ought to). Space exploration is a form of research and development (R&D), and as such, like all R&D, it is an investment in the future. We incur present costs and risks in the hope of future rewards and opportunities.[12] Interestingly, one spin-off technology promoted by space exploration, solar photovoltaic (PV) cells, is becoming quite helpful for providing electrical power to poor and remote people here on Earth. So while the R&D put into PV cells over the past few decades could no doubt have immediately helped many people in the past, without that research, people would not be benefitting from PV cells today. The **trade-offs** between immediate needs and future needs are difficult to weigh and can be a genuine source of ethical debate. However, concern for the present should not completely shut down R&D for the future because we know that, given our present technology (rooted in fossil fuels and other environmentally destructive practices), we cannot continue to survive sustainably on the planet. We need dramatic technological progress.

As a second reason, space exploration is a risky endeavor, and people have died and will continue to die as exploration proceeds. There is no way to make space exploration 100 percent safe, just as there is no way to make exploration here on Earth 100 percent safe. There will always be a risk of something going wrong, though there are certainly ways to make things safer.

Of course, life on Earth is also risky. We could be hit by a car or die of cancer, and certainly someday we will all die. Risk is not absolute but can only be relative; for example, commercial air travel is much safer than other common forms of transportation.[13] But space exploration is not yet a "safe" activity, and it may never be so due to the enormous energies and speeds involved, not to mention the effects of radiation, microgravity, lack of access to basic human health needs, reliance on meeting needs artificially (for example, atmosphere, food, and water), and so on. One perverse aspect of making space exploration safer, however, is that it may also make it more expensive, thus exacerbating the problem of cost and the concomitant problem of justice. Exactly how risk averse people are can vary dramatically between people and cultures, and exactly how risk averse people *should be* is an open question. Given that, it might be best to leave it up to individuals, assuming at least *some* reasonable safety standard. In any case, this question is not one that should shut down plans for human space exploration.

As a third consideration, the exploration of space and movement of humans into a multiplanetary political milieu could be socially and politically destabilizing. While new human settlements in space may not directly harm intelligent Indigenous inhabitants, as happened on Earth during the time of colonialism, even without disruption to Indigenous inhabitants, new and destabilizing social and political ideas can spread back to Earth. For example, shortly after the independence of the United States of America from the British empire, waves of revolutions began to sweep Europe and other parts of the world, from the French Revolution, through the Napoleonic era, through the revolutions in Latin America, and further political instability in Europe in 1830 and 1848. New ideologies developed in space could have similar effects. Furthermore, due to the intense energies necessary for space travel and other activities in space, space settlements would have a natural technological and physical "high ground" in any kind of conflict with Earth. Interplanetary war is now complete fiction, but with the advent of settlements in space it may become a distinct possibility.[14] If we are to have settlements in space, we should also learn somehow to live in peace.

In contrast to those who desire to go into space to save humanity, there are people who say that humanity is destructive and therefore either ought to stay on Earth in order to contain its destructive tendencies or actually deserves to go extinct because it is incorrigibly evil.[15] These **eco-nihilists**, as space ethicist Kelly Smith calls them, argue against space exploration because it would just spread humankind's evils to the rest of the universe.[16] Smith, Keith Abney, and others have noted that, in its extreme form, the notion that the only ethically aware creature that we know of *ought to be destroyed because it is evil* is an ethical argument for removing ethics from the universe (thus contradicting the basis of its own argument) and is therefore logically incoherent.[17] Nevertheless this position is occasionally voiced by people who do not seem to care that they are being logically incoherent and ethically self-contradictory. The less severe critique—that humans ought to do better with regard to the environment, whether on Earth or elsewhere—is clearly true and should be accepted; in fact a great deal of this book will consider this very point. But it is not a point so strong as to merit stopping space exploration, use, and settlement entirely.

Finally, we cannot completely discount the possibility that in our explorations we may discover dangerous things that we would be better off without. For example, we cannot discount that other planets may naturally harbor microbes or other organisms capable of injuring or killing humans, other Earth life, or the ecosphere. If this is the case, contamination and decontamination become foremost concerns in space exploration—however, they must be concerns *before* contact is made, otherwise it may be too late. Planetary protection currently operates with stringent guidelines for preventing forward and backward contamination from space, but the more space activity there is, the more difficult it becomes to monitor and enforce.

Continuing with the dangers of exploration, there is a nonzero probability that we could even potentially discover dangerous **extraterrestrial intelligences (ETIs)** or artifacts. While we currently have no strong evidence of ETIs, there is no reason to think that ETIs are impossible. The existence of life and humanity on Earth proves that life and intelligence are possible, and therefore we should assume that life is also possible elsewhere in the universe. As a most obvious form of danger, ETIs themselves could be hostile, perhaps leading to their harming or exterminating humanity. Even if we do not discover intelligent life forms directly, we may discover their probes or archaeological remains, and these artifacts could be dangerous as well, for example, malevolent artificial intelligences, computer viruses, or infectious biological or nanotechnological agents.

Seemingly benign discoveries involving ETIs might still be dangerous, for example, if ETIs were friendly and wanted to help us, such interactions could still end badly, as the "interstellar do-gooder" problem has described before,[18] much harm can come from wanting to help others (and this has certainly happened on Earth as well). It is also possible that humans might discover socially, culturally, psychologically, and/or politically destabilizing new information, for example, evidence of past ETI activities or ETIs with differing religious perspectives and/or morals from humanity. Some humans might want to emulate the ETI beliefs and behaviors, while others might reject them purely because they are "alien," thus causing social strife. Importantly, while venturing into space may attract the attention of aliens and thus their wrath or blessing, remaining on Earth may do so as well, and given the scenario of encountering ETIs with humans being Earthbound or spacefaring, it would seem to be a more advantageous position to be spacefaring, as that gives humanity better informational awareness, mobility, positioning (in terms of energy), and access to resources.

## So Should We Go?

By the very fact of writing this book, it might be surmised that the author implicitly thinks space exploration is something that humans should seriously consider doing. But I will be clear—I am in favor of thoughtful, ethical space exploration, use, and settlement, and I have staked a strong position on this in the past.[19] I state this here for the sake of transparency: as author of an ethics text, I believe it is important to be honest about my own positions in this debate. Readers should not take my perspective as the last word—disagreement certainly can be reasonable, and as we move into the future, knowledge and contexts will change as well. But if I felt that the response to the question of "Should we go?" was "No," I would have written a very different book, or perhaps no book at all.

I also come from a particular perspective on ethics: I am trained in Western philosophy including historically Christian and specifically Catholic ethical traditions. What I present in this book is argued philosophically and does not appeal to or rely on religion. But I do take religion seriously as a force in human psychology, society, and culture, one that should be interrogated for its ethical relevance and value. Therefore I include a few ethical tools that are historically associated with Catholic philosophy. I include these ethical tools because I think they are useful, and I hope other ethical traditions will add their tools to the "space ethics toolkit" too.

More than anything else, this book is meant to be an *applied ethics* text—in other words, *ethics that gets things done*. I offer these ethical tools in the hope that they might be the right tools for the job—and you, the reader, will ultimately make that determination. For you can decide to use any ethical tools that you want: and if you find better ones, use them and let everyone else know.

No individual can write the story of space ethics; it is something people will write together. This text is only one piece of a long and ongoing conversation; much more remains to be said. To solve the problems presented by space, we will need to search for and utilize wisdom from all the diverse traditions of humanity. This conversation between space science and human meaning and values is also an excellent way to bridge the "two cultures" gap between science and the humanities, discussed by C. P. Snow over sixty years ago.[20] Bridging these gaps helps us to unify human knowledge and experience into a more humane whole.

Space ethics is not a simple thing; indeed it may be one of the most complicated areas that human ethics will even engage, as it reaches down into the very core meaning of human existence and purpose in the universe. This book will not be a starry-eyed view of space ethics; space is not easy, and there are certainly ways to do space exploration wrongly. To align space exploration with ethics, humanity must be careful.

Humanity as a species could be content to stay here on Earth, tending to the planet and each other and avoiding the dangers and costs of exploration. However, this avoidance of space is a purely theoretical path toward the future, and one that would require active suppression of various organizations and individuals who deeply desire to explore space. Practically speaking, this suppression would not only be costly but also counterproductive. Space exploration and use could ultimately be of great benefit to those on Earth, even despite its relative expense and danger. And perhaps the foremost benefit of space exploration and use is that it makes possible the settlement of other places in space, thus providing "backups" of humanity and Earth life should disaster befall the Earth. Currently humanity's—and the entire Earth ecosphere's—eggs are literally all in one basket, and as technological dangers grow, it behooves us to pursue technologies that not only increase safety and security on Earth but also provide the ultimate "insurance policy" against disaster in the form of off-planet settlements.

## THE STRUCTURE OF THIS BOOK

The general structure of the book's chapters reflects a movement forth from Earth in both space and time. In the earlier chapters are questions people have already faced with regard to space exploration, questions such as its safety risks, health effects, military uses, and contamination risks. In later chapters are questions that are more future oriented, questions such as planetary resource utilization, long-duration spaceflight, off-planet settlement, and planetary engineering.

Every chapter will begin with an image indicative of one theme within that chapter. Chapters 3 through 14 will all follow a similar structure, beginning with a case involving space ethics. This will be followed by a rationale for the chapter, describing its significance, then the ethical content of the chapter, including space-related questions and ethical tools. There is also a section called "Back in the Box" where we will apply ideas from space ethics to contemporary problems on Earth. Finally, the chapter will close with another case, discussion questions, and further readings. Chapters 1, 2, and 15 are atypical and will not follow the same structure.

Chapter 2 will provide a brief introduction to ethics as it applies to space exploration and use. Ethics involves "should" and "ought" questions, and while chapter 1 specifically asked whether we should go into space at all, chapter 2 will break that question down into more specific pieces, not topically (which is what chapters 3 through 13 do) but analytically, in terms of what considerations ethics involves, such as approaches to ethics, perspective in ethics, centrism in ethics, sources of ethical value, and so on.

Chapter 3 will be a detailed study of the role of risk in ethical decision making as it applies to space. Risks include technical risks, such as structural failure; risks related to expertise, such as failures due to engineering error; operator risks, such as when a systems operator knowingly engages in risky behavior; and managerial risks, such as those created by management cutting corners on safety. Considerations will include causes of risks, ways to look at risks, coping with uncertainty, reducing risks, and what risks are worth taking. This chapter will not consider long-term health risks—those will be discussed in the next chapter.

Chapter 4 will consider human health risks in space. Such health risks include radiation exposure, the effects of microgravity (including on bones, muscles, digestion, etc.), nutrition, motion sickness, the dangers posed by physical particles (such as the sharp silica dust on the Moon), chemical dangers (such as perchlorate on Mars), contamination issues (biological and chemical), psychological stress and mental health, and so on. This chapter will also look at several tools for thinking about health risks such as bioethics, **informed consent**, and double-effect reasoning.

Chapter 5 will look at the dangers of space debris in the immediate space above and around the Earth, as well as in orbits beyond the Earth. Space debris is becoming a serious problem with a growing number of orbits simply being unsafe due to high-velocity clutter. Between dumping of waste, missile testing, and satellite collisions, we may be slowly closing space to human access through our own lack of care and competence. This chapter will examine the sources, ethical questions, and possible solutions to this orbital "tragedy of the commons." This chapter will begin to look at the dual-use aspects of technology as well.

Chapter 6 will cover ethical issues involving the military use of space. According to the OST, no permanent weapons platforms or weapons of mass destruction can be put in space. However, there are numerous spy satellites in orbit right now, the GPS network (controlled by the U.S. military, with more positioning systems expanding, such as the Russian Glonass), and many weapons systems that pass through space, such as long-range ballistic missiles, antiballistic missiles, and antisatellite missiles. This chapter (like the one before it) will also consider the dual-use aspects of space technology, for example, any program to clean up space junk could double as a program to gather, gain control of, destroy, or otherwise manipulate other nations' satellites and/or space assets.

Chapter 7 will discuss extraterrestrial dangers to Earth including those posed by asteroid and comet impacts as well as solar storms and less probable risks such as those posed by **gamma ray bursts**, black holes, and so on. It will consider detection, threat assessment, and solutions to such hazards, as well as the social and political difficulties of dealing with rare but catastrophic events. It will consider that human power has grown to such an extent that we could completely eliminate one class of natural disaster, namely asteroid impacts, if we simply put our attention to it. At the same time, our new power gives humanity the ability equally well to direct asteroids toward Earth, thus presenting us with a new class of disaster: human-induced asteroid impacts. This chapter will discuss in depth the ethical implications of human power, for example, adding new ethical responsibilities where once there were none, and consider how ethics ought to grow to meet our growing power.

Chapter 8 will explore astrobiology, which is the study of the origin, extent, and future of life in the universe. Astrobiology also considers the scientific, social, and cultural impacts of the potential discovery of **extraterrestrial life (ETL)**. Should we expect ETL to be rare or common? The Kepler planet detection mission has searched for potentially habitable worlds orbiting distant stars and has found that there are many. Ethically, how should we think about life that we might find in space?

Chapter 9 looks at issues of responsible exploration and planetary protection, which is the practice of trying to prevent contamination during space exploration. Planetary protection considers forward contamination from Earth on outbound journeys into space, while backward contamination returns contaminants from space to Earth. Both can be dangerous, though in differing ways: forward contamination can potentially ruin scientific research in space by contaminating it with Earth compounds or organisms; backward contamination threatens to harm Earth with unknown space compounds or organisms (likely microbes). Planetary protection deals with the unknown and as such faces complexities that we may poorly understand. How can we adequately protect what we do not know from that which we do not understand?

Chapter 10 will consider the human search for ETI. **Radio telescopes** have been searching for ETI signals. Frank Drake, via his **Drake Equation**, famously proposed that civilizations could be common in our galaxy. Was he right? The **Fermi Paradox** asks why, if life and intelligence seem like they should be so common, have we never detected any signs of them? Should we even bother looking? Why do some people feel compelled to look? Are there dangers in looking for ETI? The chapter will also examine METI: actively attempting to send messages to ETIs.

Chapter 11 examines the transformation that is occurring in space exploration as the traditional players in space—the governments of highly developed nations—are joined by more nations and private and commercial organizations. Are there any new ethical issues related to new nations exploring space? Should private interests be allowed to explore space or should they be stopped? If they should be allowed, ought they to be regulated, and how? While the OST applies to governments, and to private parties within the signatory governments by proxy, should the OST be amended to more broadly consider these issues? Considerations of rights, claims, and territoriality will be considered, as well as the role of law, politics, and economics in the exploration and use of space.

Chapter 12 looks at human biology and long-duration spaceflight, such as that required to reach other planets or stars. Right now we have the technology to reach most places in our solar system in a few years' time, which would be very taxing on human explorers. Reaching the stars with current rocket technology would take orders of magnitude longer, so long as to make such missions essentially impossible. But technology progresses and what is impossible or too expensive now may become possible and less expensive in the future. What ethical considerations are there when planning and executing long-duration spaceflight? Should humankind be engineered to make us better candidates for space flight?

After crossing space, chapter 13 will consider human voyagers who arrive at their destinations and need to learn to survive and ultimately flourish in their new environments. The first human mission to Mars might be a return mission, where the same astronauts return home after a few weeks or months on the surface, but it also might not be a return mission, instead beginning a permanent presence on the Red Planet. For those settlers who stay on Mars, survival and flourishing will require not only learning to utilize local resources but also learning how to live together in a human community in the harshest environment humans have ever tried to settle.

Chapter 14 considers planetary-scale engineering projects, such as the **terraformation** of Mars or other objects in space. Only recently have humans become aware that we are capable of changing the atmospheric composition and climate of our own planet, but such considerations for Mars have been a staple of science fiction for decades. Should humans terraform other planets? What if planets have Indigenous life, even if only microbial—should we still terraform them? Would terraformation of a very hostile planet like Mars actually help the life there, and if so should we terraform it for the benefit of the local inhabitants?

Chapter 15 is the conclusion of the book and brings the questions once again into the broadest possible perspective, that of a universe in which humans have grown in power yet retain a limited perspective and so are in need not only of greater knowledge but also greater understanding, wisdom, and ethics.

In conclusion, and harking back to the history of the exploration of Earth, as we explore space and utilize its resources, what similarities and differences are there compared to the past? What successes should we emulate and what tragedies can we learn from and improve? What ought our goals to be? How should we pursue these goals?

For all of human history, our perspective on all things, including ethics, has been limited to Earth. As we go forth into space we should think not only about how to act ethically, utilizing our admittedly limited ethical perspective, but also about how to improve our ethical perspective based on the new knowledge, understanding, and wisdom we may gain in space.

## DISCUSSION AND STUDY QUESTIONS

1. Would you choose to explore space? Why or why not?
2. Do you think that humans ought to explore space? Why or why not?
3. What arguments do you find to be the most convincing for or against space exploration?
4. What is the proper balance between current concerns (for example, military, poverty, things that should be done right now) and future concerns (for example, space exploration, scientific and technological research, things that do not need to be done now but might be helpful in the future)?
5. Continuing question 4, historically, do you think that various nations and groups have properly balanced these concerns? Use historical evidence to support your position.

## FURTHER READINGS

Jacques Arnould, "The Emergence of the Ethics of Space: The Case of the French Space Agency," *Futures* 37 (2005): 245–54.

Jacques Arnould, *Icarus' Second Chance: The Basis and Perspectives of Space Ethics* (New York: SpringerWein-NewYork, 2011).

Seth D. Baum, "Viewpoint: Cost–Benefit Analysis of Space Exploration: Some Ethical Considerations," *Space Policy* 25 (2009): 75–80.

P. Ehrenfreund, M. S. Race, and D. Labdon, "Responsible Space Exploration and Use: Balancing Stakeholders Interests," *New Space Journal* 1, no. 2 (2013): 60–72.

## NOTES

1. *Star Trek: The Original Series*, "Title Sequence," 1966, with precursors in The White House, "Introduction to Outer Space" (Washington, DC: U.S. Government Printing Office, March 26, 1958); and John F. Kennedy, "We Choose to Go to the Moon/Address at Rice University on the Nation's Space Effort," Rice Stadium, Houston, Texas, September 12, 1962, https://er.jsc.nasa.gov/seh/ricetalk.htm. This narrative, of course, can very much be critiqued: for example, Patricia Nelson Limerick, "Imagined Frontiers: Westward Expansion and the Future of the Space Program," in *Space Policy Alternatives*, ed. R. Byerly (Boulder, CO: Westview Press, 1992), 249–61; and W. R. Kramer, "To Humbly Go: Guarding Against Perpetuating Models of Colonization in the 100-Year Starship Study," *Journal of the British Interplanetary Society* 67 (2014): 180–86.

2. James S. J. Schwartz, "Myth-Free Space Advocacy Part I—The Myth of Innate Exploratory and Migratory Urges," *Acta Astronautica* 137 (August 2017): 450–60.

3. Alfred North Whitehead, *Adventures of Ideas* (New York: Free Press, 1967), 279.

4. Alfred W. Crosby, *The Columbian Exchange: Biological and Cultural Consequences of 1492* (Westport, CT: Greenwood Press, 1972).

5. This debate over terminology goes back decades. See, for example, Carl Sagan, "Comments on O'Neill's Space Colonies," in *Space Colonies*, ed. Stewart Brand (San Francisco, CA: Waller Press, 1977), 42, who says, "I think 'Space Colonies' conveys an unpleasant sense of colonialism which is not, I think, the spirit behind the idea. I prefer 'Space Cities.'"

6. United Nations Office for Outer Space Affairs, "Treaty on Principles Governing the Activities of States in the Exploration and Use of Outer Space, including the Moon and Other Celestial Bodies" (The Outer Space Treaty), United Nations General Assembly Resolution 2222 (XXI), agreed upon in 1966, signed and entered into force 1967, https://www.unoosa.org/oosa/en/ourwork/spacelaw/treaties/outerspacetreaty.html.

7. Planetary Resources, "About," 2014, www.planetaryresources.com, cited in Audra Mitchell, "Can International Relations Confront the Cosmos?" in *The Routledge Handbook of Critical International Relations*, ed. Jenny Edkins (Abingdon, Oxon: Routledge, 2019). Planetary Resources was acquired by ConsenSys Space in 2018.

8. Nick Bostrom, "Existential Risks: Analyzing Human Extinction Scenarios and Related Hazards," *Journal of Evolution and Technology* 9 (March 2002).

9. Brian Patrick Green, "Should Christians Explore Space?" The Moral Mindfield, 2012, https://moralmindfield.wordpress.com/2012/01/05/should-christians-care-about-space-exploration/.

10. Paul Levinson and Michael Waltemathe, eds., *Touching the Face of the Cosmos: On the Intersection of Space Travel and Religion* (New York: Connected Editions, 2016); and Ted Peters, with Martinez Hewlett, Joshua M. Moritz, and Robert John Russell, *Astrotheology: Science and Theology Meet Extraterrestrial Life* (Eugene, OR: Cascade Books, 2018).

11. Casey Dreier, "Reconstructing the Cost of the One Giant Leap: How Much Did Apollo Cost?" The Planetary Society, June 16, 2019, https://www.planetary.org/blogs/casey-dreier/2019/reconstructing-the-price-of-apollo.html.

12. In 1970 NASA scientist Ernst Stuhlinger wrote a letter to Sister Mary Jucunda, a nun in Africa working to save starving children, as a response to her questioning the value of space exploration, which addresses these points. Ernst Stuhlinger, "Letter to Sister Mary Jucunda/Why Explore Space? A 1970 Letter to a Nun in Africa," May 6, 1970, *Roger Launius's Blog*, February 8, 2012, https://launiusr.wordpress.com/2012/02/08/why-explore-space-a-1970-letter-to-a-nun-in-africa/.

13. Ian Savage, "Comparing the Fatality Risks in United States Transportation Across Modes and Over Time," *Research in Transportation Economics* 43 (July 2013): 9–22.

14. Dan Deudney, *Dark Skies: Space Expansionism, Planetary Geopolitics, and the Ends of Humanity* (New York: Oxford University Press, 2020).

15. Examples can be found in both ecological and space ethics literature (though it is often implicit). Alan Drengson, *The Deep Ecology Movement* (New York: North Atlantic Books, 1995); L. U. Knight, "The Voluntary Human Extinction Movement," 2019, www.vhemt.org; Lori Marino, "Humanity Is Not Prepared to Colonize Mars," *Futures* 110 (2019): 15–18; and Adam Morton, *Should We Colonize Other Planets?* (New York: Wiley, 2018).

16. Kelly C. Smith, "*Homo Reductio*: Eco-Nihilism and Human Colonization of Other Worlds," *Futures* 110 (2019): 31–34.

17. Ibid.; Keith Abney, "Ethics of Colonization: Arguments from Existential Risk," *Futures* 110 (2019): 60–63; and Brian Patrick Green, "Self-Preservation Should Be Humankind's First Ethical Priority and Therefore Rapid Space Settlement Is Necessary," *Futures* 110 (June 2019): 35–37; and other as well.

18. Brian Patrick Green, "Astrobiology, Theology, and Ethics," in *Anticipating God's New Creation: Essays in Honor of Ted Peters*, ed. Carol R. Jacobson and Adam W. Pryor (Minneapolis, MN: Lutheran University Press, 2015), 339–50.

19. Green, "Self-Preservation Should Be Humankind's First Ethical Priority and Therefore Rapid Space Settlement Is Necessary."

20. C. P. Snow, "The Two Cultures," in *The Two Cultures* (Cambridge: Cambridge University Press, 1998).

*Chapter Two*

# Questions of "Should"

## *Ethics Applied to Space*

**Figure 2.1. Lead engineer Wernher von Braun stands by the five F-1 engines of the Saturn V Dynamic Test Vehicle on display at the U.S. Space and Rocket Center in Huntsville, Alabama. Earlier in his career, during World War II, von Braun used his expertise in rocketry to build the V-2 "vengeance weapons" for Nazi Germany that killed thousands of Allied soldiers and civilians, as well as thousands more concentration camp slave laborers killed while manufacturing the weapons.**

## CLARIFYING WHAT ETHICS IS

This chapter will begin with a series of clarifications about exactly what ethics is and is not. First, ethics is the study of questions of **should and ought**. It is not the study of "could" or "must," which involve the possible application of power and necessity, respectively, although those concepts are often important for delineating the scope of ethics. If there is no choice, as in matters of necessity, there is no ethics (for example, we cannot choose to cancel the law of gravity, alter the flow of time, or adjust the number of spatial dimensions). This important fact about ethics was summed up by Immanuel Kant, who is generally attributed as being the originator of the idea that "ought implies can."[1] In other words, the material of ethics considers situations where one could act (one has the power and ability), but one could freely choose not to act or to act differently.[2] One cannot be held responsible for that which is outside one's control.

It is worth noting that "should" has two senses—one relating to behavioral prescription and one relating to prediction. "Should" in the ethical sense prescribes behavior—an agent ought to do something because it is the right or good thing to do or is at least not wrong or evil. "Should" in the predictive sense is a different matter; rather than prescribing behavior, it predicts that something ought to happen in a future descriptive sense. For example, if one is driving an automobile, then one might predict that there should be a refueling station in the next sufficiently large town. This is not an ethical "should" but rather a use of the word "should" in a predictive sense. (Note that in the case of **artificial intelligence (AI)** and **machine learning (ML)** predictive analytics, which draw predictions from past patterns in data, ethical or not, these senses can easily become confused.)

Additionally, some scholars make a distinction between "ethics" and "morals," for example, relating ethics more to philosophy and morals more to religion. Both words come from the Greek and Latin words for "customs," respectively, and in keeping with their common origin, in this book I make no distinction between the two. I will attempt to avoid confusion by predominantly using the word "ethics" and not the word "morals." In cases where "moral" is the standard word, however, I will use the word "moral" (for example, with the phrase "moral uncertainty" or "moral philosophy").

Ethics also involves several ethically freighted words such as "good," "bad," "evil," "right," and "wrong." For the sake of clarity I will define them. "Good" has multiple senses, for example, food can be good (tasty, etc.), music can be good (pleasing to the ear, meaningful), etc., but the ethical meaning of the word "good" refers to good actions (for example, actions that benefit) and good persons (for example, persons characterized by reliably performing good actions). When choosing among ethically good options, one ought to choose the best one; "better" and "best" operate as comparatives with respect to good. A fair amount of ethics involves discussing how exactly to determine what "good" is because the word is so central to ethics.

"Bad," like good, also has multiple meanings. Food and music can be bad (for example, displeasing), construction can be bad (for example, shoddy), and so on. But bad can also have an ethical meaning in reference to actions and persons. A bad action, for example, harms people, and a bad person is someone to be wary of in case he or she might perform a bad action upon you or someone you care for. Notice that bad food, music, and construction can likewise have a moral aspect to them, for example, if the food makes you sick, the music damages your hearing, or the bad construction collapses and injures people. In these cases, descriptively "bad" things have caused ethically bad outcomes. When forced to choose among bad options, one should choose the least bad option, which is the "best" option, despite still being ethically bad; that is, despite all options being bad in an absolute sense, one is the least bad in a relative sense to the options available. It is worth noting that there are situations in which only ethically bad choices are available, and decision makers should try very hard to avoid entering such situations in the first place.

"Evil" goes one step beyond "bad" in that it is a purely ethically freighted word. Typically food cannot be called evil, unless perhaps it is poisoned and intended to kill; but even then, that use of the word is a bit odd because the evil intent is not in the food but in the mind of the one who poisoned it. Similarly, but more easily, music might be called "evil" if it is spreading racism, lies, or otherwise encouraging evil activities, and construction might be called "evil" if it is, for example, designed to facilitate mass murder (for example, the gas chambers at Auschwitz). However, as with the food example, the use of the word "evil" is somewhat unusual because evil relates very strongly to intent and actions and less so to technological products, items, or places that carry the intentions and actions of others (music and structures, however, tend to carry more information content than food, hence could be better said to carry and convey "evil" as well, unless, for example, the food might contain some symbolic message). While every piece of technology has a built-in purpose and is by that reason potentially ethically freighted, truly "evil" technologies are evil because of the intent they actualize as a material object. Weapons of mass destruction, for example, biological or nuclear weapons, exist in themselves as potential facilitators of mass murder or even genocide. Yet even weapons of mass destruction are rationalized as "good" by their deterrent value (for example, their existence on opposing sides deters their use) or for reasons of dual use (for example, their research value). "Evil" is a very strong word, often with spiritual or religious connotations, and some people, even ethicists, do not like to use it, but in a book on ethics, as well as in life, it has its appropriate uses.

"Right" and "wrong" are two words that, like "good" and "bad," have both ethical and nonethical meanings. "Right" can refer to both truth, as in 2 + 2 = 4 is the right (true) answer, and ethics, as in "that was the right (ethical) thing to do." "Wrong" is the same, but in reverse: there are wrong answers as well as wrong actions. Typically, when carrying an ethical load, "right" means fair, just, virtuous, beneficial, good, etc., and "wrong" means unfair, unjust, vicious, harmful, etc. Of note is that sometimes the right course of action is not necessarily good, for example, if one has only bad options to choose from, one must choose the least bad option, which is relatively the best of the bad options. Likewise, wrong does not always mean bad; one might have multiple good options to choose from, and if one does not choose the best good option, then one has chosen a lesser good, which would be the wrong choice.

Notably, "right" also has an **objective** and **subjective** sense, as in "that is right" and "that is *my* right." The first is objective: something is either empirically or logically right or wrong (in the sense of true or false) or objectively ethically right or wrong (that is, violating fundamental ethical principles and values). A subjective right, on the other hand, is a right that belongs to a person, such as human and/or civil rights (one has a right to life, freedom of speech, fair trial, etc.) or rights gained by privilege of social role, for example, the leader of a nation typically has the right to direct that nation's military, while other people in that nation do not have that right. Note that the idea of "subjective rights" that pertain to individual subjects should not be confused with the very different assertion that "rights are subjective"—in other words, that they are morally relative or mere opinion.

It can also be helpful to limit the scope of ethics by mentioning a few additional things that ethics is not. Ethics is not merely doing what feels right, or following laws or authorities, etc. Ethics comes before those things; rightly understood, what feels right should feel that way because it is ethical, laws should be based on ethics, authority should derive from ethics (for example, Mohandas [Mahatma] Gandhi's idea of ***satyagraha***—"truth-force"), and so on.

Furthermore, ethics is not merely a strident unthinking advocacy of certain controversial political topics, nor it is a therapeutic assurance that "anything goes" if I say it does. Ethics is between those two extremes. Ethics does agree that we should advocate for the good, but we must do so with deliberation and consideration, not stridently and/or unthinkingly. Ethics does also agree that we need to be able to think for ourselves, for there are too many peculiar circumstances in life for rules or other external controls to decide them all for us in advance. But merely thinking them out for ourselves does not mean that we have come to an acceptable answer. Even very wise decision makers can still make mistakes, and therefore when decisions are made ethical debate is not over; it may have only just begun. Ethics is always open to review, and therefore everything we do in ethics must include a note of humility, both individually and socially.

It should also be emphasized that ethics is not merely a negative field that demands that we avoid doing bad things and punish those who do. This understanding of ethics reflects a punitive and/or minimalist approach that might avoid evil but never result in anyone actually achieving good. It also ignores the vast history of ethics that sees ethics as a positive endeavor, one that seeks out the best way to live one's life. In the words of the philosopher Bernard Williams, "It is not a trivial question, Socrates said: what we are talking about is how one should live."[3]

We cannot live life as a negative. There are certainly things to avoid, but there are good things to seek as well, such as helping others. This more positive perspective on ethics is vital if we are going to create a good future together for all of humankind.

Last, among the things that ethics is not, ethics is not equivalent to religion, faith, or spirituality. Certainly, most religions have ethical aspects to them, but ethical discussion is separable from religion, existing independently, on philosophical grounds. This book will examine space ethics purely secularly, though this secular approach need not be inconsistent with or antagonistic toward religious approaches. Indeed, because space ethics should be a global human endeavor, it must attempt to include all of humanity, drawing upon all of humankind's ethical resources, the sacred and the secular both. How exactly this pluralistic approach to global ethics will play out is a task for the future, though some attempts have already been partially successful, such as the United Nations *Universal Declaration of Human Rights*.[4]

## A DEFINITION OF ETHICS

A basic definition of ethics could be this: ethics is the theory and practice of how to make good decisions, perform good actions, and thereby become a good person.

Because ethics involves all of human free action, it covers an immense area of human life and therefore, throughout history, has had many definitions that variously delineate the subject. But all these definitions have the same sense: ethics is about how to live a good life and the actions one must perform in order to attain that good life. There are many ways to be a human being and live a human life. The aim of ethics is to help us determine which ways are better and which are worse and help us to choose the better ones.

But this immediately raises another question: What is better? What is good? Over two millennia ago, Aristotle defined good as "that which everything seeks."[5] Later, medieval European philosophers and theologians defined the first principle of practical reason as being that "good is to be done and pursued, and evil is to be avoided."[6] These bare-bones definitions help us know that the point of ethics is to seek good and avoid evil; however, as formal statements—words without clear content—we must then ask ourselves what *are* good and evil?

Ethicists may vary in how they understand the meaning of good, but one of the traditional and still current understandings of good is that good consists of fulfilling one's proper goals (proper in the sense of being properly natural to a being or species of beings). This dates back to Aristotle and is still regarded as a good answer by ethical systems as diverse as **virtue ethics** (which seeks to grow virtuous character, virtue being properly natural to humankind) and **preference utilitarianism** (which seeks to maximize pleasure through preference satisfaction, this being properly natural to humankind). Notice that this applies as much to humans as to other living things because all life forms have goals (even if characterized mechanistically or nonconsciously) that they strive to achieve, whether they be plants growing toward the Sun or animals seeking food.

Humans have natural goals such as seeking survival, reproduction, social interaction, and truth (and more concrete subgoals such as food, warmth, health, relationships, and education),[7] but beyond those basics, people have many understandings of the purposes of human life. Some people find purpose in helping their family, some in doing their job, and some in fulfilling their religious obligations. Because humans differ greatly in their understanding of the good, it may be easier to define good by its opposite, evil, and primarily seek to avoid that rather than to impose a rigid conception of the good onto others. In this alternative approach, most people will agree that they do not like to be in pain or suffer or to be forced to act against their will. Thus, actions that cause pain or suffering to others or force them to act in ways that they do not want should be avoided. However, as mentioned earlier, while framing ethics purely negatively leaves much room for freedom and might help us avoid evil, it does not necessarily lead to good. Both positive and negative framings are useful for ethics: a negative framing being like the outer ring of a bull's-eye target, a positive framing the center dot, and the ring in between as a zone of relatively neutral behavior.

The definition of good as flourishing (particularly for humans, but also including other living things) is ultimately what I will use in this book. Flourishing can be defined as living in a state of well-being and displaying excellence (for example, in the case of humans, exhibiting the virtues of kindness, skill, wisdom, etc.). The definition of evil is the opposite of that of good: evil harms flourishing. It damages the well-being of living things and denies them the capacity to display their excellence by either keeping their innate capacities underdeveloped or distorted toward vices. Still the precise contents of good and evil can remain elusive. Many contemporary philosophies and religions seek to provide the greater details necessary to provide the deep and vital context for flourishing. Some important contemporary theories for thinking about goodness are Amartya Sen's and Martha Nussbaum's capabilities theory,[8] John Rawls's theory of justice,[9] international human rights agreements (such as the Universal Declaration on Human Rights), and many more.

The philosophical field that examines the various basic questions of ethics is called **metaethics**, as it is just "beyond" ethics ("meta" meaning "beyond" in Greek) or perhaps underlying ethics, to use a different metaphor. Ultimately the question of the definition of good and evil is a **metaphysical** one (existing beyond the physical world, at the level of foundational principles underlying reality), which cannot be absolutely answered by science, technology, public opinion, or any empirical test. As the scholar Ian Barbour has noted, metaphysical theories (whether scientific, religious, or otherwise) can only be evaluated relative to each other by comparing four things: (1) their internal *coherence*, (2) their *correspondence* to known facts, (3) their explanatory scope (*comprehensiveness*), and (4) their pragmatic *conse-*

*quences* for believing in them (for example, if the theory is "beautiful," "fruitful" for thought, and/or beneficial).[10]

As we go forth into space our ethical theories will be tested in new ways. We will gain new insights into their coherence, factual correspondence, scope, and fruitfulness. And in that way we will hopefully be able to grow our ethical theories more toward the truth.

## FOUR APPROACHES TO ETHICS

Ethics has several main approaches (often called methods), four of which are of primary significance for space exploration and use: deontology, consequentialism, case-based analysis (also called casuistry), and virtue ethics. The first three approaches concern action ("doing") and the last approach concerns the characters/dispositions of agents ("being").

### Deontology

**Deontology** is the approach to ethics that primarily emphasizes having good intentions and dutifully following rules and fulfilling obligations ("deontology" meaning the "study of duty"). This approach is typified by Immanuel Kant, whose *Foundations of the Metaphysics of Morals* remains to this day the archetypal work describing a deontological ethical framework. Kant asserts that a goodwill is the most important thing in ethics; it is what makes our actions good. Kant presents several formulations of his categorical imperative—the one necessary rule that is the foundation of all ethical behavior. While different authors may say there are as many as five versions of the categorical imperative,[11] here I will provide three of Kant's formulations:

1. *The "universal law" formulation:* "Act only according to the maxim by which you can at the same time will that it should become a universal law."[12]
2. *The "humanity" formulation:* "Act so that you treat humanity, whether in your own person or in that of another, always as an end and never as a means only."[13]
3. *The "kingdom of ends" formulation:* "All rational beings stand under the law that each of them should treat himself and all others never merely as means, but in every case, at the same time, as an end in himself. Thus there arises a systematic union of rational beings through common objective laws. This is a realm which may be called a realm of ends (certainly only an ideal) because what these laws have in view is just the relation of these beings to each other as ends and means."[14]

There is debate in academic circles not only as to number but also as to whether these formulations are in fact equivalent to just one paramount rule of ethics. However, no matter how this debate is settled, the sense remains: ethics is about one rule, requiring our absolute adherence, with all other rules flowing from it subordinately. Importantly, the categorical imperative must be only one rule, or else it becomes vulnerable to internal conflicts where multiple rules might compete with each other. Multiple rules lead to conflicts, and this would ruin the harmony of the "realm of ends."

Whenever we talk of ethics in terms of following rules we are speaking in a deontological fashion about the obligatory duties we ought to fulfill. Legal systems, secular and religious codes of ethics, customs, taboos, sets of cultural and ritual obligations, and many other forms of guidance rely on deontology because rules are a simple and typically very effective way of directing human behavior.

With regard specifically to space exploration and use, there are two deontological ethical systems that are particularly relevant. The first is Hans Jonas's **imperative of responsibility** that he originally developed as an environmental ethic for Earth but that has profound relevance for space.[15] In this form of deontological ethics, Jonas asserts that the only truly categorical imperative is not the one that Kant asserted but rather a more fundamental one that underlies Kant's imperative, that we ought to "act so that the effects of your action are compatible with the permanence of genuinely human life."[16] Because every ethical rule is premised upon human existence, the first ethical rule logically should be that *humans ought to exist*.

The second deontological system oriented toward space is Richard Randolph and Christopher McKay's principle focused on "protecting and expanding the richness and diversity of life."[17] This rule is meant to guide future human exploration and use of space toward both protecting the diversity of life in the universe, if we happen to find extraterrestrial life, and expanding life, be it Earthly or otherwise, in the universe. In both cases life also ought to be enriched—given the opportunity to flourish, further develop, and evolve into new forms.

Both Jonas's and Randolph and McKay's ethical systems will be explored in further detail over the course of this book. There are other rule-based systems that affect space too, such as the Outer Space Treaty and human rights law.

## Consequentialism

**Consequentialism** is the approach to ethics centered on maximizing the best consequences and minimizing the worst, regardless of rules or intentions. Consequentialism in its utilitarian form is typified by the works of Jeremy Bentham and John Stuart Mill.[18] More recently, the philosopher Peter Singer has been a vocal proponent of utilitarian ethics,[19] and utilitarianism has become a sort of social movement in the form of "**effective altruism**" that seeks to maximize the beneficial impact that individuals can make on the world.[20]

As an understandable response to the suffering he saw around him, Bentham's approach to ethics was centered on decreasing the overall suffering in the world and increasing the overall amount of pleasure. Bentham summarized the "principle of utility" (or "greatest happiness principle") by saying, "it is the greatest happiness of the greatest number that is the measure of right and wrong."[21]

Bentham's ideas went on to influence Mill's utilitarianism, which he set out in his 1863 book by that name. Mill further developed Bentham's ideas by making intellectual pleasures more worthy than more animal ones[22] and by relating "the greatest good for the greatest number" to political processes and personal liberty.

The philosopher Peter Singer extended utilitarianism toward animal liberation,[23] as well as toward preference satisfaction, at one point arguing that the best level to decide the meaning of pleasure and pain is at the level of personal preferences.[24] This form of utilitarianism has become quite influential in Western thought and is aligned well with the worldview of individualism, personal independence and control, and consumerist capitalism. Singer has since moved away from preference utilitarianism and toward **hedonistic utilitarianism**.[25]

Regarding space ethics, the consequentialist theories of Nick Bostrom and Seth Baum reveal some of the intricacies of thinking expansively in space and time. Bostrom's "Astronomical Waste" paper from 2003 begins by pondering the massive loss of resources in the universe per unit time. The faster humanity could utilize these resources for the benefit of sentient beings, the sooner this waste would stop.[26] However, in the 2011 paper "Infinite Ethics" Bostrom reverses course somewhat, arguing that in an infinite universe (as our universe may be), both infinite happiness and infinite sadness (in the form of infinite numbers of happy and sad people) ought to pervade the universe, and therefore any finite increase or decrease in happiness or sadness is nothing in comparison. This makes "aggregative consequentialism," at the largest scales, ineffective as a moral system and therefore in need of careful delineation in order to still be useful.[27]

Seth Baum's 2016 chapter "The Ethics of Outer Space: A Consequentialist Perspective" returns to the idea of space having the potential for astronomically good or bad outcomes but also adds two main additional factors. First, to protect the possible astronomically good values of the future humanity needs to work very hard on preventing global catastrophes now because any catastrophe now could close off the future possibility of a bright future. Second, the discovery of extraterrestrial intelligences, if there are any, could throw the utilitarian calculus into utter turmoil because the encounter could either turn out very good or very bad, and there really is no way to predict which it will be, at least at this point in time. Therefore the greatest care is warranted in both fields.[28]

Consequentialism shall be examined further throughout the book, but for now it is worth remembering that while it is extremely useful for considering the scale of good and bad, it is also a difficult approach to ethics when the future is so unpredictable.

## Case-Based Analysis or Casuistry

**Case-based analysis** or **casuistry** (meant in the strict philosophical sense, not the derogatory sense) is the approach to ethics based on analogizing contemporary cases with past cases in order to find continuity and the most fitting solution. As new and difficult cases appear, this approach seeks to compare the new undecided cases to old decided cases and see what those comparisons may indicate about how to decide the new case.

For example, space law and treaties governing space, in general, are based analogously on aeronautical law, nautical law, and other forms of Earth law, such as the Antarctic Treaty.[29] Certainly, each set of laws is different, but each is built analogically upon the others. For another example, if life forms were discovered on other planets, people would wonder how to treat them, and they would naturally look to analogous creatures on Earth for ethical guidance. Is the organism like a bacteria? Is it more like a plant or animal? Does it seem intelligent? Is it unlike anything we have ever seen before and therefore invalid for analogy? Insofar as the organism is similar to something we have seen, we are likely to treat it in a similar way, barring of course the dis-analogies of the situation, namely the ways it is unlike what we are comparing it to, and also most importantly that it is extraterrestrial and therefore extremely scientifically interesting, as just one point.

This approach is typified by various traditions of case law on Earth: legal, religious, and ethical. In the past, "casuistry" became a term of abuse thrown at those who were perhaps splitting and joining their analogies too closely or remotely. But in more recent decades case-based analysis has regained some standing philosophical discussions, due in no small part to the book *The Abuse of Casuistry* by Albert R. Jonsen and Stephen Toulmin.[30] In their book, Jonsen and Toulmin describe six steps for casuistic analysis:

1. *Paradigm cases and analogy:* an "orderly taxonomy" of cases is kept at the ready for analogizing to current cases that may appear.
2. *Application of maxims:* moral maxims related to paradigm cases are presented for consideration.
3. *Circumstances:* the traditional "who, what, where, when, why, how, and by what means" are added to the analysis, and not just as minor details, but always with the knowledge that "circumstances make the case."

4. *Probability:* the certainty of the conclusion of the case is assessed, ranging from "certain" to "hardly probable." In more certain cases, the weight of the decision is stronger, and less so in the reverse.
5. *Cumulative argument:* or case precedent, sets the case in the context of a tradition of decisions on similar cases, typically perceived as adding innumerable small arguments into a whole rather than appealing to a single or few arguments.
6. *Resolution:* finally, a recommendation is formulated with a level of probability attached to it, for example, "In these circumstances, given these conditions, you can with reasonable assurance act in such-and-such a way. By so doing, you will not act rashly or imprudently, but can be of good conscience."[31]

Jonsen and Toulmin go into vastly more detail in their book, but it is worth noting that while case-based analysis is the bread and butter of most forms of professional ethics (in other words, ethics particular to specific professional fields, such as medicine, law, engineering, etc.), Jonsen and Toulmin's approach, which is descriptive, is often used implicitly, incompletely, or not in a recognizable form at all. These six steps are one way to do casuistic analysis, but they are not the only way. Other approaches also exist, though they can tend to resemble parts of Jonsen and Toulmin's more comprehensive approach, for example, concentrating on circumstances and case writing[32] or on maxims and cumulative argument.[33]

For space ethics the main lesson to draw from casuistry is that case-based analysis and analogy is already commonplace and will remain so. This book, for example, examines many cases studies and therefore lends itself to supporting case-based analysis. Furthermore, because space is such a huge subject, the case library to analogize against will need to be concomitantly vast. All the lessons of human history will need to be at our disposal, and casuistic analysis will require enormous human resources, and likely computational ones too, at the very least in the form of databases to help us remember what we ought not to forget. Artificial intelligence, then, might have a place in ethical analysis, perhaps particularly in the case-based approach. Casuistry also need not only use historical examples: literature can also spark the imagination, and when it comes to space, science fiction can be an especially rich source.

## Virtue Ethics

Virtue ethics is that approach to ethics centered on agents and not actions, on *being* more than *doing*. Virtue ethics aims to promote good character, good habits, and virtues and to discourage vices and other bad character traits. Unlike the other approaches, which focus on discrete choices and actions, this approach focuses on persons and their long-term predispositions to act in a certain way. This approach is strongly represented in the Western philosophical tradition typified by Aristotle and Thomas Aquinas, though the concepts of good character, virtue, and vice seem to be nearly universal among human cultures (for example, courage and wisdom are appreciated by nearly all human groups).[34] More recently, virtue ethics has become more fashionable thanks to the work of Elizabeth Anscombe, Philippa Foot, Alasdair MacIntyre, and others.[35]

Overall, virtue ethics seeks to promote individual and communal human flourishing, and it gives a more well-rounded opportunity for ethical thinking as it does not see individual actions in isolation but rather as the effects and causes of patterns in human character and culture. Every act that a human performs makes future acts of that same type easier than before—both for good and for ill. This means that everything that we do is more significant than we are likely to imagine. It also means that human beings, as components of larger cultures, contribute to the creation of each other, and our cultures to us in return. Virtue ethics is perhaps the most socioculturally engaged ethical approach because it takes very seriously the intersection of human psychology with culture and society.

With respect to space, this psycho-socio-cultural engagement might be particularly interesting given that human space exploration and settlement will involve, initially, a few very select individuals who have been carefully chosen for their talents and characters. These early founders will likely set the tone for the cultures of the settlements they create. Similarly, the vastness of space could allow for much diversity of culture, and so in various places unique cultures with their own particular notions of virtue and vice may arise.

Uniquely to virtue ethics, uncertainty can be made certain in one respect. The one thing we know about human exploration of space is that it will be *humans* doing it (even if they are behind robots and artificial intelligences, now or in the future, or are our modified descendants), and so their human characters and dispositions will be of particular importance. Similarly, virtue ethics allows for the consideration of the nonliving universe in more depth than in other ethical approaches. As a field, environmental virtue ethics has been growing recently and has gained traction with regard to several issues that are difficult for other approaches, including the value of the nonliving world, which can be valued in virtue ethics for the effects that it has upon human character, even if matters such as intrinsic value cannot be conclusively resolved.[36]

## Comparing Approaches

This text does not advocate any particular approach to ethics and at times may utilize any of them, as suits the specific situation. This is one of the benefits of applied ethics as a field: theoretical allegiances are typically weakened for the sake of achieving practically satisfactory answers. However, virtue ethics and casuistry will be prominent, given the numerous case studies that make up the substance of ethics as applied to space and the importance of human decision making (and the characters of the decision makers) to the future of humanity. Deontology and consequentialism are also prominent, however, as they are absolutely vital components of ethical analysis, particularly in applied ethics.

In the midst of this utilization of the various approaches to ethics it is worth highlighting the similarities and contrasts between argument by analogy, principles, and goals. Argument from analogy allows incremental and consistent progress. Argument by analogy builds from footpaths, to horse paths, to dirt roads, to paved roads, to superhighways. In ethics, analogous thinking might move from banning killing of those in your own family, to your own tribe, to your ethnic or religious group, to all humans, and then perhaps to certain kinds of animals, and possibly even further.

Working from principles or toward goals, on the other hand, can lead to very different types of movements. Returning to the metaphor of the path, we might ask if turning the path into a superhighway is a good idea. Will the superhighway efficiently accomplish the task that we ask of it? Perhaps a reanalysis of the situation is in order rather than just simply following the same analogical thinking. The purpose of a path is to move people and goods from one place to another. Roads and superhighways can do that too, but under certain circumstances, they may become unwieldy—too massive and jammed with traffic, thereby no longer serving their function. Instead we can return to the first principle: the purpose is to get from one place to another. What could also perform this function, besides vehicles on roads? Trains, subways, buses, ferries, and aircraft might also serve this same function. Rather than being trapped by the thinking of ever-widening roads, perhaps a more rational approach would be to construct a subway, thus moving many people without the need to pave such large swaths of the surface of the Earth.

Applied to ethics, first principle and goal thinking can be both helpful and sometimes disconcerting (though possibly disconcerting in a good way). For example, one can argue for expanding the circle of nonkilling through analogy, but one could also make the same argument through a principle, simply: respect life. But to what extent should we respect life? Ought we not permit anyone to die? Should we not use antibiotics because they kill microorganisms? And what if rules or interests conflict? From another cultural direction, the primary rule might be to enhance human freedom. But what of places where human freedom might impinge upon the rule to respect life? Here this becomes no longer an academic debate: multiple contemporary ethical issues concern this very point, with abortion, the death penalty, euthanasia, vegetarianism, alleviation of poverty, provision of health care, radical life extension, and many other issues concerning precisely this matter.

This raises the question of what to do when ethical rules conflict, because these conflicts do happen. With regard to space exploration, spending on space may preclude spending on other important ethical duties. These trade-offs between differing ethical rules and values cause difficult questions to arise that are not easily remedied by any particular formulaic solution. In fact, reasonable people can remain in disagreement on these sorts of questions. Often disagreements on trade-offs are based on differences in first principles, on differing beliefs about human nature, or on varying predictions of the future, and so need to be negotiated with great care, with sufficient input from many people.

Similarly, with **dual-use technologies** (technologies that can be used to help or harm), developing technologies for space might in fact be a corollary to enhancing political prestige or military prowess. The trade-off here is to advance spaceflight, a potentially good thing, while also enabling the creation of risky new weapons, a potentially bad thing. World War II German rocket scientist Wernher von Braun, after all, pioneered rocketry as a weapon through his V-2 ballistic missile, while later in the United States of America pioneering both rocketry for nuclear ballistic missiles and for human space exploration in the Mercury and Apollo programs. The technological overlap is so great that there is no large difference between the two: rocket technology can both expand human civilization into space and destroy it here on Earth. All that differs is the aim and the payload (and as this technology continues to progress it will become increasingly important that peaceful and military rockets are not confused).

Given the four approaches discussed here, one might legitimately ask whether all the approaches to ethics tend to either converge on the same answer or diverge from each other. The answer to this (as with many things in ethics) is that it depends on the circumstances. In many cases, the same approach can be used to argue multiple sides of an issue, and when several approaches are in play, all might serve to argue for multiple sides. The question then becomes not one of which "side" an approach tends to be on but rather how well reasoned sides can be and then be evaluated on those merits.

However, there are sometimes reasons for optimism on the convergence of ethical decisions. Despite longstanding differences on sensitive ethical-political issues and the pluralism of ethical values around the world, there has been good effort put into reaching consensus on many issues, and consensus has been reached in many cases, as various United Nations human rights treaties demonstrate. The causes of these convergences might involve human psychology, sociocultural commonalities, biology, cosmology (for example, shared laws of nature), and so on, but ultimately the causes are not as practically important as the consensus itself.[37] The ontology of ethical norms is a difficult question that is ultimately beyond the scope of this book; however, there are some clear themes among ethical norms among human groups around the world, themes such as protecting human life, protecting important objects and property, protecting the truth, and protecting human relationships. These insights are helpful for better understanding human flourishing, but while this general convergence is helpful, it is not always enough, as the particulars of the norms may prove to be divergent at the level of specifics.

## FURTHER IMPORTANT ETHICAL CONCEPTS

Other important ethical concepts include justice and fairness, human rights, the common good, narrative ethics, feminist and intersectional ethics, environmental ethics, the locus of ethical value (that is, intrinsic versus extrinsic value), "centrism" in ethics, and the expanding scope of human action and power.

### Justice and Fairness

**Justice and fairness** are highly developed senses among humans and human cultures. The concepts of justice and fairness have the status of human universals, though precisely *what* is just and fair can of course vary between cultures. These concepts can be found even among animals such as many mammals and birds.[38] Justice and fairness can be defined as giving to each what they are due, whether that be benefits, responsibilities, punishments, etc. Justice can be classified into three main forms: distributive justice, which distributes social goods (such as public education, infrastructure, police protection, etc.); commutative, contractual, or procedural justice, which governs contracts and fair process (such as government or legal proceedings); and social justice, which distributes social responsibilities (such as taxes or a military draft).

With respect to space exploration and use, many justice questions arise, beginning particularly with the justice of the costs: Is this a fair way to be spending money? Second is the justice issue of who benefits and who pays: Are the costs and benefits distributed equitably? Third are questions of access to space: Will all people have access to space or only a certain few, and by what criteria will these distinctions be made (and will that be fair)? Fourth, if humans begin to settle in space, what legal system will they develop, and will that be fair? Fifth, justice typically is defined as existing between humans, but what of the natural environment, including the nonliving environment, and the potential for justice with newly discovered life, even intelligent life? These are only a few justice questions that arise when considering space. Answering these questions will be integral to the ethical exploration, use, and settlement of space.

### Human Rights and Dignity

**Human rights** are ethical standards for behavior owed to persons simply because they are human. Often a bearer of rights may create a concomitant duty for others to treat them in a specific way, and thus (as a rule and duty-based system) rights connect to deontology. Human rights language has become increasingly important over the course of recent centuries. From the origin of the concept of subjective rights in medieval canon law through the progressive expansion of the concept from natural law and natural rights in the Western world,[39] human rights discourse has become a shared language for speaking about moral issues. International institutions have done much toward advancing human rights around the world and enforcing them.

In contrast with the idea of objective right and wrong, subjective rights are particular to *subjects*: individuals. Objectively, it is ethically right for people to have a fair trial, but subjectively *a particular person* has a right to a fair trial. Human rights also include **positive rights** (rights to something, such as a right to an education) and **negative rights** (rights not to be subject to something, such as censorship of speech). Negative rights require no actions from other people, while positive rights do require others, including the state, to help fulfill them. However, even so-called negative rights in fact require quite a bit of positive effort by the state: legislating laws, institutionalizing them, teaching them, enforcing them, etc. This is significant for space because, for example, on Earth breathing is typically a negative right, requiring no actions on the part of others, but in space breathing becomes a positive right, requiring an entire infrastructure of others to help it be fulfilled.[40] Sometimes rights are also classified as **civil-political rights** (civil rights such as voting, freedom of speech, exercise of religion, etc.) and **social-economic rights** (such as rights to food, housing, health care, etc.), which can be a helpful reformulation from the positive/negative distinction.[41]

With regard to space exploration, use, and settlement, human rights are relevant in much the same ways that deontological ethics are relevant, or that justice and fairness are relevant, but with some specific alterations of language. Rather than talking about what rules declare or what is fair, rights language tends to talk about "my rights," "our rights," or "their rights" in a possessive sense. In this terminology, rights violations can be considered unjust because the distribution of rights has been improperly disrupted, taking from individuals what they are rightly due. The harsh environment of space makes providing rights much harder than on Earth, and so we should expect both that "rights violations" will be more likely and also that people may have lower expectations about gaining access to some rights or should at least give informed consent to the dangers of rights violations that may occur while in space. If space settlements grow and disperse, we should also expect that their individual cultures may develop different expectations of rights, just as different cultures on Earth already have divergent expectations.

As one last point, human rights also connect to the idea of intrinsic human dignity. **Human dignity** is a conception of the human being as sacred, inviolable, and intrinsically worthy of respect and protection. Humans have rights because of their dignity. While dignity is typically conceived of as a transcendent value, it can also be reduced to a mere conferral of value by a group or government. This positivist version of dignity is not really dignity at all but a mere human construct. Positivist dignity has dangers in that anything that humans make up we can also decide to unmake, which is why the more transcendent idea of dignity tends to remain popular.

## The Common Good

The **common good** is another concept vital to ethical discussion. The common good seeks to protect not only individuals but also the community and society itself, through protecting such shared resources and benefits as clean air and water, quality education and health care, safe and rapid transportation, and quality relationships and social institutions. Note that this is not the same definition as the very specific economic definition of a "common good," which is rivalrous (people compete over the good because if one takes it another cannot have it) and nonexcludable (anyone can access it); the common good that is being explored here is more general and inclusive (encompassing many economic notions such as common pool resources and public goods) and derives from ethical rather than economic traditions.

"Common good" vocabulary is used when discussing the preconditions for providing benefit to everyone. In some very minimalistic understandings of the common good, the common good is merely the individual good writ large. But what is in the community's best interest may sometimes not be in the individual's interest, as the concept of a military draft in war recognizes: sometimes the individual must sacrifice their personal good for the good of the group. The common good sometimes does align with human rights, but sometimes it also does not. Property seizure practices such as "imminent domain" (in the United States, where civic plans allow for the confiscation of property, with reasonable restitution) or cases where intellectual property (such as pharmaceutical exclusivity rights) are revoked provide additional examples where the rights of some are violated in order to improve the common good (though whether such practices achieve such ends in particular cases can, of course, be open to dispute). "Common good" language can sometimes, then, be abused, especially if a leader is attempting to gain benefit for one group by imposing costs on another.

With respect to space, there are clear common good appeals to be made for the prevention of the problem of out-of-control space debris, for example, which might fill Earth's orbit with so much junk that it becomes difficult or impossible to get anything off the planet. Similarly, but at a larger scale, the existence of additional communities off of Earth protect the common good of the human species, especially against existential risks. Having "backup" human communities means that even if the vast majority of humans were lost in a catastrophic event, at least some would live on, thus preventing the waste of all of human effort in history.

On a more mundane level, people who take upon themselves the risk of exploration gain for humanity the knowledge and resources that they discover, thus benefitting the common good. Analogously to the military draft, these individuals put themselves in danger for the sake of the whole, but because these dangers are taken up voluntarily, the sacrifice is all the more heroic. Questions of the common good become vastly more complex in space where more and more groups are likely to multiply, and so the notion of "common" may become fragmented (this is to say nothing of the possibility of what the common good might mean were extraterrestrial intelligences discovered). In the future, as now, and as should have been in the past, humanity would do well to be as inclusive as possible in its notion of the common good or it risks making terrible ethical errors. Note that the common good does not always mean giving in to opposing groups; if there is a destructive force in society the common good requires opposing it, as in World War II, where the Axis powers forced war upon their neighbors and the Allies had to respond in self-defense. The more holistic "common good" in World War II was restored after the war, after the Axis powers were reduced to surrender; then the Western Allies helped to rebuild them and lifted them up through such projects as the Marshall Plan, thus benefitting the overall state of the world (though certainly there was much further work to do with decolonization, resolving the Cold War, etc.).

## Narrative Ethics, Feminist Ethics, and Intersectional Ethics

**Narrative ethics** relates stories and identity to ethics. In contemporary usage it is often related to virtue ethics and to **bioethics**, in which a patient's story can be integral to determining better outcomes—outcomes more appropriate for that individual's life narrative. Likewise, groups of people and cultures can have narratives (or more philosophically "metanarratives" that organize and interpret subordinate narratives) and draw ethical lessons from those narratives, and in a pluralistic world, these stories are important to keep in mind for their practical effects on human action.

Narrative ethics is of particular interest to the ethics of space exploration and use because while very few humans have spent much time in space, we have had many peoples and individuals writing and telling stories about space. These stories form a very large body of mythology, literature, and media, which contains innumerable ethical cases and dilemmas worthy of consideration. Importantly, behind these narratives are often metanarratives about the purpose of human life in the universe: to learn, to pioneer, to fulfill destiny. Returning to individuals, those early humans in space will no doubt have very strong personal narratives and goals with which they will identify: goals of exploration, freedom, and settlement, for example. If they accomplish their stories they are more likely to feel fulfilled, but if they are thwarted, this will negatively impact their self-image. Of course, whether they succeed or fail should align with whether their goals are good or evil, for not all goals should be assumed to be good ones.

**Feminist ethics** is the field of study that particularly examines ethics with respect to women. Feminist ethics originated with Mary Wollstonecraft's 1796 book *A Vindication of the Rights of Woman*, which proposes the simple idea that women ought to have rights similar to those of men.[42] Over time, feminist ethics has strengthened its rights claims to full equality and developed further to question the foundations of Western ethics, while suggesting alternative formulations for ethics and new solutions to social problems. It has also pondered the more specifically female aspects of life such as childbirth and motherhood, which have typically been ignored by the male philosophers of the past.

**Intersectional ethics** takes the particularization of ethics to women one step further and particularizes to the intersection of multiple categories parsed by race, sex, gender, sexuality, class, immigration status, and more. While each of these categories has its own particular difficulties placed upon it by society, the intersections of these groups are further specifically oppressed in ways unlike other groups. Intersectional ethics was founded by Kimberlé Crenshaw in 1989 and has become an important force not only in scholarly discussions of social conditions and ethics but also in social, political, and cultural activism.[43]

Feminist and intersectional ethics are quite relevant to space exploration because the space environment will likely affect males and females differently, as well as possibly affecting various intersectional groups differently. Additionally, space settlement will likely add complexity to group relationships, as well as broadening the plurality of society (thus potentially adding more possible intersections of groups). We should be aware of these ethically relevant aspects of identity or else we risk making mistakes and being oblivious to having made them.

## Environmental Ethics, Ethical Centrism, and the Imperative of Responsibility

**Environmental ethics** considers forms of ethical thinking and behavior that promote good human interactions with our natural environment. Because space is such an extreme natural environment, full of dangers and demanding significant technological intervention in order to maintain human life, environmental ethics is an absolutely crucial field of study for space ethics. For example, environmental ethics includes investigations into the ethical value, or lack thereof, of lifeless things and places, as is most of space.

Related to environmental ethics and specifically relevant for space is the question of the **locus of ethical value**—what is it that makes things ethically valuable? Is it intrinsic or extrinsic? For example, we might ask: Are other life forms valuable in themselves or only because humans deem them to be so? If we answer the first, we are attributing to those life forms intrinsic value, while if we answer the second, we are attributing to them only extrinsic or instrumental value (value that exists only as they benefit something or someone else).[44]

In general, people ought to avoid harming entities with ethical significance. However, this raises another ethical issue: How much weight should we give to the value of a whole species versus individuals of a species? It might seem like the individuals of a species might be the locus of ethical value, but if this is the case, then it would be no worse to kill any member of a species than it would be to kill the very last member of a species, thus driving it extinct (which seems to be an intuitively wrong conclusion). Weighing the interests of humans as compared to other life forms is a very sticky ethical debate; weighing the values of groups and species becomes even more so. Related to this ethical weighing is the necessary question of why weights might justifiably differ; that is, what is the source of ethical value in the world? Is it intelligence? Consciousness? Teleology? Ability to experience pleasure or suffering? Mere existence? On a spectrum, how much of those traits (or whatever the ethically relevant trait might be) must something have in order to warrant consideration? And why those traits and not something else? For example, plants, microbes, and algae do not seem to have any form of consciousness; do they therefore deserve very little ethical consideration? Some animals have very little intelligence, such as jellyfish and insects, while creatures such as elephants, great apes, cetaceans, and corvid birds have much more; does that mean they deserve more consideration? And why? Many space ethicists have been exploring these topics.[45]

Notice that this is an attempt at measurement of intrinsic value. Life forms may also have instrumental value that may make them particularly worth caring for. For example, novel life forms originating from a **second genesis** (a beginning of life not common to Earth) would have immense scientific value and therefore ought not to be arbitrarily destroyed or driven extinct, even if only microbial.[46] Life forms used for food and human industry are likewise instrumentally valuable, such as crops, food animals, and so on. Similarly, life that is particularly pleasing to humans (flowers, pets, etc.) ought also to be given more consideration because their flourishing affects the flourishing of those who are pleased by them.

We can also ask whether other life forms "own" their locations or territories. If they are intelligent and have a concept of property, does that mean they own their location? That would seem plausible, as humans often acknowledge this of each other. What if they are intelligent but lack a similar concept of property? This parallels in some ways the colonial era of human history, which certainly contained many ethical disasters involving the concept of territory and property rights. What if other life forms are semi-intelligent or not intelligent at all, like a microbe; do they still own their location, or is their location available for use by others? How, why, and who decides?

In the extremes of space, ecosystemic wholeness is also an important consideration when valuing other life forms. On Earth, we have countless other life forms living in intricate ecosystemic networks that remain relatively dynamically stable over the course of the natural history of the planet. If humans were to terraform another planet, we would need all the organisms necessary in order to construct and cultivate such Earth-analogous conditions there. If some particular organism—say a particular microbe, like nitrogen-fixing bacteria—were rare or unavailable, it would certainly be worth investing extra effort into caring for it and protecting it, even if that meant other life forms might be neglected in its favor.

Related to the question of locus of value is the concept of **centrism** in moral value and decision making.[47] Centrism asks either where the focus of ethical value lies or from what perspective ethical value should be viewed from, for example, anthropocentrism, biocentrism, ecocentrism, cosmocentrism, and theocentrism. These categories help to reframe the ideas of the locus of value. For example, we might not agree on why humans ought to be top of the ethical value chart, but if we agreed that it ought to be the case (regardless of reason), then we would be anthropocentrists. Biocentrists, by contrast, agree that living things are the locus of ethical value. Ecocentrists place the locus of ethical value in ecosystems, distributed across the life forms in it. Cosmocentrists agree that the universe ought to be the focus of ethical value.[48] Last, theocentrists believe that God ought to be the center of ethical value; this is a more common perspective among monotheistic religious believers, and nonmonotheistic religions have their own perspectives as well.[49] Typically, centrists of various sorts do provide justification for why their form of centrism ought to be considered correct. However, centrism is sometimes merely assumed, for example, most ethical systems prior to the twentieth century tended to be anthropocentric or centric in a religious way, often without much by way of rational justification.

Building upon the problem of centrism, we return to the philosopher Hans Jonas. Specifically, Jonas asserted that his imperative of responsibility—that humans ought to exist—is the case because humans are the "foothold" of ethics in the universe.[50] His ethics is "ethico-centric" or "moral-centric," so to speak, and humans, as the only known ethical creatures that humans are aware of, are therefore required as a presupposition for ethics to exist at all. Jonas argues that this has been an unnoticed and unquestionable assumption of all previous ethics, but that assumption is no longer unquestionable.[51] After the world wars and the onset of the environmental crisis, with the continuing existence of humanity now in question, we now had to ask: Ought humans to exist? To this he answered yes, for we are the precondition for the concept of "should" and "ought" to be asked at all. Without us, ethics ceases to exist. Therefore the first rule of ethics, the first good that ethics must seek before it can do any other good, should be to preserve its own precondition, the existence of humanity. Only when threatened by nonexistence could we come to see the assumption of and importance of our own existence.

Jonas also asked: What is it that is threatening humanity with nonexistence? And the answer is: our own technology. Technology has changed the **nature and scope of human action and power**.[52] Whereas formerly human actions were spatially and temporally limited, now our actions and their effects are expanding across space and time, affecting places beyond Earth and far into the future. The answer to this question is now even more relevant than it was in Jonas's time, particularly with respect to the ethics of space exploration and use. The realm of human action is now leaving planet Earth for the reaches of space. The energies involved and the scale of these actions are vastly beyond what ancient human ethical systems were designed for. Therefore we need a new ethical system based on the imperative of responsibility. It is technology that has forced this change, a gradual change in quantity of power has now become a qualitative change in power, one that also requires a qualitative change in ethics. This expanding scope of human power necessitates an expanding scope of human ethics.

Many more ethical concepts will be found in the following pages of this book, but these basics are important for laying out the initial framework and tools for thinking about ethical issues in the exploration and use of space.

## DISCUSSION AND STUDY QUESTIONS

1. Everyone has their own tendencies to view ethics in particular ways, and it is beneficial to know our own tendencies and blind spots. Of the ethical concepts discussed in this chapter, which are you most familiar with? Which are you most likely to use when thinking about or discussing ethics? Which are you least familiar with? Which are you least likely to use when thinking about or discussing ethics? Why?
2. Building upon question 1, once you have considered this about yourself, can you see how this might lead to agreements and disagreements with others on ethical issues? What ways might there be to use this knowledge to enhance or reduce disagreements between people?
3. With regard to space ethics, how might these ethical concepts affect how space ethics is considered in particular contexts, such as in a classroom, at a corporation, or in a governance meeting? How might these ethical concepts affect discussions across cultures (both for good and for bad)?

4. Why should people be ethical? (Note: This can be construed as asking "why should people follow 'should' statements?") Why?

## FURTHER READINGS

Constance M. Bertka, *Exploring the Origin, Extent, and Future of Life: Philosophical, Ethical, and Theological Perspectives* (Cambridge: Cambridge University Press, 2009).

Eugene C. Hargrove, ed., *Beyond Spaceship Earth: Environmental Ethics and the Solar System* (San Francisco: Sierra Club Books, 1986).

James S. J. Schwartz and Tony Milligan, eds., *The Ethics of Space Exploration* (Switzerland: Springer International, 2016).

## NOTES

1. See, for example, Robert Stern, "Does 'Ought' Imply 'Can'? And Did Kant Think It Does?" *Utilitas* 16, no. 1 (2004): 42–61, citing various passages in Kant, including *Religion Within the Boundaries of Mere Reason*, trans. George di Giovanni, in *Religion and Rational Theology*, trans. and ed. Allen W. Wood and George di Giovanni (Cambridge: Cambridge University Press, 1996), 92, and *Critique of Pure Reason*, trans. Norman Kemp Smith (London, 1933), 637.
2. Questions concerning the reality of counterfactual free will (whether we can really choose to do other than we do) will not be addressed in this book other than to say that while counterfactual free will can certainly be disputed, it is not a scientifically testable hypothesis either way, neither in favor nor against it (we cannot replay time in order to conduct the experiment over and over again). What we can justifiably say we know is that humans have agency (we can make things happen, whether we are free or not) and therefore have physical responsibility for our actions. Insofar as ethics is capable of influencing our actions for the better, ethics is crucial for creating a better world.
3. Bernard Williams, *Ethics and the Limits of Philosophy* (Cambridge, MA: Harvard University Press, 1985), 1, quoting Plato, *Republic*, 352d.
4. United Nations, *The Universal Declaration of Human Rights*, 1948.
5. Aristotle, *Nicomachean Ethics*, trans. Terence Irwin, second edition (Indianapolis, IN: Hackett Publishing Co., 1999), 1 [I.1].
6. Thomas Aquinas, *Summa Theologiae*, trans. Fathers of the English Dominican Province, Complete English Edition in 5 Volumes (New York: Benziger Bros., 1947, republished by Notre Dame, IN: Ave Maria Press, 1981), 1009 (I–II, 94.2).
7. Ibid.
8. Amartya Sen, "Equality of What?" in *Tanner Lectures on Human Values*, ed. Sterling M. McMurrin (Cambridge: Cambridge University Press, 1979), 197–220; Amartya Sen, "Rights and Capabilities," in *Morality and Objectivity: A Tribute to J. L. Mackie* (London: Routledge and Kegan Paul, 1985), 130–48; Martha C. Nussbaum, "Non-Relative Virtues: An Aristotelian Approach," *Midwest Studies in Philosophy* 13 (1988), 32–53; and Martha C. Nussbaum, *Women and Human Development: The Capabilities Approach* (Cambridge: Cambridge University Press, 2000).
9. John Rawls, *A Theory of Justice* (Cambridge, MA: Harvard University Press, 1971).
10. Ian Barbour, *Religion in an Age of Science: The Gifford Lectures*, Volume I (New York: Harper Collins, 1990), 34–39.
11. H. J. Paton, *The Categorical Imperative: A Study in Kant's Moral Philosophy* (Philadelphia: University of Pennsylvania Press, 1947), 129–97.
12. Immanuel Kant, *Foundations of the Metaphysics of Morals and What Is Enlightenment?* trans. Lewis White Beck, second edition (Upper Saddle River, NJ: Prentice Hall, Inc., 1997), 38.
13. Ibid., 46.
14. Ibid., 50.
15. Hans Jonas, *The Imperative of Responsibility* (Chicago: University of Chicago Press, 1984).
16. Ibid., 11.
17. Richard O. Randolph and Christopher P. McKay, "Protecting and Expanding the Richness and Diversity of Life, an Ethic for Astrobiology Research and Space Exploration," *International Journal of Astrobiology* 13, no. 1 (2014): 28–34.
18. Jeremy Bentham, *A Fragment on Government* (London, 1776), preface, accessed August 27, 2014, http://www.constitution.org/jb/frag_gov.htm; and John Stuart Mill, Utilitarianism *and* On Liberty*: Including Mill's "Essay on Bentham" and Selections from the Writings of Jeremy Bentham and John Austin*, ed. Mary Warnock (Malden, MA: Blackwell Publishing, 2003).
19. See, for example, Katarzyna de Lazari-Radek and Peter Singer, *Utilitarianism: A Very Short Introduction* (Oxford: Oxford University Press, 2017).

20. See, for example, Effective Altruism, https://www.effectivealtruism.org/.
21. Bentham, *A Fragment on Government*, 1.
22. Mill, Utilitarianism *and* On Liberty, 186–87.
23. Peter Singer, *Animal Liberation*, new revised edition (New York: Avon Books, 1990 [1975]).
24. Peter Singer, *Practical Ethics*, second edition (Cambridge: Cambridge University Press, 1993), 12–14, 94–95.
25. "I have moved from preference utilitarianism to hedonistic utilitarianism." Robert Wiblin, Arden Koehler, and Keiran Harris, "Peter Singer on Being Provocative, EA, How His Moral Views Have Changed, & Rescuing Children Drowning in Ponds," interview with Peter Singer, *80,000 Hours Podcast*, December 5, 2019, https://80000hours.org/podcast/episodes/peter-singer-advocacy-and-the-life-you-can-save/.
26. Nick Bostrom, "Astronomical Waste: The Opportunity Cost of Delayed Technological Development," *Utilitas* 15, no. 3 (2003): 308–14.
27. Nick Bostrom, "Infinite Ethics," *Analysis and Metaphysics* 10 (2011): 9–59.
28. Seth Baum, "The Ethics of Outer Space: A Consequentialist Perspective," in *The Ethics of Space Exploration*, ed. James S. J. Schwartz and Tony Milligan (Switzerland: Springer, 2016), 109–23.
29. Margaret S. Race, "Policies for Scientific Exploration and Environmental Protection: Comparison of the Antarctic and Outer Space Treaties," in *Science Diplomacy: Antarctica, Science, and the Governance of International Spaces*, ed. Paul Arthur Berkman, Michael A. Lang, David W. H. Walton, and Oran R. Young (Washington, DC: Smithsonian Institution Scholarly Press, 2011), 143–52.
30. Albert R. Jonsen and Stephen Toulmin, *The Abuse of Casuistry: A History of Moral Reasoning* (Berkeley: University of California Press, 1988).
31. Ibid., 250–57.
32. Brian Patrick Green with Irina Raicu, "A Template for Technology Ethics Case Studies," Markkula Center for Applied Ethics, March 5, 2019, https://www.scu.edu/ethics/focus-areas/technology-ethics/a-template-for-technology-ethics-case-studies/.
33. Martin Peterson, *The Ethics of Technology: A Geometric Analysis of Five Moral Principles* (Oxford: Oxford University Press, 2017).
34. See, for example, Shannon Vallor's *Technology and the Virtues: A Philosophical Guide to a Future Worth Wanting* (Oxford: Oxford University Press, 2016), which takes a multicultural perspective on virtue ethics and the relevance of those virtues to technology ethics.
35. G. E. M. Anscombe, "Modern Moral Philosophy," *Philosophy* 33 (1958); Philippa Foot, *Natural Goodness* (Oxford: Clarendon, 2001); Alasdair MacIntyre, *After Virtue: A Study in Moral Theory*, second edition (Notre Dame, IN: University of Notre Dame Press, 1984).
36. See, for example, Geoffrey B. Frasz, "Environmental Virtue Ethics: A New Direction for Environmental Ethics," *Environmental Ethics* 15, no. 3 (Fall 1993): 259–74; Rosalind Hursthouse, "Environmental Virtue Ethics," in *Working Virtue: Virtue Ethics and Contemporary Moral Problems*, ed. Rebecca L. Walker and Philip J. Ivanhoe (Oxford: Clarendon, 2007), 155–72; Simon P. James, "For the Sake of a Stone? Inanimate Things and the Demands of Morality," *Inquiry* 54, no. 4 (2011): 384–97; Ronald L. Sandler, "Environmental Virtue Ethics," in *International Encyclopedia of Ethics*, ed. Hugh LaFollette (Blackwell Publishing Ltd., February 1, 2013), 1665–74.
37. Brian Patrick Green, "Convergences in the Ethics of Space Exploration," in *Astrobiology: The Social and Conceptual Issues*, ed. Kelly C. Smith and Carlos Mariscal (Oxford: Oxford University Press, 2020), 179–96.
38. See, for example, Mark Bekoff and Jessica Pierce, *Wild Justice: The Moral Lives of Animals* (Chicago: University of Chicago Press, 2009).
39. Brian Tierney, *The Idea of Natural Rights: Studies on Natural Rights, Natural Law, and Church Law 1150–1625* (Grand Rapids, MI: Eerdmans, 1997); and Thomas Mautner, "How Rights Became 'Subjective,'" *Ratio Juris* 26, no. 1 (March 2013): 111–33. Mautner presents a thorough history, but critiques the idea of "subjective rights" as redundant.
40. Keith A. Abney, personal correspondence, February 10, 2021.
41. See, for example, Jack Donnelly, *Universal Human Rights in Theory and Practice*, second edition (Ithaca, NY: Cornell University Press, 2003), 30–31; and Henry Shue, *Basic Rights: Subsistence, Affluence, and U.S. Foreign Policy*, second edition (Princeton, NJ: Princeton University Press, 1980), 35–64, 153–55.
42. Mary Wollstonecraft, *A Vindication of the Rights of Woman: With Strictures on Political and Moral Subjects* (London: J. Johnson, 1796).
43. Kimberlé Crenshaw, "Demarginalizing the Intersection of Race and Sex: A Black Feminist Critique of Antidiscrimination Doctrine, Feminist Theory and Antiracist Politics," *University of Chicago Legal Forum* (1989): 139–67.
44. See, for example, Kelly C. Smith, "The Trouble with Intrinsic Value: An Ethical Primer for Astrobiology," in *Exploring the Origin, Extent, and Future of Life: Philosophical, Ethical, and Theological Perspectives*, ed. Constance M. Bertka (Cambridge: Cambridge University Press, 2009), 261–80; and Charles S. Cockell, "The Ethical Status of Microbial Life on Earth and Elsewhere: In Defense of Intrinsic Value," in *The Ethics of Space Exploration*, ed. Schwartz and Milligan, 167–80.
45. Eugene C. Hargrove, ed., *Beyond Spaceship Earth: Environmental Ethics and the Solar System* (San Francisco: Sierra Club Books, 1986); Holmes Rolston III, "The Preservation of Natural Value in the Solar System," in *Beyond Spaceship Earth*, ed. Hargrove, 140–82; Christopher P. McKay, "Does Mars Have Rights? An Approach to

the Environmental Ethics of Planetary Engineering," in *Moral Expertise*, ed. D. MacNiven (New York: Routledge, 1990), 184–97; Christopher P. McKay, "Planetary Ecosynthesis on Mars: Restoration Ecology and Environmental Ethics," in *Exploring the Origin, Extent, and Future of Life*, ed. Bertka, 245–60; Kelly C. Smith, "Manifest Complexity: A Foundational Ethic for Astrobiology?" *Space Policy* 30, no. 4 (November 2014): 209–14; Kim McQuaid, "Earthly Environmentalism and the Space Exploration Movement, 1960–1990: A Study in Irresolution," *Space Policy* 26 (2010): 163–73; Alan Marshall, "Ethics and the Extraterrestrial Environment," *Journal of Applied Philosophy* 10 (1993): 227–36; Erik Persson, "The Moral Status of Extraterrestrial Life," *Astrobiology* 12 (2012): 976–84; and Mark Williamson, "Space Ethics and Protection of the Space Environment," *Space Policy* 19 (2003): 47–52; and so on.

46. Christopher P. McKay, "The Search for a Second Genesis of Life in Our Solar System," in *First Steps in the Origin of Life in the Universe*, ed. J. Chela-Flores, T. Owen, and F. Raulin (Dordrecht: Springer, 2001), 269–77.

47. Michael N. Mautner, "Life-Centered Ethics, and the Human Future in Space," *Bioethics* 23 (2009): 433–40.

48. Mark Lupisella, *Cosmological Theories of Value: Science, Philosophy, and Meaning in Cosmic Evolution* (Cham, Switzerland: Springer, 2020); Mark Lupisella, "Cosmological Theories of Value: Relationalism and Connectedness as Foundations for Cosmic Creativity," in *The Ethics of Space Exploration*, ed. Schwartz and Milligan, 75–92; Mark Lupisella and John Logsdon, "Do We Need a Cosmocentric Ethic?" paper IAA-97-IAA.9.2.09, presented at the International Astronautical Federation Congress. American Institute of Aeronautics and Astronautics, Turin, Italy, 1997, 1–9; and Anna Frammartino Wilks, "Kantian Foundations for a Cosmocentric Ethic," in *The Ethics of Space Exploration*, ed. Schwartz and Milligan, 181–94.

49. James M. Gustafson, *Ethics from a Theocentric Perspective: Theology and Ethics* (Chicago: University of Chicago Press, 1983).

50. Jonas, *The Imperative of Responsibility*, 10.

51. Ibid., 139.

52. Ibid., 1.

*Chapter Three*

# Risk and Safety

**Figure 3.1. The Space Shuttle *Challenger* disintegrates seventy-three seconds into flight after the failure of an O-ring seal in one of its solid rocket boosters on January 28, 1986.**

## THE SPACE SHUTTLE *CHALLENGER* DISASTER

At 11:38 a.m. on January 28, 1986, the Space Shuttle *Challenger* lifted off from Cape Canaveral in Florida. The day before, the entire solid rocket booster engineering team had agreed that the launch should not proceed because the temperature was too low to ensure pliability in the solid rocket booster O-rings. If the O-rings were insufficiently pliable, then hot exhaust gases could escape between the joints in the booster segments, leading to catastrophic failure. Despite the booster team's unanimity, NASA leadership was determined to launch, with or without the team's approval. They found someone to approve the launch despite the team's objections.

The space shuttle did not explode upon booster ignition, as some on the engineering team had worried. Instead a slow leak of exhaust gases burned through the side of the booster, and, seventy-three seconds into flight, the solid rocket booster catastrophically failed, blowing the entire space shuttle assembly into pieces. The seven astronauts aboard the shuttle, including the first teacher in space, all perished. Millions of schoolchildren watched the disaster live on television, including the author of this book.

The Space Shuttle *Challenger* disaster was preventable, just as so many disasters are. But despite the potential for prevention, space exploration, due to the energies and environments involved, will always remain extremely hazardous to human life. This chapter will explore these dangers, and how we ought to think about them ethically.

## RATIONALE AND SIGNIFICANCE: THE RISK OF SPACE EXPLORATION

Space exploration is an intrinsically risky activity. The amount of energy required to get off of the surface of the Earth and into orbit is staggering, the space environment itself is deadly, and human bodies are simply not capable of withstanding these forces without robust technology protecting them. When this technology fails, people die, and because no technology is perfect, in the future these technologies will fail again and people will die again. And yet, despite this risk, there are many people who think that going into space is something that humans should do, and they are willing to take those risks upon themselves.

Space exploration relies upon blending complex scientific understanding with highly sophisticated technology. Technical issues, as well as particular managerial, engineering, and operator decisions, coupled with the realities of the dangers of the space environment, cause many safety concerns worthy of ethical reflection, as the case of the *Challenger* exemplifies.

This chapter will focus on acute, immediate, and short-term safety and risk concerns, such as those presented by abrupt failures of technology, while chapter 4 will focus on longer-term or chronic health risks such as those presented by exposure to radiation and microgravity.

## ETHICAL QUESTIONS AND PROBLEMS

### What is safety?

**Safety** is an individual and collective perception about what is not dangerous and what is acceptable in terms of health and harm. The social conception of safety is collective social decision, also called a social construction.[1] That does not mean it is not real or merely arbitrary, far from it: the concept of safety is just as real as the concepts of money and government, which are also social constructions. A social construction just means that it is a social judgment based on the changing contexts of the human social and natural environment.

A social judgment means that no individual person is allowed to make the decision for everyone. Instead judgments of safety are made by society: the combination of engineers, businesspeople, lawyers and judges, politicians and lawmakers, and of course individual members of the public.

Judgments of what qualifies as adequate safety can change over time as willingness or unwillingness to take risks—**risk aversion**—wax and wane. For example, travel in all forms used to be much less safe than it is now. And yet many people still traveled—it was not considered to be unacceptably safe. But how many people today would be willing to go back in time to join a sixteenth-century ocean voyage knowing the deadly risk one would be taking? However, we know some today who while perhaps unwilling to take such risks if transposed into the past would be willing to travel between objects in space and settle truly "new" worlds. Safety is not an objective judgment and notions of safety are highly context dependent. Some people find riding motorcycles to be an acceptable risk and others do not. Some people find smoking to be an acceptable risk, while others do not. And yet society tries to influence these decisions through sometimes requiring helmets and/or taxing tobacco products. There are many societal responses to risk ranging from banning certain risks (for example, private ownership of military weaponry) to mandating others (for example, children must go to school, thus exposing them to the risks of the outside world such as bullying and viruses).

## What is risk?

As defined by Mike W. Martin and Roland Schinzinger, "**Risk** is the probability that something unwanted or harmful may occur."[2] It is often expressed as an equation, known as the **risk equation**, where risk = harm × probability. Risks are typically taken for the sake of some opportunity: the probability that something wanted or good might occur.

While the risk equation might look like a straightforward method of assessing risk, it is actually quite complex. One problem is that getting numbers that adequately quantify harm and probability is not always possible. Indeed, especially with new technologies and activities, there may be no reliable way of determining probabilities of failure, and with certain types of failure, the consequences may range from insignificant to catastrophic. How does one quantify a harm?[3] How does one assess probabilities for untested new technologies? Often harm is reduced to monetary cost, thus allowing a number to be placed into an equation, but how does one quantify the worth of the fruit of exploration or a human life?

Another reason is that not everyone evaluates risk in the same way. As noted earlier, to some people some activities seem risky, to others not. Risk can be measured in some ways (for example, known rates of failure) but not in others (for example, how much one values one's own safety versus achieving a risky aim), and so there is always an element of the personal when it comes to risk, thus making informed consent all the more important. Individuals do not need to agree on risks, but societies together do need to come to some level of agreement so that reasonable social expectations are set and met.

Life is intrinsically risky. Every day we face risks that may damage our health or kill us, for example, living in a location that is subject to earthquakes, being exposed to environmental pollutants, driving an automobile, or eating unhealthy foods, yet we do them anyway. So the first ethical question with regard to risk is: Beyond the inescapable risks, what additional risks should we take?

## When is it legitimate to elect to take a risk beyond that which is "normal"?

If risk is intrinsic to life, then we often may not have a choice about what risks to take. Humans have always been exposed to natural risks such as those presented by accidents, natural disasters, and disease. We have also always been exposed to risks associated with other humans, including human-associated accidents, neglect, emotional and physical abuse, interpersonal violence, and organized violence. There is nowhere on Earth (or off it) that is completely safe from all risks. Above this baseline of risk, however, there come risks that we can choose or reject voluntarily, such as choosing to drive a vehicle and then driving recklessly or taking up a risky hobby such as high-altitude mountain climbing. There is an ethical distinction between normal risk and elective risk.

Typically people voluntarily take risks upon themselves in order to receive some sort of benefit. For example, people drive automobiles in order to have freedom of movement, to be able to get to places that they want to go on their own schedules. The benefit—freedom of movement—seems to be worth the risk—breakdown or an accident resulting in injury or death. Crucial to this judgment is the balance between the likelihood and the severity of the negative outcome versus the likelihood and the benefit of the positive outcome. Ways to think about determining these judgments will be discussed shortly, but in general, a risk is worth it if the probable benefit outweighs the probable negative outcome; that is, there is a high enough probability of a good outcome relative to a low enough probability of a harmful outcome. Note this echoes the risk equation but with consideration of benefits as well.

## What makes a risk worthy of being accepted and why? How do we decide and who gets to make these decisions?

Whether it is the near certainty of a slightly good outcome (such as traveling to visit a friend) or a low probability of a very good outcome (starting one's own business and becoming wealthy), these must be deemed worthy relative to their opposite and concomitant risks, namely, the small probability of a very bad outcome (an accident while traveling) or the fair probability of a rather bad outcome (the business fails and one has wasted time and effort and now must find other work). Many people will choose to travel to visit friends because they judge the risk to be acceptable, while many people will not choose to start their own business because they judge the risk to be not worth it. Notice that in these examples it is the decision makers themselves who are choosing to put themselves at risk. This is not always the case.

There are some types of risks that legitimate authorities do not think private individuals ought to take. When legitimate authorities intervene to make decisions on behalf of others it is called **parentalism**, and associated terms include such negative and gender-freighted words as "paternalism" and "nanny state." Parentalism is a more neutral term because we can generally think of both good and bad examples of parenting. Parents try to protect those in their care, but sometimes they are too protective and sometimes not enough. Contrarily, when authorities decide instead to let individuals make these choices for themselves, they are recognizing personal **autonomy**. The tension of **parentalism versus autonomy** is a major ethical consideration in engineering design, law, educations, health care, family life (of course), and many other forms of human interaction.[4]

For example, in many nations there are certain types of drugs that are illegal. While individuals might judge the risk of taking these drugs to be acceptable, lawmakers do not—these risks are therefore made much riskier by making them illegal on top of their already risky pharmacological effects on their users. Similarly, many places require motorcycles riders and/or bicyclists to wear helmets to protect their heads in the case of an accident. Helmets demonstrably lower the risks to these riders, yet many riders would prefer not to wear them, so other decision makers intervene to try to force this lower-risk behavior. Numerous other examples exist for this type of parentalistic decision making, including rules about smoking tobacco, sanitation, food safety, product safety requirements (for example, seat belts), building codes, etc. Notice that there can be no "blanket" judgments for or against parentalism or autonomy; these are social judgments that can legitimately vary between places and times depending on the specific circumstances.

### When are people legitimately allowed (or not) to accept risks for themselves or to foist risks upon others?

Besides the unavoidable risks of life and the regulated risks subject to social control, there is still much room for individual free choice. In societies that emphasize liberty, individuals are generally allowed to choose what risks to accept, with major exceptions such as in time of war when drafts may be instituted for the sake of collective safety or in times of pandemics when public health needs can override individual autonomy. The **libertarian** perspective, typified by John Stuart Mill in *On Liberty*, is "that the only purpose for which power can be rightfully exercised over any member of a civilised community, against his will, is to prevent harm to others. His own good, either physical or moral, is not a sufficient warrant."[5] But is individual danger really of no relevance to the rest of society? Clearly not, because harms very often ripple outward through society, hence attempts to control risk parentalistically.

In contrast, societies that emphasize community often are more collective and/or hierarchical in their decision making, and the group may choose whether to put individuals at risk for the sake of group benefit. The **utilitarian** perspective, typified by John Stuart Mill in his *Utilitarianism*, can actually be a balance against the libertarianism of *On Liberty*. In this perspective, happiness maximization takes priority, and therefore actions that offer that end should be encouraged, while those that endanger happiness should be avoided. The socialist perspective, often typified by cost-controlling welfare states, can illustrate this position because the social costs of risky behaviors are often subject to forms of control (as noted earlier in various restricted and banned behaviors) and particularly when it comes to national health care and those behaviors that can damage health (and thus drive up costs, which must be paid by the rest of society).

### Why should we spend money on space if such expenses also risk disasters, especially when that money is so needed for other programs here on Earth with more certainty of positive results?

Even if there are people willing to endanger their own lives for the sake of space exploration, perhaps society ought not let them because that pursuit harms the rest of us, even if it is just by misallocation of resources. Perhaps we can do much more to reduce risk by spending money on Earth rather than in space.

This is particularly relevant for "backup Earth" arguments against existential risk because it is, for example, almost certainly more cost-effective to reduce nuclear arsenals and other such actions than it is to maintain such weapons *and* start a backup Earth on another planet. The same for stopping climate change—it is certainly more cost-effective to mitigate future risk now than it would be to pay for the damages such future risks would bring.

And yet ethical judgment must always work with the reality it has, not the reality it would like to have. These actions to reduce existential risks are possible but have been tried for decades without success because powerful forces—governments and corporations—do not want these risks to be removed. They perceive the existence of these risks to actually be of benefit to them, or they believe that the coordination and trust problems associated with removing these risks are insurmountable. Given these realities, the more obvious option—to simply reduce risks on Earth and not move into space—is not a real option. It is an ideal we could work toward, indeed should work toward, but its implementation is currently vastly more difficult than even the settling of space.

## ETHICAL CONCEPTS AND TOOLS

### Informed Consent, Part I

One of the first ethical tools to consider when facing risky activities is informed consent. Informed consent is one of the major rules in biomedical ethics and legal ethics and can be found in such foundational documents as the 1947 Nuremberg Code and the 1979 Belmont Report.[6] More mundanely, the concept of informed consent can be found in every end-user license agreement (EULA) for smartphone apps and other computer programs and devices, as well as every legal contract. In brief, informed consent says that, under normal circumstances, all people (1) ought to fully know the risks of participating in an activity and (2) ought to fully consent to the activity before engaging in it or being subjected to it.

For example, informed consent provides a helpful perspective on the Space Shuttle *Challenger* disaster. All the astronauts knew that flying the space shuttle was not without risk. They knew that rockets could explode, there was no atmosphere to breathe in space, and a multitude of other risky aspects of spaceflight. By training for and agreeing to ride on the shuttle the astronauts seemed both to have full knowledge and to give full consent—but did they really? Because while the astronauts knew of the many general dangers of spaceflight, they did not know of the specific danger presented by the cold rubber O-rings crucial to the safety of the solid rocket boosters the way the engineers at Morton Thiokol knew them. These engineers wanted to delay the flight of the shuttle but were overruled. If the astronauts had known that Morton Thiokol's solid rocket booster engineers felt that it was unsafe to fly the shuttle under those particular circumstances, would they still have agreed to go? This lack of knowledge violated the principle of informed consent: the astronauts did not have full knowledge of the risks of the activity they were participating in and so their consent could not be said to have been fully informed.

### Risk Ethics, Part I: Coping with Probability and Uncertainty

Sociologist Charles Perrow coined the term **normal accident** or **system accident** for accidents that are based purely on the probabilities of failures of parts within complex systems. As he defines the term, "*normal accident* is meant to signify that, given the system characteristics, multiple and unexpected interactions of failures are inevitable."[7]

In 1996, author Malcolm Gladwell drew on Perrow's ideas and managed to predict a future space shuttle loss "for the most mundane of reasons." Indeed, in 2003 the Space Shuttle *Columbia* was lost upon reentry for the mundane reason of a block of foam hitting and compromising the leading edge of the orbiter's wing during ascent. Gladwell expressed that if we are not willing to take such risks, "then our only option is to start thinking about getting rid of things like space shuttles altogether."[8] He is indeed correct—spaceflight cannot be made completely safe; it will always include a risk of total failure. But if we are unwilling to give up on spaceflight, we can at least try to reduce the risk as much as humanly possible.

Conducting ethical analysis under the conditions of uncertainty has been called **risk ethics**, which combines risk analysis with ethical decision making.[9] But such a form of ethics is not actually new; ethical decision makers have always had to consider uncertainty, hence the inclusion of the criteria in such traditional ethical tools as cased-based analysis (as noted in the previous chapter) and in questions surrounding noncooperation with evil (chapter 6).

Moral uncertainty is a kind of uncertainty directly related to uncertainty about what good and bad are in a particular case: "uncertainty that stems not from uncertainty about descriptive matters, but about moral or evaluative matters."[10] One historical example of coping with specifically moral uncertainty in Western tradition is the **probabilism** debate in sixteenth-century Catholic moral theology. In this debate, there was a spectrum of positions on how to deal with moral uncertainty, ranging from the laxest position to the most rigorous, based on how many respectable opinions affirmed the position:

**minus probabilissimus:** one opinion affirms, but no others
**probabilism:** a few opinions affirm
**aequiprobabilism:** there are equal opinions on each side
**probabiliorism:** most opinions affirm
**tutiorism:** follow the safest (most morally conservative) option[11]

This scale of certainty about moral matters allowed for gradations in treatment of moral problems. For issues that were extremely grave, stricter options made more sense, but on issues that were less grave, laxer options might be permissible; note that this presages the risk equation in terms of harm and probability. It also connects to current debates on expert consensus, in which, for example, scientists may differ on how to most effectively respond to climate change, and policy makers must make decisions based on these differing interpretations. Note that in free societies today, very often expert scientists and technologists can accomplish something if they simply have the money and expertise to do it themselves: this reflects a *minus probabilissimus* approach to risk, as they only need themselves to approve their actions. This is good for innovation but a very weak approach to risk. Echoes of these historical positions can still be found in the approaches to risk described shortly.

In addition to moral uncertainty, there is also descriptive uncertainty (uncertainty about the facts of a case) and predictive uncertainty (uncertainty about what the future will bring), among others. Each of these uncertainties alone is difficult enough, but piled on top of each other the complexity can be quite difficult.

Currently there are several **risk principles** for dealing with the uncertainties posed by new technologies. The most well-known of these is the **precautionary principle**:

> The precautionary principle . . . relates to an approach to risk management whereby if there is the possibility that a given policy or action might cause harm to the public or the environment and if there is still no scientific consensus on the issue, the policy or action in question should not be pursued. Once more scientific information becomes available, the situation should be reviewed.[12]

The precautionary principle encapsulates some common pieces of folk wisdom such as "look before you leap" and "better safe than sorry." It shifts away from older approaches to damage control that emphasized civil liability and toward anticipatory damage control. It is suitable when damages are potentially massive and/or irreversible or when offending parties are unlikely to be held accountable but can be prevented from acting beforehand.[13]

The **prevention principle** takes a more cautious approach than the precautionary principle but specifically relates to situations with more certainty of negative outcomes. It follows the general rule that "prevention is better than cure" and historically appeared soon after it became clear that cause and effect were occurring in new cases of risk, such as with pollution or new consumer products. This principle is generally uncontroversial because cause and effect are clear and certain; it is when it moves toward uncertainty that more controversy appears, and the precautionary principle must be invoked instead.[14]

The **polluter pays principle** is an approach to risk that permits risk-taking behavior and then, if something goes wrong, requires restitution from those who created the problem.[15] It is assumes that if risk takers make a mistake they will then make restitution for it. This principle makes sense for situations in which harms can be tolerated and the agents of that harm held accountable. Environmental dumping is a classic example of this principle; if the dumping does not cause harm, then the principle does not come into play, but if it does become harmful then the polluter is held accountable. This approach to risk is not suitable for truly disastrous and irreversible mistakes, nor for situations in which the polluter cannot be held accountable.

Another approach is that taken by the U.S. Presidential Commission for the Study of Bioethical Issues. This principle is known as **prudent vigilance**, which counsels going forward watchfully and only slowing down if definite troubles appear.

> Responsible stewardship calls for *prudent vigilance*, establishing processes for assessing likely benefits along with safety and security risks both before and after projects are undertaken. A responsible process will continue to evaluate safety and security as technologies develop and diffuse into public and private sectors. It will also include mechanisms for limiting their use when indicated. Prudent vigilance does not demand extreme aversion to all risks. Not all safety and security questions can be definitively answered before projects begin, but prudent vigilance does call for ongoing evaluation of risks along with benefits. The iterative nature of this review is a key feature of responsible stewardship.[16]

Prudent vigilance makes sense for technologies that are relatively harmless in their effects, but for technologies with more catastrophic potential it would seem to be inadequate and more stringent principles called for.

Philosophers of technology Michael Davis and Hans Jonas have both advocated the **gambler's principle**, which counsels us "don't bet more than you can afford to lose"[17] and not to "play a '*va banque*' ["go for broke" or "all in"] game with humanity."[18] In other words, for any elective scenario with a finite probability of an intolerable outcome, the scenario should not be chosen. This applies particularly to catastrophic scenarios in which human extinction (an infinite loss) is at risk due to elective human activities and technologies. The gambler's principle resembles the prevention principle and precautionary principle in that it seeks to control damage before it occurs, but it hews more closely to the risk equation in that for any version of the risk equation in which the harm is intolerable, no finite probability, no matter how low or uncertain, can be acceptable. It therefore grants the strongest preventive tolerance, but only for intolerable disasters, not for more minor cases of pollution or harm to individuals. Its appropriate use is for preventing catastrophes. In this way, the gambler's principle takes proportionality seriously: the more harmful an action is, the more it ought to be prevented. With regard to global catastrophic and existential risks, this rule might be summarized as "reasonable gamblers should not risk infinite losses on a finite win, even if the probability of that loss is quite low."[19]

Last, the transhumanist philosopher Max More advocates a **proactionary principle** wherein technology is developed as quickly as possible because the direness of our current situation of assured individual death and eventual planetary environmental destruction means we must make all haste.[20] This principle is not widely held in a theoretical sense, that is, few ethicists would claim to follow it as their preferred perspective on risk; however, it is notably in practice among many who are pushing technology very quickly, such as technology developers and entrepreneurs (for example, Facebook's former motto of "move fast and break things"[21]). It can also be seen in such historical circumstances as World War II, in which the scientists and engineers of the Manhattan Project considered the military value of creating the atomic bomb to be worth the risk of igniting the atmosphere and destroying all life on the surface of the Earth.[22] In this case, the risk of Nazi Germany gaining the bomb without the Western Allies also having it was deemed so unacceptable that risking destroying the world was acceptable by comparison. While this is an extreme example, the proactionary principle is also implicit whenever a choice is made to take a risk, such as taking a new job, moving to a new town, or embarking on a journey.[23]

Importantly, implicit in each of these perspectives on technological risk is an opinion of the contrasting brightness of the future with the brightness of the now. More cautious principles see our current situation as fairly good and future changes as risking that good for something worse. They are optimistic about the present and pessimistic about the future, and in this way can tend to be **techno-conservative** or **techno-pessimistic** in their approach ("conservative" not necessarily in the political sense but in that they do not desire change). Less cautious principles see our current situation as inadequate or poor, while the future is very bright, so the sooner we get into this bright future the better. They are pessimistic about the present but optimistic about the future, and in this way tend to be **techno-optimistic** or **techno-progressive**, that is, seeking progress (not to be confused with techno-progressivism in the political sense, that is, political progressivism enabled by technology).

All of these principles can be seen as trying to balance protecting the goods that we already have with obtaining the future goods that we do not yet have. In this way they are trying to direct human actions and technological progress toward good and away from evil. This harkens back to classical ethical theory, where the first principle of practical reasoning is that "good is to be done and pursued, and evil is to be avoided,"[24] which has illuminated some ethical traditions to this day.[25] More recently in his book *Superintelligence*, Nick Bostrom has advocated for directed technological development that prioritizes technologies that he believes are more likely to help control future risks.[26] Likewise, technologist Bill Joy[27] and philosopher Leon Kass[28] have both considered whether certain types of technology ought to be relinquished or forbidden. The simple answer would seem to be to direct technological development toward technologies that facilitate "good" actions and away from technologies that facilitate "evil" actions,[29] but as we will see over the course of this book, such judgments (and, even more so, governance) are not easy.

It is worth remembering that ethical perfection cannot be computed, and there will always be some measure of uncertainty to our actions. Ethics is messy and typically does not allow us to optimize for every relevant moral value or good. Ethics is not a "solvable" problem, and there are several reasons why. For example, in the *Nicomachean Ethics* Aristotle implies that perfect practical wisdom would require infinite experience, and on this point he is correct: what seems right at this point in history cannot really be known without knowing its full context throughout time and space.[30] Similarly, in *Quaestiones Disputate de Veritate* Thomas Aquinas states that the reason that humans have a general "appetite" for good, and not for every specific instance of good, is that reality is too varied to allow for intelligence to specify all particular goods but only, rather, an abstract large class of "good" within which specification is possible based on context.[31] More recent thinkers such as the utilitarian Gustaf Arrhenius and computer scientist Peter Eckersley agree that consequentialism cannot maximize for all goods, and choosing one puts at risk the others, therefore an uncertain and flexible pluralism of goods is necessary.[32] Last, in literature we find the wizard Gandalf (perhaps speaking for J. R. R. Tolkien) saying that "even the very wise cannot see all ends," meaning that we must have uncertainty, as well as humility and mercy, in what we do now because we cannot foresee how our actions will affect the future.[33] Given these uncertainties, we can only do what seems to be correct now given our best information and reasoning abilities.

This is not to say that we should give up on ethics and do whatever we want, because there are certainly better and worse futures that we can clearly see in our choices now. The standards of ethics exist because they generally lead us toward better choices and outcomes. It is only to say that among the choices we imagine, we still might be missing some crucial detail that could lead to unexpected consequences later on. Stochasticism and lack of knowledge keep the universe always on unexpected paths, for better or for worse.

## Risk Ethics and Time Horizons

**Risk-benefit analysis** is a form of **cost-benefit analysis** that understands that costs may be not only determinate but also probabilistic.[34] Of importance when considering risk is its connection to consequentialism. Positive and negative consequences are the primary considerations in risk-benefit and cost-benefit analyses, though both can tend to reduce these ideas to quantities of capital. When goods are not so easily quantized, this simplification can lead to morally repellent outcomes, as in the case of the notoriously unsafe Ford Pinto.

In the early 1970s, when the Ford Motor Company was calculating whether to make the Pinto automobile any safer, the company decided that it would be monetarily worth it to sell the unsafe car because it would cost more money to fix the problem than it would likely need to pay in damages for deaths and injuries.[35] This was a bad decision for both ethical and business reasons. When investing in courses of action, primary costs are paid up front, while secondary costs come later on. Ford Motor ended up paying vastly more in damages than it would have taken to simply solve the problem in the first place, and as additional damage, its reputation will retain this mark indefinitely. This case painfully demonstrates (not just for Ford but for those killed or injured in Pinto accidents and their loved ones) that when making ethical decisions, both **short-** and **long-term consequences** should be considered. **Short-termism** tends to make decisions mostly with a short time horizon, while **long-termism** makes ethical decisions with a longer time horizon (such as the **seven generation sustainability rule**[36]). In general, better ethical decisions are made when one is thinking longer term, with exceptions for exigent emergencies: things that simply must be done now or there will be no future decisions to be made. Such emergencies can range from the individual scale ("I need food or I will starve") to the global ("We must win this war or we will all die"). Hans Jonas quotes the German playwright Bertholt Brecht on the importance of being able to think past the short term with this phrase: "first comes the feed, and then morality."[37] If ethics involves thinking long term, then, for the sake of helping everyone in society, individuals' short-term concerns should be reduced (for example, by making sure everyone has food, safe water, shelter, etc.) in order to make longer-term concerns more thinkable.

## Risk Tolerance at the Individual and Social Levels

Individuals tend to have an "accepted" level of risk that we return to when situations become more or less safe. This accepted level of risk is called **risk tolerance**. For example, if someone knows that their car has the best safety features, then they may drive more recklessly.[38] If we know that our car is very unsafe, we may drive more carefully to compensate. This behavior is called **risk compensation** and it generates **risk homeostasis**: the tendency to return to a set level of risk tolerance.[39]

With regard to space, safety and risk tolerances may vary with circumstances: government versus private spaceflight, human versus cargo-only spaceflight, military versus tourist spaceflight, etc. For example, with respect to cargo versus human risk tolerance, crewed missions have much more stringent safety requirements. This difference can be seen, for example, in the long approval process for space capsules such as SpaceX's Dragon, the cargo version of which was approved years ago, while the crewed version underwent years of further testing for safety.

In general, humans can mitigate or adapt to risk or both. **Risk mitigation** can be defined as "avoiding the unmanageable," in other words, reducing the likelihood of scenarios that are beyond control. **Risk adaptation**, on the other hand, can be defined as "managing the unavoidable," in other words, coping with those risks that are inescapable.[40] These terms are often found in literature on climate change but are broadly applicable across all risk literature. In general, it is easiest to remember that mitigation reduces risks (for example, trying to mitigate climate change by reducing greenhouse gas emissions), while adaptation tolerates risk and changes human behavior in order to deal with that risk (for example, building seawalls in response to sea level rise). With respect to space exploration, risk mitigations could include, for example, not trying overly risky missions that we are not capable of conducting safely, such as flying humans to Mars with current technology (this is technically possible, but the danger is significant because these technologies are still under development). Risk adaptations, on the other hand, consist of all the measures taken to reduce damage from radiation, system failures, nutrition, etc.

Considerations of global catastrophic risks and existential risks are also important when approaching the subject of risk. Risk is not only an individual phenomenon; entire societies can be put at risk and even potentially large regions of the world, the entire planet, or large swaths of space. Nick Bostrom and Milan Circovic define global catastrophic risks as ones that "might have the potential to inflict serious damage to human well-being on a global scale."[41] Bostrom defines existential risks as ones that "threaten the entire future of humanity."[42] These risk are real and are worthy of serious response.

These sorts of risks include both natural and anthropogenic risks. Natural risks originating on Earth include natural climate change, **oceanic anoxia**, pandemics, and large volcanic eruptions (including **supervolcanoes and flood basalt eruptions**). Natural risks of extraterrestrial origin include **asteroid** and **comet** impacts, gamma ray bursts and **supernovae**, **solar flares** and storms, and the gradual expansion of the Sun over geologic time scales.[43]

Anthropogenic risks are risks brought on by human technology and can be further divided into accidents or malicious attacks (in most cases, weaponized technologies used for malicious purposes are the riskiest). These include nuclear technologies, technologies for biological manipulation, computer control technology over **cyber-physical infrastructure** (physical infrastructure controlled by computers, such as power plants or rocket/missile control systems), industrial and chemical technologies inducing environmental degradation (including general **environmental toxicity**, climate change, **ocean acidification**, and oceanic anoxia), artificial intelligence technology, nanotechnology, **climate engineering and geoengineering** technologies, and space technologies (such as directing asteroids toward Earth).[44]

Hans Jonas's imperative of responsibility provides one form of response to global catastrophic and existential risks. As noted in the previous chapter, Jonas asserted that the only truly categorical imperative (the one ethical directive that is always in force) is that humans ought to exist because human existence is a precondition for ethics[45] (at least as far as we know[46]). Therefore we should take great care to protect the preconditions for our own existence and mitigate and adapt to risks as well as we can.

Because failure is inevitable when dealing with complex systems, it is wise to plan for **safe exits** from failures.[47] Safe exits provide ways to escape when technology fails or disaster strikes. For a simple example, large passenger ships carry lifeboats, even though they are very unlikely to need them, so that if the ship sinks passengers can safely disembark and avoid drowning. Likewise, passenger aircraft are equipped with emergency slides to facilitate exit after an accident, and military aircraft often have ejection seats for rapid escape in dire situations. For a more complex example, a typical fossil fuel power plant, when it loses power, will safely cease operation, in contrast to a nuclear power plant, which has a nuclear reactor needing active cooling even when the power plant is no longer functioning. A fossil fuel power plant has a natural safe exit, while a nuclear power plant requires much more careful planning to provide a safe shutdown scenario.

Shifting examples into space, some space launch systems have emergency separation capabilities for the passenger compartment in the case of disaster, rocketing the crew capsule away from the rest of the launch vehicle. Unfortunately, the Space Shuttle *Challenger* did not have these emergency escape systems; when the solid rocket booster attached to the shuttle failed, there was no safe exit. It is worth noting that a safe exit might have been possible: the crew compartment of the shuttle apparently survived long enough for some of the astronauts to deploy their personal emergency safety systems.[48] In other words, the crew did not die in the explosion but rather on impact with the ocean. If the crew compartment had been designed with a safe exit—such as a system to eject the whole compartment and deploy parachutes—perhaps the crew could have survived even such a horrendous disaster as *Challenger*.[49]

## Technology Ethics

To tie together this chapter, it might be helpful to think of risk in terms of the technology that has given us these risks. These new technological issues in ethics are appearing now because in the past humankind was involuntarily limited in its power due to our collective weakness. We simply could not do things because we lacked the power. Now we are growing in power and we should learn to be voluntarily restrained by our own judgment and ethics. Weakness is no longer a barrier against choosing disaster, and so we should build our own barricades against the disasters we could choose.[50]

In plain terms, if technology enhances our ability to act efficiently (as well as increasing our scope of action), then we should work to be efficient at good and inefficient at evil. If instead we do the reverse, we will create a terrible world.[51] So we should ask these questions:

> "How can technology help make good efficient?"
> "How can technology help make evil inefficient?"

It is no exaggeration to say that our answers to these questions will help to shape the future. Unfortunately, these questions are not asked often enough—but we can learn to ask them more.[52]

It is absolutely critical to note that technology is not value neutral; it is value laden. Every technology has a purpose for which it exists, and that purpose is inextricable from its nature. There are values embedded into every artificial object: a hammer is for hitting and assumes that hitting is good, which, with respect to nails, it often is. Technology is a form of power and it is therefore intrinsically related to ethics. Technology must serve humanity, not the reverse.[53]

## BACK IN THE BOX

What do these space-associated risks tell us about our own lives here on Earth? Do we take too many risks or too few? Or just different kinds of risks?

Unlike in the harshness of space, here on Earth we have a very robust and fertile ecosphere, with immense quantities of breathable air, abundant water resources, many sources of food, access to various energy resources, and so on. Given the availability of these things on Earth, the harsh desolation of space might make us want to reconsider how often we unwisely squander these resources and even outright pollute and destroy them. Such abundance as we are used to on Earth is utterly rare in space. Perhaps the experience of space will help us to better understand the importance of preserving our home planet and help us to reconsider our unwise activities causing environmental damage, wasting resources, use of resources for weaponry instead of human development or risk reductions, lack of direction in technological development toward mitigating risks, and so on.

Contrarily, where might thinking about the risks of space exploration and settlement indicate that we take too few risks? For example, much of Earth is quite uninhabited and yet still vastly easier to explore, develop, and settle than space (for example, remote places such as the oceans or Antarctica). Analogously, perhaps it means that we should risk more on the development of risky or dual-use but highly beneficial technologies such as artificial intelligence or genetically modified organisms. Overall, this becomes a question of when to take a risk and what makes something worthy of taking a risk. As we have been discussing, the benefit should be deemed worth it by those put at risk, thus indicating the need for informed consent, even at a societal level. This concept of **societal informed consent** has generated a small but important literature and will be examined more in later chapters.[54]

## CLOSING CASE

As a counterexample to the "spaceflight is too risky" narrative, we can ask if there are risks to not exploring space as, for example, Elon Musk has stated.[55] The Earth will inevitably become uninhabitable as the Sun expands, but long before that, other disasters could make Earth-based civilization very difficult or impossible: nuclear war, rogue nanotechnology, use of bioweapons, unfriendly artificial intelligence, natural pandemics, supervolcanoes, asteroid strikes, and so on.[56] There is an existential risk to being a one-planet species. This could be solved by establishing a "backup Earth" where humankind and Earth life could continue to exist in the event of global catastrophe. Given this risk situation—that it is both risky not to settle space and risky to settle space—which risk should we take? The first option avoids the short-term and individual risks of space exploration but at the cost of risking the entirety of human civilization (as well as potentially every other living thing on Earth) in the long term—a probability that increases over time toward 100 percent. The second choice poses immediate threats to individual persons via repeated and terrible disasters associated with spaceflight and space settlement while offering a long-term reduction in existential risk; individual particular humans would risk their lives to preserve humanity as a whole. Individuals who choose to save many others while endangering themselves are often called heroes. For many reasons, this more heroic path is the one that I think humanity should take.

## DISCUSSION AND STUDY QUESTIONS

1. Should humans take these risks in order to explore space? Is space exploration and settlement worth it? Why or why not?
2. In your own mind, what would be the trade-off point between the risks of space exploration and the risks of not exploring space? For example, if the risk of space exploration and settlement was a 1 percent chance of one person's death each year for fifty years while the risk of remaining on Earth was a 1 percent chance of the collapse of civilization within fifty years, which risk would you choose? How might your mind change as the numbers change? Why?
3. Are there ways that we can make reliable systems out of unreliable parts? Think not only in terms of technological objects but also in terms of humans participating in decision-making systems.
4. Of the ethical concepts and tools discussed in this chapter, which do you find to be the most useful and why? Which are least useful and why?

## FURTHER READINGS

William MacAskill, Krister Bykvist, and Toby Ord, *Moral Uncertainty* (Oxford: Oxford University Press, 2020).
Mike W. Martin and Roland Schinzinger, *Introduction to Engineering Ethics*, second edition (New York: McGraw-Hill, 2010).
Toby Ord, *The Precipice: Existential Risk and the Future of Humanity* (New York: Hachette, 2020).
Charles Perrow, *Normal Accidents: Living with High Risk Technologies* (New York: Basic Books, 1984).
Shannon Vallor, Brian Green, and Irina Raicu, "Ethics in Technology Practice," Markkula Center for Applied Ethics, June 19, 2018. https://www.scu.edu/ethics-in-technology-practice/.

## NOTES

1. Carl Mitcham, "A Philosophical Inadequacy of Engineering," *The Monist* 92, no. 3 (2009): 349; and Mike W. Martin and Roland Schinzinger, *Introduction to Engineering Ethics*, second edition (New York: McGraw-Hill, 2010), 77–129.
2. Martin and Schinzinger, *Introduction to Engineering Ethics*, 108.
3. There are good tools for this, for example, Microsoft Corporation has developed a harms modeling tool specifically to try to prevent harms that can be generated by technology: Mira Lane, Josh Lovejoy, Arathi Sethumadhavan, Katherine Pratt, Neil Coles, Karen Chappell, and Harmony Mabrey, "Foundations of Assessing Harm," Microsoft Azure, May 18, 2020, https://docs.microsoft.com/en-us/azure/architecture/guide/responsible-innovation/harms-modeling/, and "Types of Harm," Microsoft Azure, May 18, 2020, https://docs.microsoft.com/en-us/azure/architecture/guide/responsible-innovation/harms-modeling/type-of-harm.
4. Jason Millar, "Technology as Moral Proxy: Autonomy and Paternalism by Design," IEEE International Symposium on Ethics in Engineering, Science, and Technology, Symposium Record, *IEEE Xplore*, May 2014.
5. Mill, Utilitarianism *and* On Liberty, 94–95.
6. "The Nuremberg Code," Trials of War Criminals before the Nuremberg Military Tribunals under Control Council Law No. 10, Nuremberg, October 1946–April 1949 (Washington, DC: U.S. Government Printing Office, 1949–1953), accessed December 30, 2020; https://www.ushmm.org/information/exhibitions/online-exhibitions/special-focus/doctors-trial/nuremberg-code, and the National Commission for the Protection of Human Subjects of Biomedical and Behavioral Research, *The Belmont Report*, Department of Health, Education and Welfare (DHEW) (Washington, DC: U.S. Government Printing Office, April 18, 1979), https://www.hhs.gov/ohrp/regulations-and-policy/belmont-report/read-the-belmont-report/index.html.
7. Charles Perrow, *Normal Accidents: Living with High Risk Technologies* (New York: Basic Books, 1984), 5.
8. Malcolm Gladwell, "Blowup," *The New Yorker*, January 22, 1996, http://gladwell.com/blowup/.
9. Neelke Doorn, "The Blind Spot in Risk Ethics: Managing Natural Hazards," *Risk Analysis* 35, no. 3 (2015); G. Ersdal and T. Aven, "Risk Informed Decision-Making and Its Ethical Basis," *Reliability Engineering and System Safety* 93, no. 2 (2008): 197–205; E. Vanem, "Ethics and Fundamental Principles of Risk Acceptance Criteria," *Safety Science* 50, no. 4 (2012): 958–67; Terje Aven, "On the Ethical Justification for the Use of Risk Acceptance Criteria," *Risk Analysis* 27, no. 2 (2007): 303–12.

10. William MacAskill, Krister Bykvist, and Toby Ord, *Moral Uncertainty* (Oxford: Oxford University Press, 2020), 1.

11. John Mahoney, *The Making of Moral Theology: A Study of the Roman Catholic Tradition* (Oxford: Clarendon, 1987), 135–43.

12. "Precautionary Principle," *Glossary of Summaries, EUR-Lex: Access to European Union Law*, accessed July 6, 2016, http://eur-lex.europa.eu/summary/glossary/precautionary_principle.html.

13. World Commission on the Ethics of Scientific Knowledge and Technology (COMEST), "The Precautionary Principle" (Paris: United Nations Educational, Scientific and Cultural Organization [UNESCO], 2005), 7–8.

14. Ibid.

15. Ibid.

16. Presidential Commission for the Study of Bioethical Issues, *New Directions: Ethics of Synthetic Biology and Emerging Technologies* (Washington, DC, December 2010), 27, 123, http://bioethics.gov/sites/default/files/PCSBI-Synthetic-Biology-Report-12.16.10_0.pdf.

17. Michael Davis, "Three Nuclear Disasters and a Hurricane," *Journal of Applied Ethics and Philosophy* 4 (August 2012): 8.

18. Jonas, *The Imperative of Responsibility*, 38.

19. Brian Patrick Green, "Emerging Technologies, Catastrophic Risks, and Ethics: Three Strategies for Reducing Risk," IEEE International Symposium on Ethics in Engineering, Science, and Technology, *ETHICS 2016 Symposium Record*, *IEEE Xplore*, Vancouver, British Columbia, May 13–14, 2016, http://ieeexplore.ieee.org/abstract/document/7560046/; Brian Patrick Green, "Little Prevention, Less Cure: Synthetic Biology, Existential Risk, and Ethics," Workshop on the Research Agendas in the Societal Aspects of Synthetic Biology, Arizona State University, https://cns.asu.edu/sites/default/files/greenp_synbiopaper_2014.pdf.

20. Max More, "The Proactionary Principle, Version 1.0," Extropy.org, 2004, accessed April 10, 2020, http://www.extropy.org/proactionaryprinciple.htm.

21. Hemant Taneja, "The Era of 'Move Fast and Break Things' Is Over," *Harvard Business Review*, January 22, 2019, https://hbr.org/2019/01/the-era-of-move-fast-and-break-things-is-over.

22. Emil Konopinski, Cloyd Margin, and Edward Teller, "Ignition of the Atmosphere with Nuclear Bombs," *Classified US Government Report*, declassified 1979, August 14, 1946, https://fas.org/sgp/othergov/doe/lanl/docs1/00329010.pdf; Daniel Ellsberg, "Risking Doomsday I: Atmospheric Ignition," in *The Doomsday Machine: Confessions of a Nuclear War Planner* (New York: Bloomsbury, 2017), 274–85.

23. Brian Patrick Green, "Six Approaches to Making Ethical Decisions in Cases of Uncertainty and Risk: The Principles of Prevention, Precaution, Prudent Vigilance, Polluter Pays, Gambler's, and Proaction," Markkula Center for Applied Ethics, November 14, 2019, https://www.scu.edu/ethics/focus-areas/technology-ethics/resources/six-approaches-to-making-ethical-decisions-in-cases-of-uncertainty-and-risk/.

24. Aquinas, *Summa Theologiae*, 1009.

25. Brian Patrick Green, "The Catholic Church and Technological Progress: Past, Present, and Future," *Religions* 8 (June 2017), 10, https://www.mdpi.com/2077-1444/8/6/106.

26. Nick Bostrom, *Superintelligence: Paths, Dangers, Strategies* (Oxford: Oxford University Press, 2014), 281.

27. Bill Joy, "Why the Future Doesn't Need Us," *Wired*, April 2000, http://archive.wired.com/wired/archive/8.04/joy_pr.html.

28. Leon R. Kass, "Forbidding Science: Some Beginning Reflections," *Science and Engineering Ethics* 15 (2009): 271–82.

29. Brian Patrick Green, "Are Science, Technology, and Engineering Now the Most Important Subjects for Ethics? Our Need to Respond," paper presented at the 2014 IEEE International Symposium on Ethics in Engineering, Science, and Technology, Chicago, Illinois, USA, *IEEE Xplore*, May 23–24, 2014; and Green, "Emerging Technologies, Catastrophic Risks, and Ethics."

30. Aristotle, *Nicomachean Ethics*, trans. W. D. Ross, in *The Basic Works of Aristotle*, ed. Richard McKeon (New York: Random House, 1941), 1027–30 (lines 1141a17–1142a31).

31. Thomas Aquinas, *Quaestiones Disputate de Veritate*, trans. Robert W. Mulligan, James V. McGlynn, and Robert W. Schmidt, html edition, ed. Joseph Kenny (Chicago: Henry Regnery Company, 1952–54), 25.1 (Responso, para. 4), accessed December 29, 2020, https://isidore.co/aquinas/english/QDdeVer25.htm.

32. Gustaf Arrhenius, "An Impossibility Theorem for Welfarist Axiologies," *Economics and Philosophy* 16 (2000), 247–66; and Peter Eckersley, "Impossibility and Uncertainty Theorems in AI Value Alignment or Why Your AGI Should Not Have a Utility Function," arXiv.org, 2019, accessed December 29, 2020, https://arxiv.org/abs/1901.00064.

33. J. R. R. Tolkien, *The Fellowship of the Ring* (New York: Houghton Mifflin, 2004 [1954]), 74.

34. Martin and Schinzinger, *Introduction to Engineering Ethics*, 118–19.

35. Ibid., 58.

36. A rule attributed to the Constitution of the Iroquois Confederacy that states that the living generation ought to be concerned for the welfare of future generations, stating, for example, "Look and listen for the welfare of the whole people and have always in view not only the present but also the coming generations, even those whose faces are yet beneath the surface of the ground—the unborn of the future Nation." In "The Constitution of the Iroquois Nations," IndigenousPeople.net, August 5, 2016, http://www.indigenouspeople.net/iroqcon.htm.

37. Jonas, *Imperative of Responsibility*, 172, citing Bertholt Brecht, "The Threepenny Opera," 1928.

38. Sam Peltzman, "The Effects of Automobile Safety Regulation," *Journal of Political Economy* 83, no. 4 (August 1975): 677–726.

39. Gerald J. S. Wilde, "Risk Homeostasis Theory: An Overview," *Injury Prevention* 4 (1998): 89–91.

40. Robin Mearns and Andrew Norton, "Equity and Vulnerability in a Warming World: Introduction and Overview," in *Social Dimensions of Climate Change: Equity and Vulnerability in a Warming World*, ed. Robin Mearns and Andrew Norton (Washington, DC: World Bank, 2010), 8, citing the Scientific Expert Group on Climate Change, *Confronting Climate Change: Avoiding the Unmanageable and Managing the Unavoidable*, Report to the United Nations Commission on Sustainable Development (Research Triangle, NC, and Washington, DC: Sigma Xi and the United Nations Foundation, 2007).

41. Nick Bostrom and Milan M. Circovic, "Introduction," in *Global Catastrophic Risks*, ed. Nick Bostrom and Milan M. Circovic (Oxford: Oxford University Press, 2008).

42. Nick Bostrom, "Existential Risk Prevention as Global Priority," *Global Policy* 4 (February 2013): 15.

43. Green, "Emerging Technologies, Catastrophic Risks, and Ethics," 1–2.

44. Ibid.

45. Jonas, *Imperative of Responsibility*, 43.

46. After all, other ethically capable beings may exist, whether animals, extraterrestrial intelligences, artificial intelligences, unknown intelligences, and so on.

47. Martin and Schinzinger, *Introduction to Engineering Ethics*, 127–28.

48. Mike Mullane, *Riding Rockets: The Outrageous Tales of a Space Shuttle Astronaut* (New York: Simon & Schuster, 2006), 245; Joseph P. Kerwin, "Letter, Joseph P. Kerwin to Richard H. Truly," July 28, 1986, https://history.nasa.gov/kerwin.html; Jay Barbree, "Chapter 5: An Eternity of Descent: Evidence Hints That Astronauts Were Alive During Fall," *NBC News*, January 1997, accessed April 16, 2020, http://www.nbcnews.com/id/3078062#.XpiDw_1Kjcs.

49. *Report of the Presidential Commission on the Space Shuttle Challenger Accident* (Rogers Report), "Chapter IX: Other Safety Considerations" (Washington, DC: U.S. Government, June 6, 1986); https://history.nasa.gov/rogersrep/v1ch9.htm; and "Implementation of the Presidential Commission Recommendations: Recommendation VII" (Washington, DC: U.S. Government, June 1987), https://history.nasa.gov/rogersrep/v6ch6.htm.

50. Green, "The Catholic Church and Technological Progress," 10.

51. Brian Patrick Green, "The Technology of Holiness: A Response to Hava Tirosh-Samuelson," *Theology and Science* 16, no. 2 (2018): 223–28.

52. Brian Patrick Green, "Ethics Is More Important Than Technology," Markkula Center for Applied Ethics, August 10, 2020, https://www.scu.edu/ethics/all-about-ethics/ethics-is-more-important-than-technology/.

53. Shannon Vallor, Brian Green, Irina Raicu, "Ethics in Tech Practice: An Introductory Workshop," slide deck, Markkula Center for Applied Ethics, June 22, 2018, slide 14, https://www.scu.edu/media/ethics-center/technology-ethics/Updated-Sample-Slide-Deck-for-Ethics-in-Tech-Practice-materials.pdf.

54. Laurence R. Tancredi and Arthur J. Barsky, "Technology and Health Care Decision Making: Conceptualizing the Process for Societal Informed Consent," *Medical Care* 12, no. 10 (October 1974): 845–59, https://www.jstor.org/stable/3763494; see also concepts in Mitcham, "A Philosophical Inadequacy of Engineering," 351–52, referring to Mike W. Martin and Roland Schinzinger, *Ethics in Engineering*, fourth edition (New York: McGraw Hill, 2005), 92; see also Martin and Schinzinger, *Introduction to Engineering Ethics*, 77–103, esp. 81–84.

55. "One path is we stay on Earth forever, and then there will be some eventual extinction event. I do not have an immediate doomsday prophecy, but eventually, history suggests, there will be some doomsday event." Elon Musk, "Making Humans a Multi-Planetary Species," *New Space* 5, no. 2 (June 1, 2017): 46–61, https://www.liebertpub.com/doi/abs/10.1089/space.2017.29009.emu?journalCode=space.

56. Toby Ord, *The Precipice: Existential Risk and the Future of Humanity* (New York: Hachette, 2020).

*Chapter Four*

# Space and Human Health

**Figure 4.1. Identical twin astronauts Mark Kelly (*left*) and Scott Kelly (*right*) before Scott's nearly yearlong stay on the International Space Station**

## THE CASE OF THE TWO KELLYS

At Baikonur, Kazakhstan, on March 28, 2015 (March 27, GMT), American astronaut Scott Kelly was launched into space, along with two others, to reside on the International Space Station for nearly a year. Scott was an extremely unique astronaut in that he had an identical twin brother, Mark, who was also an astronaut. But Mark would not be on this flight; Mark would remain on Earth as the control subject in an experiment on the effects of long-term spaceflight on the human body.

After 340 days in orbit, Scott Kelly returned to Earth on March 2, 2016 (Baikonur time). Many categories of data had been collected over the course of the experiment and continued to be collected upon return to Earth: biochemical, cognitive, **epigenomic**, gene expression, immune function, **metabolomic**, **microbiome**, **proteomic**, physiological, and **telomeric**.[1] The study discovered that there were "significant changes" in many categories of data during spaceflight, including "changes in telomere length, gene regulation measured in both epigenetic and transcriptional data, gut microbiome composition, body weight, carotid artery dimensions, [eye health], and **serum** metabolites." Other metrics changed significantly upon return to Earth, including inflammation response, immune gene networks, and cognitive performance. There were even examples of long-term changes six months after returning to Earth, "including some genes' expression levels, increased DNA damage from chromosomal inversions, increased numbers of short telomeres, and attenuated cognitive function."[2]

The case of the two Kellys illustrates the significance of spaceflight on the human body. Spaceflight strongly affects many bodily systems, and the effects can last for months and perhaps even years.

## RATIONALE AND SIGNIFICANCE: SPACE IS BAD FOR YOUR HEALTH

Space is an extremely hostile environment for most life forms evolved for conditions on Earth. Beyond technical safety and decision making, even an otherwise very safe space mission could still yield poor health outcomes, whether astronauts, tourists, explorers, or settlers. This chapter, in contrast to chapter 3, will concentrate on long-term medical concerns for human health in space. Relatedly, it should also be noted that many organisms besides humans will be involved in space exploration, use, and settlement, and these other organisms can be expected to also face serious impediments to their flourishing.

## ETHICAL QUESTIONS AND PROBLEMS

### Is health a scientific, technological, or ethical question?

Often when doing ethics of technology there is a question of whether a particular problem is scientific, technological, or ethical in nature. For example, health problems in space concern all three categories. When an astronaut becomes ill, there is a scientific question as to the cause of the illness. There is a technological question of how to resolve the illness—either to cure it or control it—and how to prevent similar situations in the future. And there is an ethical question of whether we should put people in such extreme situations that risk their health in new and poorly understood ways. We should acknowledge from the beginning that all health problems in space are problems that we are choosing to create and then solve. There is currently no immediate *need* to go into space (nothing is dragging us there against our will), though there may still be an ethical reason to do it: an *ought*.

## What are the health dangers of space travel?

**Radiation exposure** is one of the most well-known and researched aspects of space travel. The Earth is a relatively low-radiation environment when compared to much of the universe. On Earth we are protected not only by our thick atmosphere but also by the planet's **magnetosphere**, which diverts **ionizing radiation** into specific areas such as the atmosphere near the Earth's poles (causing auroras) and the **Van Allen radiation belts**.

Space is full of electromagnetic radiation from one end of the spectrum to the other, and it also contains **cosmic rays** (atomic nuclei traveling at a significant fraction of the speed of light) that can be very dangerous to human health. The more intense forms of radiation are called ionizing radiation, named so because they are capable of stripping electrons from atoms. When this type of radiation hits human tissue, it can harm important biological molecules. In particular, DNA, if hit by intense ionizing radiation, can sustain damage that can lead to mutations and/or cell death. While on Earth we might worry about getting sunburned (a radiation burn from ultraviolet rays) or getting cataracts in our eyes, we do not have to worry as much about cosmic rays or others of the most high-energy forms of radiation found in space, even though some do make it to the surface. Because humans evolved on Earth, we have repair mechanisms that can fix much of the damage caused by normal amounts of radiation exposure on Earth's surface. However, in space these higher energy forms of radiation, in large doses, can lead to radiation poisoning and death.

Radiation exposure can cause both acute and chronic health concerns, and both are problems in space. There are certain places in space where the radiation flux is intense enough to kill humans in relatively short order; for example, Jupiter has intense radiation belts that would make human missions to the inner **Galilean moons** essentially impossible. Earth also has its own radiation belts—the Van Allen Belts—that are dangerous to human health but are not acutely lethal except under conditions of prolonged exposure or a **geomagnetic storm**.

There are indications that even a few days of radiation exposure beyond **low Earth orbit (LEO)**[3] can have lasting and dramatic consequences for cardiovascular health. Delp et al. claim that the Apollo astronauts have died at a four to five times higher rate of cardiovascular disease than astronauts who stayed on Earth or who never left LEO[4] —although these findings are disputed.[5] This unexpected finding may indicate not only that space is dangerous but also that it is dangerous in unexpected ways, and the controversy over these findings themselves adds to the uncertainty of the situation. Thus, while there is much that we do know about the effects of space on human physiology, there is also much that we still do not know. Therefore we should plan for large margins of error in our risk assessments and exercise proportionate caution.

**Microgravity**, colloquially called "weightlessness," is one of the first things that people often think of when they think of being in space. The "weightless" effect is caused by being in a state of free fall while in orbit. There is still gravity, but it is balanced due to "falling" at orbital velocity so that the net force feels like zero. Microgravity is associated with many health problems, some short and some long term. Short-term effects include **space adaptation syndrome or "space sickness,"** a condition related to motion sickness that involves nausea induced by the effects of microgravity upon the vestibular system and is typically associated with the first few days of transition into space. Other short-term effects include "moon face" caused by the upward redistribution fluids within the body, increased eye pressure due to this fluid redistribution, etc.

Space sickness may not seem like a very serious medical condition, but it can become one in extreme cases. The most extreme case of space sickness observed to date was that of U.S. senator Jake Garn who went to space as a congressional observer in 1985. Despite being an accomplished military pilot, Garn experienced severe space sickness for his entire space shuttle mission, rendering him unable to perform his assigned duties.[6]

Long-term effects of microgravity include muscle, bone, and cardiovascular deterioration. Needless to say, these are very serious, and much is done to try to prevent them, such as regular exercise; however, even with these measures, long-term astronauts typically return to Earth weakened and in need of both medical and physical support.

Additionally, there are relatively minor yet strange effects of microgravity such as the sloughing off of calluses on the feet. Again, there is much still to learn about the physiological effects of microgravity upon humans and what measures may be taken to mitigate or adapt to the risks posed by those effects.

**Environmental control**, known more colloquially as "life support," consists of everything environmental that humans need in order to stay alive. As creatures evolved for the Earth's surface, humans require atmospheric conditions with certain specifications, including chemical composition, temperature, humidity, pressure, acceptable loads of microorganisms and pathogens, acceptable levels of allergens, etc. Each of these concerns has its own particular needs. For example, carbon dioxide scrubbers can keep oxygen and carbon dioxide in balance given sufficient resources, but depending on conditions and problems that may develop, resources could be cut short. The Apollo 13 mission is a classic example of this problem, when an unexpected malfunction in an oxygen tank caused it to explode and damage various spacecraft systems, leading to dangerous levels of carbon dioxide.

However, other aspects of environmental control are vital as well. For example, evaporated sweat and exhaled breath contain water, which may condense and collect in unexpected areas of the spacecraft, leading to possible problems such as electrical shorts or growth of microorganisms. This problem affected the Soviet/Russian space station Mir, where large amounts of condensation were discovered behind panels and were later found to contain potentially harmful microbial life.[7] The International Space Station is vulnerable to similar problems.[8]

Associated with environmental control are also a range of issues associated not just with maintaining life but also comfort. For example, while humans might be able to live in environments that are, relative to Earth's surface average, very hot or cold or very humid or dry, we have a preferred range for these environmental conditions. Outside of the preferred range, people can become stressed both physically and psychologically. Another example is that air does not flow in a spacecraft in the same way that it does on Earth. Heat does not rise under conditions of microgravity, nor do gases tend to move about except by passive diffusion, unless actively propelled with fans. This can have negative effects on carbon dioxide scrubbing and temperature control; ventilation is therefore very important for maintaining a healthy environment.

As the interior of a spacecraft must form a complete system, human metabolism must also be taken into account in environmental control. The food that we eat becomes carbon dioxide and water. Humans produce urine and feces that must be appropriately dealt with, whether through urine recycling or by sending it "overboard"—which in the cases of gases may just be venting to space, but solid waste requires more elaborate disposal, typically sending it to be burnt up upon reentry into Earth's atmosphere.

**Nutrition** is a vital concern for humans in space. Astronauts obviously need adequate nutrition, including proper quality and quantity of food, which must be very carefully planned ahead of time. Additionally, certain types of food are not suitable for consumption in free fall. For example, carbonated sodas are completely unsuitable for consumption in space: not only are they nutritionally empty, but they are also heavy, their gas content would unduly tax carbon dioxide scrubbers, and they would create gas problems in astronauts' stomachs. This is not a serious concern on Earth because on Earth gravity will separate the bubbles of gas from the liquid in a person's stomach and lead to burping, but in the absence of a strong gravitational field the gas bubbles and liquid do not separate, thus leading not to an urge to burp but to vomit, with associated health consequences.

Likewise because food residue and disposal is a problem for astronauts, food containing too much fiber is not appropriate. Heavy foods and foods without nutritional value are likewise not suitable for space as they are expensive to lift off of the Earth and therefore wasteful.

**Micrometeoroids** are a potential health problem for astronauts, though they tend to be exceedingly rare. Micrometeoroids may be quite small, but their very high velocity makes them dangerous projectiles. While the direct effects of a micrometeoroid on a human body would be obviously catastrophic, the effects of micrometeoroids on spacecraft are dangerous too. Vital equipment can be damaged and hull breaches can even occur (though depressurization would not necessarily be rapid, it would need to be patched quickly). The jet of gas from a hull breach also might act as a thruster on the spacecraft, thus pushing it off course or causing it to begin to tumble. Space debris and the threat it poses will be discussed further in chapter 5.

**Physical particle danger** is another problem in space. Apollo astronauts on the Moon noted, upon returning to their lander and removal of their helmets, that the dust stuck to their space suits was extremely fine, very clingy, and smelled like gunpowder.[9] While most astronauts experienced no negative effects from exposure to Moon dust, Apollo 17 astronaut Jack Schmitt developed "hay fever"–like symptoms and severe nasal congestion, which, however, dissipated rapidly and were not so severe with further exposures.[10] The Apollo missions give us data from the only direct human exposure to the material of another world. Bad smells and nasal congestion notwithstanding, the results were fairly mild, but we can envision worse scenarios in which dust, perhaps sharp glassy particles or filaments, injures astronauts' respiratory and/or digestive systems, skin, eyes, and other exposed areas or damages equipment.

**Chemical toxicity** is another concern in space travel. While spacecraft themselves are highly complex and carry potentially dangerous chemicals, other extraterrestrial surfaces may carry dangerous chemicals as well. For example, in 2008 the Mars Phoenix lander discovered very high levels of perchlorate on the Martian surface, and while perchlorate can be useful as rocket fuel (serving as a powerful oxidizer), it is also hazardous to human health, strongly inhibiting thyroid function, and as such has been used for decades as a medicine to treat hyperthyroidism. It is not currently known whether perchlorate will cause a problem for human exploration or settlement of Mars, however, the fact that perchlorate was only recently discovered, and that Mars represents only one of several extraterrestrial environments that humans have even begun to test, should indicate to us that there are many unknowns, and potentially dangerous ones, waiting to be discovered. If explorers had been sent to Mars with no knowledge of the perchlorate contamination of the surface, they could have quickly experienced hypothyroidism and, tens of millions of miles from Earth, would have no ability to diagnose it or determine the cause, much less treat it.

Contamination can come from both chemical and biological sources. With regard to planetary protection (chapter 8), both **forward and backward contamination** is considered, but with regard to human health it is mostly backward contamination that is of concern. Certainly, bringing biological or chemical contamination to a new environment from the Earth could be bad, but threats from Earth are mostly known, while threats from space could be largely unknown. Chemical contamination will certainly be an issue for explorations of Mars and possibly too for other bodies in the solar system. Biological contamination is another matter, however. The Moon appears to be completely lifeless, and likely so are many other moons, planets such as Mercury and Venus, and the asteroids. However, places such as Mars and the subsurface oceans of Europa and other large icy moons may harbor life, and possible future explorers to these places will need to take precautions against contamination.

**Sleep disturbance** is common among astronauts. The absence of gravity as well as lack of airflow, noise, exposure to cosmic rays (causing light flashes in closed eyes), and other phenomena generally reduce the amount of sleep that astronauts can get. This leads to numerous negative health effects associated with sleep deprivation, including physiological, cognitive, and psychological effects.

**Weakening of the immune system** is a serious problem for astronauts, who face it particularly upon return to Earth. While a weakened immune system might be of no concern were there no pathogens in the environment, this is wishful thinking, even in space. Everywhere that humans go we bring a microbiome with us containing trillions of microorganisms, some of which can exhibit pathological behavior under certain conditions. Additionally, other microorganisms will no doubt "stow away" on equipment and other materials on board the spacecraft, thus making it nearly certain that some types of pathogens will accompany astronauts. Immune system health is a major concern for spaceflight.

**Psychological stress** due to isolation, lack of social diversity, lack of freedom of movement, and other psychological stressors is a potentially serious problem for humans in space. While astronauts are currently very carefully selected, in the future if more people travel in space the numbers of psychological problems will no doubt rise, especially on long-term space missions.

There are many ideas for how to alleviate many of the health problems associated with spaceflight. For example, radiation shielding, ways to simulate gravity, means to improve nutrition, ways to reduce the dangers of toxicity and contamination, ways to reduce psychological stress, etc., all exist. However, they take knowledge, effort, and resources to implement.

Given these health concerns, we should finally ask the question of whether long-term human space exploration, use, and settlement can be both technically and ethically feasible. We can conclude that while it is very difficult, there are ways that it can be.

## ETHICAL CONCEPTS AND TOOLS

### Bioethics and Medical Ethics, Part I

Whenever we talk about human health we enter into one of the relatively more well-explored areas of applied ethics known as medical ethics or, more broadly, bioethics. Medical ethics has to do with ethical issues surrounding medical care, and bioethics expands upon this scope to additionally include biological research, biotechnology, genetics, the role of animals in research, and so on. Medical ethics might, for example, consider whether and when it could be legitimate to discontinue medical treatment for a terminal patient. Bioethics might consider, for example, whether it is right to alter the genes of an animal so that it will produce organs for human use.

In relation to space, bioethical and medical ethical issues are bound to appear. For example, would it be right for settlers in space to try to have children? Certainly, part of settling a place is to settle it permanently, with a reproducing population, rather than having to constantly rely on immigration. However, is the high-risk environment of space any kind of world to bring children into? Radiation damages DNA and could lead to birth defects. Low gravity or microgravity weakens bones and muscles and might make it impossible for humans born away from the Earth to ever return. Relatedly, would it be ethically legitimate to biologically manipulate human embryos so that they would be better adapted to space? These are huge ethical questions.

At the other end of life, there are many medical ethical issues surrounding death, such as the ethics of when to stop life-sustaining treatment, physician-assisted suicide, and euthanasia. In space, where medical resources will likely be in short supply, it is easy to envision possible scenarios in which treatments are impossible to maintain, where **triage** is necessary, or where chronically severely disabled patients might have others wish they were not a burden on the health care system. While these are analogous to problems we already face on Earth, the new environment of space may make these problems all the more complex and difficult.

## Parentalism versus Autonomy

As discussed in the previous chapter, there can be a tension between those who want to take risks upon themselves and those who want to control risks for others. For example, there comes a time in every growing child's life when they tell their parent, "I'm old enough to make my own decisions now!" and the parent must then decide whether the child is or is not correct. This marks the beginning of the transition from a status marked by external authority and external decision making by parents to one marked by internal and autonomous authority. Parents at this transition point can only hope that the child has developed proper decision-making capabilities.

This, in essence, is the parentalism versus autonomy question. It appears not only in familial situations but also in social and political calculations and in technological design, as in the difference between products that restrict freedom yet protect the user from failure, in contrast to products that allow the user to have more freedom yet may permit a user to make damaging choices.[11] Parentalism versus autonomy is a question of freedom in decision making. For those able to make good decisions for themselves, they should be allowed to do so. For those unable to make good decisions for themselves, whether because of immaturity, lack of intellectual ability, and so on, then someone should decide for them, with their best interest in mind. Of course, embedded in this debate is precisely the question of who can make a good decision and on what basis that is judged; this has been and remains a root of many kinds of ethical problems. For example, the history of democracy has acutely demonstrated this problem: it was formerly thought that only people of a certain status (race, sex, wealth, etc.) should be allowed to vote because only such people would be reasonable and competent to make political decisions; we now look back and find these criteria to be completely unethical. However, these types of restrictions still exist; for example, children are not allowed to vote, but in that case the "parentalism" argument is essentially made for them—children need parents to make decisions for them, in the child's best interest.

With respect to space, the parentalism versus autonomy tension remains an active one. As in the previous chapter, individuals are taking risks upon themselves, and these risks could conceivably burden society with their costs, even if (or perhaps particularly if) the harms are chronic rather than acute (for example, continuing medical bills, backward contamination from space, etc.). Can or should society or the government request, demand, or force people not to go into space for their own good? There are laws governing the exploration of space, such as the Outer Space Treaty, which already imposes a certain parentalistic imperative on nations to police their own citizens with regard to space exploration, but these laws and treaties are likely to develop and grow in complexity in the future. Deciding exactly how much autonomy can be allowed while still maintaining order and protecting the common good is a matter for serious ethical debate, especially because the possible harms of space exploration are so high.

## Informed Consent, Part II

Informed consent for acute risks is a slightly different matter than informed consent for more chronic health issues. It is clear that a spacecraft exploding or depressurizing is a serious problem that ought to be avoided because it immediately risks the survival of the crew and any mission they might be on. But long-term issues such as radiation exposure are different. For example, inside a properly constructed spacecraft, radiation issues will at least be tolerable, even while radiation damage may begin to slowly accumulate in a person's body. It may take years of chronic exposure for this damage to start producing negative effects. The person may be able to continue and complete their mission, but they might also become chronically or permanently disabled, or even terminally ill. And there are many long-term issues in space, as noted earlier, besides just radiation, which may all be active and interacting in unpredictable ways to harm human health.

Informed consent, then, becomes particularly difficult in this situation because while we know that harm is likely, we do not know the exact likelihood nor extent of these harms. Humans in space will know they are placing significant risks upon themselves, but they will not be able to have full information about those risks, thus making *truly informed* consent impossible. Instead, what is possible will be only a partially informed consent, with knowledge that serious harm is likely, but without precise knowledge of what exactly that harm might be.

## Positive and Negative Human Rights

Building upon the discussion of human rights in previous chapters, many philosophers have noted that there is a distinction between positive rights and negative rights. Negative rights are those conceived of as needing nothing external in order to protect and enable them—the rights holder simply needs to not be interfered with. Common negative rights are those such as freedom of speech, freedom of association, free exercise of religion, freedom of assembly, and so on. For each of these rights, the right is inherent and it requires outside interference to suppress. For example, people naturally have freedom of speech. Stopping a person from speaking requires active measures, such as covering their mouth, injuring them, coercively threatening them, forcing them somewhere they cannot be heard, etc. Another example would be freedom of movement. People naturally can walk from place to place or use other means of transportation, as available. In order to prevent people from ending up where they should not be, we have to build barriers, threaten them, physically stop them, etc.

Positive rights are those that require additional outside inputs in order to be exercised. For example, the right to a fair trial requires numerous governmental and social structures in order to function, including law enforcement officers and judges (and/or juries) who are competent and not corrupt, legislators and legislative bodies to create the laws to be enforced, social systems to place legislators where they are, widespread social values of fairness and impartiality, etc. Similarly, in some nations there are considered to be positive rights to safe food, drinking water, health care, education, and so on, which require massive social institutional structures that produce and distribute these resources, such as sufficiently advanced agriculture, reservoirs and pipelines for water, hospitals, and schools, and everything further that those institutions require.

Obviously positive rights are more complex than negative rights, but some scholars have noted that even negative rights may require positive actions and institutions to protect them. For example, if social forces conspire to suppress unpopular views, the government may be required to intervene in order to protect holders of those views from this discrimination. If social forces conspire to, for example, burn down a religious gathering place, likewise the government should actively protect the rights of this group.[12]

What of space travel? Is there a right to space travel (or a right to access to space), and is it more like a positive or a negative right? Could one justifiably be forced to or stopped from going into space? While we have freedom of movement here on Earth, we do not have utter freedom of movement. We cannot naturally place ourselves 100 kilometers underground any more than we can 100 kilometers above, and we cannot travel to any place we desire by any means that we desire either. At this time it cannot be said that there is a right to space travel; however, our right to freedom of movement may at some point in the future make such a right to space travel at least a relative possibility—not a right to go anywhere but a right to go places within reason. Freedom of movement in space would require immense expenditures of resources that could not reasonably be provided to all and so cannot be considered a right. If a government wanted to prevent its own citizens from going into space, it might cite these and related parentalistic concerns. On the other hand, there have been points in the past where governments have forced small or large numbers of people to move, whether through such legal means as imminent domain, through forced relocation (legal or otherwise), or as punishments for crimes (such as the forced settlement of Australia with British convicts). If a government wanted to offload convicts into a space colony, it would not be outside the scope of what humans have done before, despite the risk, expense, and ethical issues.

## Virtue Ethics, Part I

Virtue ethics is an important ethical approach when considering human health and activity in space. A virtue, as defined by Aristotle, is any human excellence. For example, Aristotle says that the excellence of the human eye is to see well. Analogously to the body, the human mind has excellences too, manifesting in various virtues of thought and character, such as wisdom, courage, temperance, prudence, diligence, and various particular skills such as sports or music. With respect to human activity, in Aristotle's philosophy, becoming strong in virtue and of good moral character seems to be *what humans are for*. An excellent human is one who displays the virtues, and a bad human is one who lacks virtues or, even worse, manifests vices.

When considering who should become an astronaut, the original astronauts and cosmonauts had to have what was colloquially known in the United States as "the right stuff." They had to be healthy, skilled, and brilliant, often being both tremendously skilled operators of vehicles (such as test pilots) as well as very academic (with doctoral degrees). Neil Armstrong and Buzz Aldrin, the first two men to walk on the moon, typified this level of excellence, being extremely skilled not only physically but also intellectually. They were virtuous, thanks to innate human talent refined by decades of hard work in the context of a society with opportunities to express that talent in heroic ways.

Space flight does have an element of heroism to it. Only the very most skilled and dedicated individuals are capable of it, and ultimately, while reaping rewards for themselves in the form of fame and adventure, they also put themselves at risk, including death, for the sake of pushing the boundaries of human capabilities. In this way they lend their lives to society for the sake of a greater good, even risking sacrificing their own lives for this good. They are in this way heroic and worthy of the appreciation and emulation of others in society.

Because spaceflight is a difficult task that facilitates the development of human virtues, it is ethically valuable just in itself. This presents a positive side to the risks discussed in this chapter and the last because overcoming risk for the sake of helping others, to push forward the boundaries of human knowledge, technology, and civilization, is a worthy task.

## Double-Effect Reasoning

Given that space exploration, then, both risks terrible harms and offers tremendous benefits, we should ask ourselves if there might be a way to weigh these benefits and burdens against each other. Certainly, utilitarianism is one way to do this, with complexity levels ranging from a simple cost-benefit analysis all the way to vastly detailed and complex analyses of actions into the far future.

However, utilitarianism and consequentialism in general do have some drawbacks, especially in relation to the intentions of the agent. Virtue ethics, in particular, is very concerned about the intentions and characters of the agents making decisions and performing actions. Therefore in order to help prevent the utilitarian calculus from sometimes encouraging straight barbarism (such as killing innocent people when proportionately beneficial) and thus damaging people's characters, thinkers in the tradition of virtue ethics developed tools for overcoming some of these drawbacks. **Double-effect reasoning** is one such tool, dating back over seven hundred years to the thinking of the philosopher and theologian Thomas Aquinas. Often found in bioethics, but also as a component of just war thinking, environmental impact studies, and elsewhere, the principle of double effect has four main parts.

1. *The action itself must be good or neutral*; the action must not be ***malum in se***, that is, evil in itself. An act that is evil in itself cannot be done in a right way. While this terminology seems old, and indeed is rooted in medieval Catholic ethics, *malum in se* is still used today in legal terminology to refer to acts such as murder, sexual assault, and torture. Those acts are *always wrong by definition*. Notice that this distinction arose in American foreign policy after the events of September 11, 2001—American officials made it U.S. policy to allow for "enhanced interrogation techniques" but could not call it "torture," an intrinsically evil act. (This is one vulnerability to ethical reasoning in general: people of bad will can attempt to redefine terms in order to make evil appear good.)
2. *The intention behind the action must be authentically good*. In other words, the objective of the action is the good effect, not the inextricable bad effect. Note that the negative effect can be *foreseen* but not *intended*. We know, for example, that taking certain kinds of powerful anticancer drugs will cause horrible side effects. These side effects are *foreseen* but not *intended*. Indeed, if those side effects could be eliminated, then they should be, which is one of the major ways that pharmaceuticals are improving over time. What would it mean to intend the bad effect of an act rather than the good? Consider surgery: a surgeon should do surgery for the sake of healing people, with the unfortunate side effect of also injuring them in the process, for example, cutting, sawing, cauterizing, etc. Surgery should not be performed because of the surgeon's secret sadistic glee in causing injury, with healing the patient as a mere side effect (for example, to make the surgeon's sadism legal). That would be habituating the surgeon to evil rather than good, and likewise with all other cases of double effect—the objective must be the good effect, not the evil one. Recognize that if you met a surgeon who stated that causing injury was their motivation for surgery, you would be justifiably disturbed.
3. *The action must be proportionately beneficial, generating more good than evil*. This step is similar to other forms of consequentialist ethics in that it merely seeks to maximize good and minimize harm. However, the other criteria of double-effect reasoning act as boundaries on what qualifies as acceptable proportionate reasoning. Proportionate reasoning is thus limited to only certain specific cases where it is less likely

to damage the character of those involved with deciding and carrying out an action. There is a common temptation to try to reduce all double-effect reasoning to just the proportionality criteria. However, while such a thing is possible, it is, then, no longer double-effect reasoning—it is consequentialism, with the concomitant risks to character.

4. *The evil of the act cannot be the means to the good of the act; rather, one action must produce two effects inextricably.* If the two effects of an action *can* be separated, for example, gaining the benefits of an anticancer drug without the harms, then these effects *should* be separated, for the sake of avoiding harming the patient. In other words, *the ends cannot justify the means*, that is, future good cannot be used to justify present evil. **Ends-justify-the-means reasoning** actually contains two actions: first, an evil one, which then, possibly, brings about a good one. A classic example of this is any use of innocent lives as bargaining chips or extortion for some "greater" good, for example, taking hostages in order to achieve some political end or destroying a city in order to shorten a war. Even if the good ends were ethically worthy, they should not be pursued by evil means because, notice, *there is no intrinsic reason for any of these means to actually bring about the desired end.* There is no direct logical reason that hostage taking, destruction, or other evil actually will cause the desired good effect: the good effect is a separate thing depending on separate decisions and actions by different people. One reason ends-justify-the-means reasoning is bad (in addition to rationalizing bad behavior) is because it is fundamentally bad reasoning—it relies on trying to manipulate others into producing a desired effect through an evil effect. But the manipulated other might just not cooperate, in which case the only thing that ends-justify-the-means reasoning achieves is a guaranteed evil now, with no possible good from it in the future. Double-effect reasoning insists upon strenuously avoiding these flaws. However, double-effect reasoning does involve inescapable evils because there is only one action, and the two effects, good and bad, are inextricable. If the effects could be extricated, then they should be, as any well-intentioned agent would agree. The distinction between ends-justify-the-means reasoning and double-effect reasoning may seem subtle, but it is quite significant.

It is worth asking whether these questions about character are actually important enough to warrant placing such delineations around consequentialism as are found in double-effect reasoning. Some of the examples discussed here have started to hint at that, but this should be made clear. *Humans* make ethical decisions. Actions do not appear out of thin air but rather from the decisions of agents, and agents tend to continue to act in ways in which they have acted previously: this is why we can designate some people as trustworthy, kind, clever, brave, etc. If we know that someone has a kind disposition, always helps others, and intends the best, then we know that that person is likely to continue to be this way, even if doing evil suddenly became permissible. (Note that to misjudge a person's character is a dangerous thing and we should expect that over millions of years of evolutionary history humans ought to be fairly skilled at it.) On the other hand, if you know that someone has evil intentions and is only acting in a beneficial way for some external reason—which may not always be there to control them—then you know that you cannot really trust that person. Wariness would be warranted, and if many people in society are like this, that lack of trust erodes society like acid.

Society utterly depends on trust: we trust banks to hold our money, we trust legal systems to uphold the law, we trust our neighbors not to harm us when we sleep, and we trust our friends or family. Losing these forms of trust destroys society, potentially quite rapidly. We have examples of this in history: when people run to the bank because they no longer trust it to have their money or when people flee nations as refugees because they no longer trust in their safety. Even in quite orderly societies, we can see a bit of this barbarian awareness when we see people who succeed by exploiting legal loopholes in order to enrich themselves at the expense of others. We know they cannot be trusted, and their success illustrates that something in society has gone deeply wrong, which then erodes trust. If people in society suspect that there are always others out there waiting "to get them," then society begins to degrade into a cold war, which sometimes even erupts into civil disturbances and worse.

Double effect, then, in its concern for character seeks to help maintain *trust* in society by maintaining *trustworthiness* among those tasked with making ethical decisions. The only way to really be *worthy of trust* is to literally be *trustworthy*: reliable, sincere, honest, having integrity.

Additionally, one could ask if double-effect reasoning is just a part of a consequentialism more broadly construed, so that it also includes consequences upon human character. The answer is *yes*, but if this is the case, then that form of consequentialism is called virtue ethics. Consequentialism and virtue ethics are both teleological forms of ethics in that they care particularly about the outcomes of decisions made and less on the inputs, such as rules. It is just that consequentialism in the narrower utilitarian sense is more concerned with material outcomes and effects on the well-being on sentient creatures, while virtue ethics includes among its consequences those that affect moral agents and their characters, including their intentions, goals, and purposes, and how those might be affected by performing evil actions. Every evil action performed facilitates future evil actions, and conversely, every good action performed facilitates future good actions. Therefore if consequentialism is expanded beyond merely material concerns and into the psychology and character of the agents performing the actions, then it becomes apparent that this more stringent standard is actually more likely to have good consequences in the end. Actions are always performed by agents, and the habitual dispositions of those agents to act in certain ways will influence what actions are performed in the future, for good or for ill.

Last, how does double-effect reasoning relate to space ethics? As mentioned earlier, spaceflight is risky. One action, which can be good (for example, gaining knowledge or making humanity safer against catastrophic risks), also intrinsically puts the lives of those same agents at risk. Many efforts are made to separate these inextricable possible effects from each other: safety systems and research and everything done for the sake of caring for astronaut health. But ultimately, space is dangerous. Humans did not evolve for the void; we cannot be made to be naturally at home there—we are suited to Earth. And therefore spaceflight will never be completely separated from its side effect of possible injury and death because spaceflight is just one action with inextricable risks and benefits.

## BACK IN THE BOX

Being alive is a terminal condition. We will die, whether on Earth or elsewhere. With the time we have, how, then, should we live? This is the fundamental question of ethics: *How should we live?* And more individually: *How should I live?* What is a good life? Who is a good person and why? What things will a good person have done during their life? And related to space: Might a good life include leaving Earth and doing good actions elsewhere?

These questions of a good life are inescapable. Many people do try to escape these questions, through work, distraction, and entertainment. If someone manages throughout their entire life to completely avoid ever thinking about what it means to be a good person, can they be said to have been a good person? In response to that we might remember that there are very good people in the world who do not know philosophical ethics at all (though they will have learned good behaviors from their parents, society, etc.). What role, then, does ethics play if one can be good without studying it? At the very least, the study of ethics allows for us to more fully understand our lives and face the ever-growing complexities of our world so that we can face new challenges better than if we had no understanding at all.

Which returns us to the complexities of life on Earth. What should we think about the standard risks some people accept on Earth, like driving a car, participating in a military, eating unhealthy food, smoking tobacco, drinking alcohol, engaging in risky sexual behavior, and so on? One response to the risks of life is to "eat, drink, and be merry," which is certainly what many humans have done throughout history (and perhaps the effects of that have led to our world being not as good as it could be). The more ethical approach is to take care of oneself in order to preserve one's own life long enough so that through skill and labor one might make the world a better place. To take a risk for no reward, or a purely selfish reward, is a foolish and short-sighted endeavor. A culture that valorizes such behavior will not last long. But to take a risk for the sake of benefitting oneself and others is a logical and even heroic purpose. A culture such as this will thrive and grow.

## CLOSING CASE

From the previous discussion it is obvious that humans are not at all well adapted to life in space. Therefore we might ask: Should humankind genetically manipulate itself to become better adapted to space? This is not a merely academic question; some scholars have asked it and answered in favor, while others are more cautious.[13] This "synthetic biology in space" question extends to the manipulation of other life forms as well,[14] but for this case we will limit ourselves to humans.

The case for genetic modification is strongest if its intent is to do good, especially for the purpose of alleviating medical conditions. For example, it might be the case that radiation in space will systematically harm the humans living there unless better DNA repair mechanisms can be developed for the human genome. These new genetic alleles could be inserted into embryos, resulting in new humans who might be more hardened to radiation. There are numerous other modifications that might be of use as well, such as resistance to the negative effects of microgravity, resistance to malnutrition (for example, the ability to innately produce necessary vitamins), resistance to low pressure or even vacuum, resistance to toxic chemicals, resistance to psychological traumas associated with space (limited social interaction, sleep deprivation, etc.), and so on.

However, there would of course be both technical and ethical risks as well. Ignoring the technical side, which might find that these problems are not easily tractable, the ethical side is fraught. This process would, for example, completely ignore informed consent, which could only be given as **proxy consent** by a parentalistically minded authority figure. The research process itself could easily harm or kill thousands of humans at various stages of development, not to mention the harms that might be incurred in the implementation phase. Even if nominally successful, engineered offspring could potentially face social stigma, perhaps not be able to interbreed with the rest of the human race, or even damage the genetic integrity of humanity as a species. In other words, if it worked, it might have benefits, but if it failed, it could be disastrous as well. Last, this adaptation of humans to space might make "regular" humans seem disabled relative to these enhanced individuals, and this differential suitedness to space could result in discrimination and exclusion.[15] In other words, even if it worked, the results could still be disastrous.

## DISCUSSION AND STUDY QUESTIONS

1. Do you think that the health risks of space exploration make it worth it or not? Why?
2. Obviously humans should be cautious about space exploration, but given that some dangers are inescapable, what is the best justification for taking these sorts of risks?
3. Thinking more multiculturally, how might other cultures respond to the health risks of space exploration, use, and settlement? What justifications might other cultures find to advocate taking these sorts of risks and what justifications for not taking these risks?
4. Of the ethical tools and concepts discussed in this chapter, which do you find to be the most useful and why? Which do you find to be the least useful and why?

## FURTHER READINGS

Marco Durante, "Space Radiation Protection: Destination Mars," *Life Sciences in Space Research* 1 (April 2014): 2–9, https://www.sciencedirect.com/science/article/abs/pii/S2214552414000042.

Francine E. Garrett-Bakelman et al., "The NASA Twins Study: A Multidimensional Analysis of a Year-Long Human Spaceflight," *Science* 364 (April 12, 2019): 1, https://science.sciencemag.org/content/364/6436/eaau8650.full.

Ann R. Kennedy, "Biological Effects of Space Radiation and Development of Effective Countermeasures," *Life Sciences in Space Research* 1 (April 2014): 10–43, https://www.sciencedirect.com/science/article/abs/pii/S2214552414000108.

David Williams, Andre Kuipers, Chiaki Mukai, and Robert Thirsk, "Acclimation During Space Flight: Effects on Human Physiology," *Canadian Medical Association Journal* 180, no. 13 (June 23, 2009): 1317–23, https://www.cmaj.ca/content/180/13/1317.short.

## NOTES

1. Francine E. Garrett-Bakelman et al., "The NASA Twins Study: A Multidimensional Analysis of a Year-Long Human Spaceflight," *Science* 364 (April 12, 2019): 1.
2. Ibid.
3. "Low Earth orbit" is defined at orbits below 2,000 kilometers; see Inter-Agency Space Debris Coordination Committee, Steering Group and Working Group 4, "IADC Space Debris Mitigation Guidelines," IADC Action Item number 22.4, September 2007, 6.
4. Michael D. Delp, Jacqueline M. Charvat, Charles L. Limoli, Ruth K. Globus, and Payal Ghosh, "Apollo Lunar Astronauts Show Higher Cardiovascular Disease Mortality: Possible Deep Space Radiation Effects on the Vascular Endothelium," *Scientific Reports* 6 (July 28, 2016).
5. Francis A. Cucinotta, Nobuyuki Hamada, Mark P. Little, "No Evidence for an Increase in Circulatory Disease Mortality in Astronauts Following Space Radiation Exposures," *Life Sciences in Space Research* 10 (August 2016): 53–56.
6. NASA, "Oral History 2 Transcript: Robert E. Stevenson Interviewed by Carol Butler," Houston, Texas, May 13, 1999, 13–35, http://www.jsc.nasa.gov/history/oral_histories/StevensonRE/RES_5-13-99.pdf.
7. C. M. Ott, R. J. Bruce, and D. L. Pierson, "Microbial Characterization of Free Floating Condensate Aboard the Mir Space Station," *Microbial Ecology* 47 (2004): 133–36, http://science.nasa.gov/media/medialibrary/2007/05/11/11may_locad3_resources/Ott%202004.pdf.
8. Trudy E. Bell, "Preventing 'Sick' Spaceships," *NASA Science: Science News*, May 11, 2007, http://science.nasa.gov/science-news/science-at-nasa/2007/11may_locad3/.
9. Tony Phillips, "The Mysterious Smell of Moondust," *NASA Science: Science News*, January 30, 2006, http://science.nasa.gov/science-news/science-at-nasa/2006/30jan_smellofmoondust/.
10. Ibid.
11. Millar, "Technology as Moral Proxy."
12. See, for example, Donnelly, *Universal Human Rights in Theory and Practice*, 30–31; and Shue, *Basic Rights*, 35–64, 153–55.
13. See, for example, Konrad Szocik and Tomasz Wójtowicz, "Human Enhancement in Space Missions: From Moral Controversy to Technological Duty," *Technology in Society* 59 (November 2019); and James S. J. Schwartz, "The Accessible Universe: On the Choice to Require Bodily Modification for Space Exploration," in *Human Enhancements for Space Missions: Lunar, Martian, and Future Missions to the Outer Planets*, ed. Konrad Szocik (New York: Springer, 2020).

14. M. S. Race, J. Moses, C. McKay, and K. J. Venkateswaran, "Synthetic Biology in Space: Considering the Broad Societal and Ethical Implications," *International Journal of Astrobiology* 11 (2012): 133–39.
15. Schwartz, "The Accessible Universe."

*Chapter Five*

# The Dangers of Space Debris

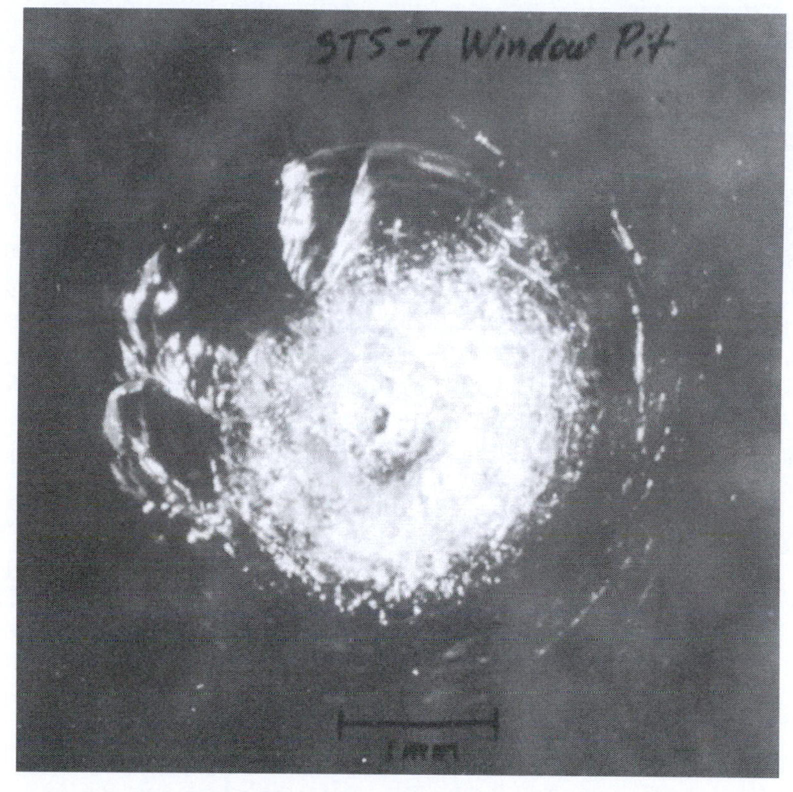


**Figure 5.1. A pit in Space Shuttle *Challenger*'s windshield from a high-velocity paint fleck in 1983**

## THE SPACE SHUTTLE *CHALLENGER* STS-7 DEBRIS IMPACT

In June 1983, on the third morning of its mission (STS-7), the crew of the Space Shuttle *Challenger* discovered that one of their windows had a five-millimeter pit on the outside. The shuttle window had been hit and damaged by a tiny object traveling at high velocity. This pitted the glass, and upon return to Earth, the window had to be removed and replaced. Upon further examination, the object was determined to have been a fleck of paint, obliterated upon impact but leaving chemical traces consistent with spacecraft paint. The fleck was estimated to be 0.2 millimeters in size and traveling between 3 and 6 kilometers per second.[1] At some point in the past, the fleck had fallen off of another space mission and had been orbiting at high velocity until it finally crossed paths with the *Challenger*.

This incident was not uncommon. In 1995, Karen Edelstein estimated that impacts of this type (ones necessitating the replacement of a window) hit a shuttle window every 10.8 days in space, on average, for a total of forty-five replaced windows over the thirteen-year history of the program until that point (December 1994).[2] This was not only expensive but also highlighted the fact that impacts—due both to artificial debris and natural objects—were a significant risk to spaceflight. While none of these impacts threatened the destruction of the shuttle, if they had hit a spacewalking astronaut or a vital shuttle system, the outcome could have been worse. Luckily space debris has not caused a serious accident yet.

## RATIONALE AND SIGNIFICANCE: IT'S GETTING CROWDED UP THERE

Space debris is a lesser-known but very important ethical issue in space exploration and use because it threatens the very possibility of future interaction with space. Already there are some orbits around the Earth that are unsafe due to debris traveling at kilometers-per-second velocities; orbital structures, vehicles, and satellites have been impacted and some even destroyed. In fact, one of the reasons humans in space vehicles and stations tend to reside in low and deteriorating orbits is to avoid space debris, which tends to fall out of such orbits (thus making them emptier and safer). This chapter will also consider the proper disposal of waste in space, including near Earth orbits, beyond Earth orbits, and on the surfaces of extraterrestrial objects.

## ETHICAL QUESTIONS AND PROBLEMS

### Why is space debris a problem?

When humans first began to explore space, Earth's orbits were mostly empty, as, over the lifetime of the Earth, most orbits cleared (a major exception being the Moon, and any natural objects that might wander into Earth orbit). But once humans started launching rockets into Earth orbit, pieces of debris ranging from minuscule to house sized started to fill the orbits around our planet. Given time, these orbits will eventually once again become clear, but humanity has begun to fill these orbits much faster than they naturally empty themselves, hence the problem: Earth orbit is getting more and more crowded with hypervelocity projectiles. Due to the incredible speeds involved, anything in orbit is a potentially destructive or deadly impactor, even a tiny fleck of paint. Because kinetic energy equals one-half mass times velocity squared, energy increases much faster due to velocity than due to mass, thus even tiny particles can have devastating effects. Additionally, because all orbital objects are moving at kilometer-per-second velocities, the closing angle of the impact can greatly affect the relative speed, for example, getting hit head on could be twice the relative velocity (or more, depending on other factors) of getting hit from the side, and getting hit from behind, by an object on the same orbital path, might not be that serious. But as a general rule, anything traveling at about 3 kilometers per second is effectively as energetic as the detonation of a bomb of the same mass, and as velocities increase, projectiles can rapidly become much more dangerous than even the most powerful conventional explosives.

### Might we someday (or already) face the "Kessler Syndrome," in which pieces of debris begin hitting each other and form a cascade of increasingly smaller particles, thus filling Earth's orbits with a cloud of debris and closing off space to further human activity?

The worst-case scenario for space debris would be the **Kessler Syndrome**, named for the NASA scientist, Donald Kessler, who described it, along with Burton Cour-Palais, in 1978. The Kessler Syndrome is a collision cascade in which pieces of space debris begin to collide and break into smaller and smaller pieces, eventually creating a cloud of debris around Earth that makes access to space too risky, uneconomical, or even impossible.[3]

One version of the worst-case scenario could be sparked by a war, in which antisatellite weapons destroy tens or hundreds of satellites, turning each one into a debris field that then continues on to destroy further satellites. Several nations possess antisatellite weapons, including China, India, Russia, and the United States, and each of these nations possess both enough weapons and targets to significantly alter the orbital debris environment. For example, just one test of an antisatellite weapon by China in January 2007 created a ring of debris around the Earth, which may remain unsafe for centuries (see the first case in chapter 6).[4]

Of note is that the Kessler Syndrome need not be a rapid phenomenon. It could be a slow-motion disaster, taking years, decades, or even centuries for enough satellites and other debris to collide and form a debris cloud. In fact, we could be at the beginning of a debris cascade already and just not realize it; after all, the first few hits of a debris cascade are likely to seem minor, but due to the exponential growth of impactors created by each collision, the situation could rapidly degrade in a few years, months, or even days. As with other such disasters that grow exponentially, such as pandemic diseases, actions taken before disaster strikes may seem too much and too early but actions taken after disaster strikes will certainly be too little and too late. In other words, some exponential-type disasters can really only be prevented; solving the disaster after the fact is orders of magnitude more difficult, if not impossible.[5]

### Orbital space around the Earth is a common area we share; how ought humanity protect this resource so that all humans, including future generations, gain the greatest benefit from it?

High-speed orbital debris threatens humanity's ability to access space, so the decisions we make now will have repercussions generations into the future. Given this situation, then, how ought humankind go about protecting this space from a **tragedy of the commons** (in which a shared resource, which could be used sustainably, is instead exploited until destroyed)? Some sort of governance is necessary. There are organizations that keep track of space debris, such as the United States Space Surveillance Network. And there are properly designated "graveyard orbits" around the Earth, such as approximately 235 kilometers above geostationary orbit, as designated by the Inter-Agency Space Debris Coordination Committee (IADC).[6] Yet space debris creation is not governed under major international treaties; that is, the IADC proposes "guidelines" —which are widely followed—but not laws.

The 1967 Outer Space Treaty (OST) does not govern the creation or disposal of space junk, although in Article VII it does mention liability for damage done by space objects, and the 1972 Space Liability Convention goes further in assigning the specifics for this liability.[7] Of course, assigning liability relies on knowing who the responsible parties are, and after several collisions, this will be difficult or impossible to determine, and accountability will likewise degrade. Returning to our discussion of risk in chapter 3, we might remember that when actions can result in harms that are truly disastrous and/or irreversible, or when it is difficult to hold parties accountable for their actions, it is better to prevent the harm from occurring (using standards such as the precautionary principle or prevention principle) rather than rely on an approach to risk such as the polluter pays principle, which assigns liability after the fact. In other words, to directly answer the earlier question, we ought to take direct and immediate action to begin efforts to protect our access to space by using a stringent approach to risk, enforced through updated international treaties, in cooperation with as many nations of the world as will cooperate.

## How, in terms of governance, should the problem of space debris be solved, and by whom?

The "tragedy of the commons" was first described (without using the term itself) by William Forster Lloyd in 1832, later published as "Two Lectures on the Checks to Population."[8] In 1968 Garrett Hardin brought the idea of the "tragedy of the commons," and the phrase itself, to prominence in an eponymous essay in the journal *Science*.[9] While both authors had a pronounced **Malthusian** perspective, assuming that overpopulation would soon cause disaster, the fundamental principle of the tragedy of the commons is useful in many cases that are not related to population.

In the case of orbital debris, the common resource is access to space in and through clear orbits around the Earth. While it might seem to be in the individual's interest to abandon debris in various orbits around the Earth, thus avoiding the immediate costs of proper disposal, this is only short-term self-interest, as in the long term much worse results will occur due to collisions. If many actors all engage in this same poor behavior, many orbits may become unsafe, thus increasing risks to expensive property and lives. In the worst-case scenario, the Kessler Syndrome makes the orbital space around Earth completely inaccessible, even to pass through. And so what once seemed like reasonable cost-saving practices in the past become a dire emergency in the future.

There are three basic solutions to the tragedy of the commons: (1) privatization of the resource, (2) self-regulation of the resource users, and (3) external regulation of the resource users. Each can work in some circumstances, but with respect to orbital debris it is difficult to see how any could work besides the third path, though even the third path does present some further choices (for example, **internalizing economic externalities**: making orbital polluters pay).

1) While privatization can often lead to good regulatory outcomes and reasonable protections on resource use, the privatization of specific orbits does not make sense in this case. While there are authorities who coordinate where satellites go into orbits, etc., because orbits can cross in an infinite number of ways it does not really make sense to think of them as being a piece of space that someone can "own" in a realistic sense, and that, if others enter them, those outsiders are "trespassing" upon. Just as one cannot really "own" a specific open piece of Earth's atmosphere, it does not make sense to think that pieces of empty space can be privatized.

2) While self-regulation is sometimes an ideal way to protect resources and encourage ethical autonomy, integrity, and responsibility at the same time, there are some cases where self-governance approaches fail. So far the history of the space debris problem has not been a complete success for self-regulation. For example, while Kessler and Cour-Palais published in 1978 and the IADC was working diligently in 2004, in 2007 China, via its antisatellite weapon test, created an enormous long-term debris field. Guidance clearly existed but was ignored. Self-governance, while perhaps more than 99.9 percent effective, can fail, and in high-harm situations those few failures can be disastrous because debris can quickly end up in other orbits, potentially initiating a cascade.

3) External regulation makes the most sense for solving the problem of orbital debris. First, because the beginnings to the solution to the problem consists of needing extremely expensive systems, such as tracking radars, satellites, space launches, and so on, only very large organizations are capable of engaging in solving the problem. Second, external regulatory frameworks and organizations already exist, they just need to extend their legal regime into this regulatory area. Then they need to coordinate better and gain the resources needed to police and clean up orbital space. Third, only a powerful external regulator can hope to enforce governance in space, which is so vast and risky. Individual actors in a self-governance framework cannot hope to do this; only a legitimate overarching authority with significant regulatory reach can. Fourth, because the approach to risk necessary for orbital regulation is preventive rather than reactive, an external regulatory agency is more likely to be able to properly analyze the situation in a way that is lower risk and more long term.

While the OST and Space Liability Convention are currently inadequate for humanity's needs with respect to preventing and cleaning up orbital debris, they do set a precedent and a framework for eventually solving the problem. These governance approaches, developed by the United Nations and signed and ratified by member states, offer a strong beginning for producing a solution to the problem of space debris, and the UN continues to work on these issues, even occasionally with a specifically ethical focus.[10]

### What should be the guiding principles for controlling access to and use of orbital space and for preventing the use of space as a dumping ground that gradually constricts our continued access?

Two clear guiding principles could be sustainability and cost efficiency. Humanity ought to maintain sustainable access to space so that future generations have the same access to space, or better, than we have today. There are multiple ethical reasons for this, including intergenerational justice, good stewardship of resources, various forms of **altruism**, and so on.

But—as with all matters of sustainability—humanity also needs to consider the cost of this endeavor because maintaining or improving access to space by controlling space debris will be expensive. However, cost needs to be considered in two ways—short-term costs and long-term costs. While it might seem less costly to abandon debris in space, it is only less costly in the short term. In the long term, as more debris accumulates, the risk of damage to property and human lives will increase.

For example, satellites might require more shielding to deal with space debris (for example, layered and spaced **Whipple shielding**), thus increasing their weight and volume and therefore costs to get into space. For another example, costs will begin to increase as more space objects are destroyed after shorter lifespans in orbit. If a satellite's lifespan is decreased by half, then it needs to double the rate at which it recuperates its costs, which may mean that certain types of satellites become less economical. Notice that the creators of the space debris and those affected by it may well not be the same people, thus unfairly rewarding the perpetrators and unfairly punishing the victims. As a third example, it is much more difficult to clean up messes than it is to prevent their creation, and cleaning up a few thousand large satellites on a few thousand trajectories is much simpler than cleaning up millions of tiny particles on millions of trajectories. Last, we must consider the danger to human life. If the Earth becomes surrounded by a cloud of dangerous debris, fewer people will be willing to risk going into space at all; indeed, it may become so unsafe that humans will never be able to leave Earth again, and perhaps even robotic missions could not be sent. The long-term costs of this—trapping ourselves on Earth without access to space—is not easily calculable. At the very least it would mean that space resources would not be accessible, that we would be cut off from scientific knowledge gained by space exploration (and perhaps geosensing and other Earth-directed satellite functions as well such as communications, GPS, and weather/climate), and that we could not make humanity a multiplanetary species and reduce the threat of existential risks.

### What should be the proper way to dispose of rubbish in space? What of the problem of waste and pollution on other planets and celestial objects?

As mentioned before, graveyard orbits can be a reasonable solution to the problem of space debris. Boosting geostationary satellites to a higher orbit or moving lower-orbiting satellites into less-used orbits (such as within the Van Allen radiation belts) can get these objects out of the way and make them much less likely to become dangerous debris. Similarly, and already in common use, satellites in low Earth orbit (LEO) can be deorbited and burned up in the atmosphere over the oceans so that any debris surviving reentry is unlikely to affect anyone.

Looking farther into the future, one of the important lessons for space exploration is that while some sorts of resources are plentiful in some places (for example, sunlight in the inner solar system), others will be rare due to the distances and energies necessary to acquire them (depending on location, for example, water, metal, etc.). Relatively rare resources will become more precious than they might otherwise be on Earth, and so, as humankind expands its operations in space, old satellites and other space junk might start to be hungrily eyed by humans seeking resources. This is actually a good solution to the problem, if done right, because orbital debris would be recycled and put to good use. Note also that in this scenario it is much more economical to track and collect larger debris than smaller debris, so it would be good to avoid debris collisions as much as possible. It is also worth noting that if this scenario is done wrongly it could actually increase the amount of space debris if satellites are ripped apart and unwanted pieces are dumped back into orbit. This sort of dumping ought to be banned.

With respect to pollution deeper into space or on extraterrestrial objects, there are several considerations. First, any free-floating orbital debris is dangerous, whether in Earth orbit or elsewhere, and its generation ought to be banned. Debris that is created on an extraterrestrial surface needs further consideration. Is the surface inhabitable? If so, then two standards need to be in play, with the stricter one always taking precedence. First, the junk or pollution could be treated as it would be on Earth, in a manner consistent with reducing harm to human and other life. Second, the junk or pollution could be treated with a level of care suitable for planetary protection—preventing forward and backward contamination that risks scientific knowledge (in accordance with the OST) or potential environmental damage (which currently exceeds the standards of the OST). (Of course, planetary protection consideration should come into play much before the disposal phase, before initial contact with the extraterrestrial surface.) For either standard, remediation may need to include sterilization (if biological contamination is an issue), incineration of other types of chemical reaction (if there are dangerous chemicals that need to be destroyed), or perhaps burial (if the objects are dangerous from being sharp, etc.). Of course, rather than these approaches, the best approach would likely be recycling, so that the value of the junk is contributed forward into future useful objects.

What if the extraterrestrial surface is not inhabitable? Then scientific planetary protection standards should be followed so as to avoid reducing an object's research value. Additionally, ethically speaking, overall damage should still be minimized, keeping debris in one place, perhaps buried, minding that toxic chemicals or biological contamination are not released. Even if the extraterrestrial surface is itself harsh, even toxic, that is no excuse to leak chemicals onto it or otherwise damage it. For larger planet-sized objects with atmospheres, it might be reasonable to burn debris up in the atmosphere. But for more vulnerable space objects, human interference ought to be limited, unless there is a good reason for it not to be.

As a general rule, **noninterference** should be one of humanity's instinctive reactions when dealing with natural places. This is not to prevent humans from interfering with places, only to presume in favor of noninterference and require a good reason to allow interference.

### In addition to halting the creation of more debris, should humankind actively collect, deorbit, or shoot down space debris? If so, who should fund and govern this effort?

While the clearest path to sustainability is to simply stop putting more debris in orbit, we may not be given the choice to merely passively protect our access to space. If a debris cascade begins—and perhaps a slow-motion one has begun already—then active removal of debris may be the only option for preserving access to space. Active removal could consist of several approaches.

First, objects could simply be actively collected, for example by something like a "garbage-collecting" robot that would grab bits of junk and then carry them away. At the end it could deliver the junk for recycling, drop the junk into the Earth's atmosphere to burn up, or park the junk, bound up, in a graveyard orbit. Second, objects could be actively destroyed, for example, via lasers or focused sunlight that either burn the target into vapor or burn a patch on the side of a larger object (the ejected gas acting like a rocket engine) in order to alter its orbit and thus guide it into the Earth's atmosphere or a graveyard orbit. The danger with this approach is that it might lead to many smaller objects breaking off and thus exacerbate the debris problem rather than remedy it. Third are passive approaches. This could involve giant blobs of elastic foam or other material that, while extremely light, would be able to catch smaller pieces of debris and trap them, while at the same time not itself breaking apart. Eventually, when the junk-collecting blob becomes full or otherwise reaches the end of its lifespan, it could be burned up, parked in a graveyard orbit, or recycled.

This effort should be funded by an organization with legitimate authority, such as the United Nations, based on fees or other income from those who have either contributed to the debris problem in the first place (whether national, corporate, or individual) or who are continuing to benefit from having orbital space clear of debris—which is ultimately all of humanity.

As a last note, all debris collecting or destroying platforms have a clear dual-use problem: they could be used to destroy good satellites and therefore act as weapons. The OST bans weapons platforms in space, but a "garbage-collecting" platform could have the same tools on it just directed at peaceful purposes—until they are not, for example, if a war breaks out. This dual-use problem is one that will be examined in more detail later in this chapter and in the next, but suffice it to say that it makes the neutrality of the governing agency over such cleaning platforms all the more crucial.

## ETHICAL CONCEPTS AND TOOLS

### The Tragedy of the Commons

The abuse of space as a dumping ground is a case of the tragedy of the commons, where an unregulated common resource is unsustainably exploited. Is there a way for this concept to be used to enlighten our thinking in ways that allow it to be a positive ethical tool?

As mentioned earlier, there are several solutions to the tragedy of the commons: privatization, self-regulation, and external governance. Each works best under certain circumstances. For example, privatization works well for things like housing and vehicle ownership. In both cases, the owner is responsible for the resource and has an incentive to take care of it because otherwise they must live with the negative consequences of their own actions. In places where public housing exists, depending on the cultural context, sometimes the housing is not cared for and common areas in particular are likely to become run down; public transportation can be similar. In both cases of public ownership, there must be a constant input of public effort and money to maintain the resource by an external regulator.

Self-governance works best where individuals and organizations trust each other and are incentivized to maintain that trust. Returning to the example of housing and vehicles, borrowing a friend's car or temporarily living in their home would be prime examples of self-governance. If a whole group of friends collaborated to share houses and vehicles, this would be the same system of trust at a larger scale. There would always be an expectation that everything would be returned in good condition so that trustworthiness would be maintained and the system would continue, to everyone's benefit.

Slightly removed from this friendship model, and therefore one step away from self-governance and toward external governance, several technology companies, such as Airbnb, Lyft, and Uber, have built business models on sharing housing and transportation. These corporations have done this by utilizing ways to develop trust between strangers who may never interact again. This governance regime involves stringently rating users and facilitating trust through this rating system, including by penalizing, even banning, those who fall below a certain threshold. In some ways this model might be seen as external governance, so it depends on the level of perspective. From the perspective of the corporation they are self-governing their own system (which exists because of their labor) and everyone involved is there voluntarily: both suppliers of housing and transportation and consumers of it. From the perspective of the suppliers and consumers, however, they might come to feel alienated from the corporation and feel as though it is pushing them around and not acting in their best interest. In this case, the corporation seems to act less like a self-regulator and more like an oppressive external regulator. If the employees and consumers of these corporations were also the main shareholders, in a cooperative arrangement, that would make the system truly self-governing, but if ownership and labor are not united, then discord can appear.

External governance seems to make the most sense in the case of space debris, but there are several ways to go about externally governing. First is by economic means, similar to the more external perspective version of the case of the technology companies. In this case, participants in the system of services are incentivized toward good behavior by economic rewards and punishments, including access to the system of services itself. With regard to space debris, this might not make sense because the "service" being provided by the regulation is merely an absence of something bad, which makes it appear, in the short term, as though one is paying for no service at all and thus wasting money. Of course this is not the case, but it is difficult to justify unless everyone comes to the realization that it is important.

Another method of external regulation that is economically based involves internalization of externalities—in other words, making users pay the full cost of their use of the common resource. In the case of space debris, this might take the form of paying an up-front clean-up fee to place satellites in orbit or "renting" an orbit from a regulatory agency, thus providing money to clean up debris if necessary. (Currently the International Telecommunications Union allocates orbits for free on a first come, first served basis.) An agency could even keep charging for nonfunctioning "junk" satellites occupying orbital space, and if the junk broke up then the agency could charge rent for more orbits, thus strongly disincentivizing keeping junk in orbit (note that this is the polluter pays approach to risk, which does assume an organization will keep paying and not just declare bankruptcy).

The last and most heavy regulatory regime for external governance is one that is legal and/or political in nature. Rather than focusing solely on monetary and economic incentives and disincentives, legal and/or political rewards and punishments can be harsher. For example, organizations could face criminal penalties for misbehavior: heavy fines, leadership going to jail, or the organization could be disbanded. Contrarily, the rewards of a legal and/or political system might be stronger as well, with important contracts only going to the organizations with the best operational record or other honors such as public recognition for achievements, etc.

While more recent formulations of the tragedy of the commons utilize concepts from **game theory**, machine learning, and other sophisticated techniques that may provide deep insights in the future, for now it can be helpful to recognize that humans have been solving these governance issues for millennia, and often quite well. The orbital tragedy of the commons can be solved, but before that can be done there needs to be the economic and political will to solve it.

## Consequentialism

Consequentialism is the ethical theory that judges the goodness of actions by their effects. Because the amount of good possible if humans settle space is literally astronomical (verging on the multiplication of near infinite numbers of people by near infinite amounts of time by near infinite amounts of space), advocates of consequentialist ethics and space exploration have good reason to express concern about access to space.[11] If human access to space is cut off, we will be trapped here on Earth and unable to fulfill the enormous amount of good we could achieve if humans could settle space. Even worse, if trapped on Earth we could be driven extinct by an existential catastrophe and thus close off all future value that humans might experience.

It is worth pointing out that despite the stark utopianism and bleakness of these scenarios, because the future is unpredictable, there are other scenarios as well. Adventuring into space, for example, could bring extremely negative experiences, such as ongoing interplanetary war, encounters with malevolent intelligences (artificial or natural), and so on. A happy future is not the only option, but neither is a terrible future of suffering—it also might just be on average neutral, despite just being at a much larger scale. Similarly, even if restricted to Earth by orbital debris, humankind might not fare so badly, and indeed might be quite satisfied despite only being able to look at space and not go there. And in fact, over time, Earth orbit would clear out and give humanity the opportunity again (if humanity still exists) to explore space.

Consequentialism facilitates some very important insights about ethical actions with respect to space debris because it is very much a problem of shortsightedness. Because consequentialism thinks longer term, it sees past these short-term concerns. But because prediction can also be difficult, and infinities present problems for calculations, it can also experience some complexities in its approach to solving these sorts of problems.

## Social Justice

**Social justice** expresses the idea that there is a responsibility of the strong to care for the weak. It has been captured in such religious phrases as "to whom much has been given, much will be required" (Luke 12:48) and in comic book heroes such as Marvel's Spiderman, whose Uncle Ben says, "With great power comes great responsibility."[12] It can also be found in derivative forms in chivalrous concepts as *noblesse oblige* (nobility obligates), though this can also be critiqued as a justification for entitlements and privilege.

The basic idea of social justice is that when people need help, those who are able to help should in fact do so. For example, in a natural disaster, people who need help turn to those who are actually capable of helping them—trained medics, police officers, etc.—and not to toddlers, the infirm elderly, etc., who are themselves more likely to need help than to be able to help others. Relatedly, those assigned to render help are obliged to do so. Authorities who shirk their duties are justifiably punishable for dereliction of duty. Note: this does not mean that being strong or wealthy is an ethical necessity or that not being strong is therefore morally blameworthy. The elderly and toddlers may have once been or one day will be able to help others in need. But at the moment of emergency, those with the capacity to help are more responsible than those who are not (and if we can become more prepared for such situations, it is responsible to do so). Note that these justice obligations are not merely about professional social roles such as with emergency responders; social justice can obligate anyone who is able to help, and in this way sometimes can be perceived as a dangerous and uncomfortable socioethical principle.

Space debris is a social justice issue because those who are capable of doing something about the situation—the current spacefaring nations—are in a position of responsibility for all of humanity. If the spacefaring nations abuse their power and ruin our access to space, then all humanity pays the price. If, rather, the spacefaring nations use their power wisely and maintain safe access to space, someday all may benefit from this protected resource, including those who are unable to act at this time: current nonspacefaring nations and future generations. Nonspacefaring nations cannot protect the resource themselves because that is not within their realm of action; only the spacefaring nations can protect against orbital pollution, and so "with great power comes great responsibility."

## Intergenerational Justice

Building on the idea of social justice, **intergenerational justice** is the concept that justice not only extends spatially between all humans currently alive but also extends temporally in how we treat the past and the future. Obviously, we cannot do anything to change the past, but we might do wrong to the past by lying about it, defaming those worthy of praise or praising those worthy of infamy, by failing to learn the lessons of the past, and so on. We should look on the past with a sympathetic eye, knowing that "the past is a different country," the people of former times often lived under harsh circumstances, and they were fallible, just as we are today. We should not assume their mistakes, or their victories, are so foreign to humankind that we are unable to make the same or worse mistakes today or achieve the same or greater victories than they did. This is cause both for caution and for hope.

The future, on the other hand, we can affect and will affect by the choices we make now. If "the past is a different country" then so is the future, and we are all immigrants to that country.

If we make good choices now, we will create a better future, and if we make bad choices now, we will create a worse future. What duties do we owe to the future? This is a difficult question to conceptualize, but we can think of the Golden Rule, the rule of positive reciprocity: do unto others as you would have them do unto you. What good things did our ancestors do for us or do we wish that they had done for us? What bad things did our ancestors do to us or do we wish that they had not done? When we consider the past, we should then think of whether we are ourselves now doing good or bad things to our descendants through our contemporary choices.

With respect to orbital debris, if we destroyed humankind's ability to access space, thereby leaving us trapped on Earth, we would remain in a precarious situation where settlements in space are no longer possible and yet our destructive power is growing while our resources and environment health are diminishing. If events, whether natural or human made, then drove humanity extinct, we would become, in the words of Toby Ord, "The worst generation that had ever lived."[13] If the people of the future are, or will be, as important as people alive now, then intergenerational justice is a crucial consideration in ethics.[14]

## The Common Good

Social justice and intergenerational justice consider how ethics engages larger numbers of people across space and time, but "people" are not just aggregated individuals but communities with social structures, relationships, institutions, and other shared goods. To seek the ethical common good is to seek to develop, improve, and strengthen goods that provide broad social benefit, such as social institutions, community organizations, transportation infrastructure, education, health care, public safety, clear air and water, and good governance.

With regard to space debris, the shared good of space access is at risk, and it is space governance in particular that needs to be developed in order to reduce that risk. For the sake of the common good we not only need some sort of governance regime to control the production of space debris but also, increasingly, mechanisms for removing debris that is already clogging orbits. For the same reasons as the tragedy of the commons, shared goods tend to be underfunded by societies because while all benefit from them, no one in particular is necessarily in charge of making sure that the good remains properly cared for, nor is it necessarily possible to force people to pay for the good when they use it, thus potentially leading to uncontrolled exploitation. Economists have studied these phenomena in some detail, and that research is quite relevant to the problem here; economic thinking can fruitfully engage ethical thinking on the common good.

## Environmental Ethics

Humans have made a mess of Earth's environment, and now as we extend ourselves into space we risk doing the same beyond Earth as well; indeed, we have already begun to do so in Earth orbit. As environmental ethics has developed it has gradually helped to enlighten us to such previously less attended to moral values such as the values of nonhuman life and the abiotic environment more generally.

And yet in the case of orbital debris this is hardly so complex an environmental situation as to necessitate appeals to such sophisticated environmental debates as those on instrumental versus intrinsic value. In later chapters of this book we will engage such issues, but in the case of orbital debris mere rational self-interest ought to be a sufficient motivator to do the right thing. And yet as many environmental issues on Earth demonstrate (that is, climate change, environmental toxicity, etc.), rational self-interest is often not sufficient for solving problems in a timely manner. Due to various limitations in human psychology, culture, and national and international governance, long-term, complex, and unpopular problems (even if extremely important) tend to be poorly attended to, while short-term, simple, and/or popular problems (even if quite unimportant) tend to take up most attention.

The only point to make here, then, is to remember to prioritize problems according to their true magnitude. The social lives of famous movie stars are not as important as the survival of humanity, which makes movie stars possible in the first place. We should remember the preconditions for our own existence—in other words, for humans to have ethical debates, humans must exist. For humans to gossip about movie stars, humans must exist. For humans to have ethical debates about the environment, humans must exist.[15]

Backing up one more step, we do know that humans do not need to exist in order for the natural environment to exist—it certainly preceded us and did fine without us until we arrived. It could certainly endure without us if humankind went extinct. And we know humanity has and continues to inflict massive damage upon the Earth's biosphere. But the biosphere without us would also be at a loss. Not only could it be heavily damaged by any existential crisis to humanity, whether natural or artificial, but eventually the Earth's biosphere will die as the Sun expands, overheats the planet, and eventually destroys the Earth and every precious thing on it, whether alive, intelligent, or not. Spacefaring humanity could prevent that final cataclysm of omni-extinction (the extinction of all life on Earth), carry Earth life to new and safer places, and create new environments that can flourish and evolve in their own ways. Humans need a healthy environment, and space demonstrates how rare and vital a good environment is. But space also will teach us how to create new environments ourselves, and within those environments we bring other life forms, starting with microbes but eventually including the most complex forms of life on Earth. The environmental ethics of space is one that must include humanity, for we are the only way that Earth life—the only life that we know of—can get there. And returning to space debris: that means that we need to keep Earth's orbit clear.

## The Ethics of Dual-Use Technologies, Part I

As mentioned briefly earlier, if it is necessary at some point in the future to take active measures in cleaning up orbital debris, then these trash-collecting or -destroying measures themselves would require significant regulatory control because they are effectively weapons. Article IV of the OST prohibits weapons of mass destruction or stationing weapons in space, but what if these "stations" were merely "collecting" or "incinerating" trash? As a silly example, we are pleased (or should be pleased) when garbage trucks carry away our garbage on collection day, but if these trucks instead scooped up our car, crushed it, and drove away with it, we would be justifiably upset. A satellite is the same. A trash-destroying platform could "collect" or destroy a derelict satellite or a fully functioning one—perhaps one belonging to an enemy. Governance needs to anticipate these possibilities and plan accordingly.

Dual-use technologies are ones that have potential for both peaceful and violent use. Any means of collecting or destroying orbital debris has a dual use as a weapon, and therefore international law becomes a concern. Dual-use technologies are not intrinsically bad, but they are always cause for concern and should be governed in a manner proportionate to their risk and benefit. In chapter 6 we will continue to explore questions of dual-use in more depth, but for now this point should be recognized because it complicates solving the problem of space debris.

## BACK IN THE BOX

The tragedy of the commons was formulated for situations here on Earth, situations such as oceanic fisheries management, climate change, atmospheric pollution, build-up of toxicity in land and water, chores among roommates, "the office refrigerator," and so on. While it might be nice if we only had to face these problems as hypothetical ones in a space-aged future, we are in fact already living in that future, and we have needed to deal with tragedies of the commons for millennia. Some of our greatest tragedies of the commons today, such as climate change, could be solved, but instead some nations choose not to cooperate in this endeavor.

One of the greatest tragedies of the tragedy of the commons is that even when we recognize that we are experiencing this type of problem, we still do not necessarily try to solve it. Jared Diamond, in his book *Collapse*, spends a chapter considering why some societies fail and others succeed when faced with challenges that are existential for their civilizations. Many apparently choose to fail—and by this, Diamond is not holding people responsible for things that are outside of their control. For example, many civilizations had no way to anticipate the problems they would face, to perceive or understand what those problems were, or to know how to solve them. No, what Diamond discovered was that many societies could recognize that something was wrong, could figure out how to solve it, but instead *chose not to try to solve the problem*. Reasons for this are complex, including both "rational" (in an economic sense) and "irrational" causes.[16] But no matter the specifics, these cases should give us pause and provoke in us an ever-stronger determination to do the right thing in the right way and for the right reasons, even if many powerful people are opposed. We solve these problems by determined, purposeful action for the sake of creating a better world.

## CLOSING CASE

On February 10, 2009, Iridium 33—an active commercial communications satellite—and Kosmos-2251—a derelict Russian military communications satellite—collided at an altitude of 789 kilometers and at a closing speed of 11.7 kilometers per second. This was the first satellite-to-satellite collision in space; before that satellites had only been hit by space debris or weapon strikes. The strike generated over two thousand pieces of debris, much of which remains in orbit to this day, in rings around the Earth approximating where the satellites previously orbited. These pieces will remain a danger to other satellites until their orbits eventually decay.[17] What do incidents like this tell us about our orbital environment? What should humanity do to address these problems?

## DISCUSSION AND STUDY QUESTIONS

1. After Edelstein's study in 1995, space shuttle missions began to spend more time flying "tail first" so that orbital debris would tend to hit areas of the space shuttle that were less sensitive to damage for the reentry phase of the mission. This simple solution reduced risk significantly. What other creative solutions might there be to reducing the risk of space debris?
2. Note that low Earth orbital velocity is approximately 7 to 8 kilometers per second. If the closing velocity between the Space Shuttle *Challenger* and the paint fleck in STS-7 was only 3 to 6 kilometers per second, this means that both objects were actually *going predominantly the same direction*. At maximum closing velocity for two objects in LEO, a head-on collision would be at approximately 16 kilometers per second. Does

this suggest any courses of action to attempt to reduce orbital impact speeds or otherwise reduce the risk of space debris?

3. What should people do about space debris? Justify your answer.
4. Of the ethical concepts and tools discussed in this chapter, which do you find to be the most useful and why? Which do you find to be the least useful and why?

## FURTHER READINGS

Jacques Arnould, *Icarus' Second Chance: The Basis and Perspectives of Space Ethics* (New York: SpringerWeinNewYork, 2011).

Donald J. Kessler, "Collisional Cascading: The Limits of Population Growth in Low Earth Orbit," *Advances in Space Research* 11, no. 12 (1991): 63–66.

National Research Council, Division on Engineering and Physical Sciences, Commission on Engineering and Technical Systems, Committee on Space Debris, *Orbital Debris: A Technical Assessment* (National Academies Press, July 7, 1995).

## NOTES

1. Donald J. Kessler, "Earth Orbital Pollution," in *Beyond Spaceship Earth: Environmental Ethics and the Solar System*, ed. Eugene C. Hargrove (San Francisco: Sierra Club Books, 1986), 57.
2. Karen S. Edelstein, "Orbital Impacts and the Space Shuttle Windshield," *Proceedings of SPIE 2483 Space Environmental, Legal, and Safety Issues*, Symposium on OE/Aerospace Sensing and Dual Use Photonics, Orlando, Florida, June 23, 1995.
3. Donald J. Kessler and Burton G. Cour-Palais, "Collision Frequency of Artificial Satellites: The Creation of a Debris Belt," *Journal of Geophysical Research: Space Physics* 83, no. A6 (June 1, 1978): 2637–46.
4. Craig Covault, "Chinese Test Anti-Satellite Weapon," *Aviation Week & Space Technology*, January 17, 2007, https://web.archive.org/web/20070128075259/http://www.aviationweek.com/aw/generic/story_channel.jsp?channel=space&id=news%2FCHI01177.xml.
5. Alexey Turchin, "Processes with Positive Feedback and Perspectives on Global Catastrophes," *Social Sciences and Modernity*, no. 6 (2009) (in Russian).
6. Inter-Agency Space Debris Coordination Committee, "Report of the IADC Activities on Space Debris Mitigation Measures," presented to the 41st Session of the Scientific and Technical Subcommittee of the United Nations Committee on the Peaceful Uses of Outer Space, 2004, https://web.archive.org/web/20150402103645/http://www.iadc-online.org/Documents/IADC-UNCOPUOS-final.pdf.
7. United Nations Office for Outer Space Affairs, "Convention on International Liability for Damage Caused by Space Objects," General Assembly Resolution 2777 (XXVI) in 1971, entered into force 1972.
8. William Forster Lloyd, "Two Lectures on the Checks to Population," 1832, republished as "W. F. Lloyd on the Checks to Population," *Population and Development Review* 6, no. 3 (September 1980): 473–96, https://www.jstor.org/stable/1972412.
9. Garrett Hardin, "The Tragedy of the Commons," *Science* 162, no. 3859 (December 13, 1968): 1243–48.
10. Alain Pompidou, *The Ethics of Space Policy*, Working Group on the "Ethics of Outer Space," UNESCO World Commission on the Ethics of Scientific Knowledge and Technology, UN, 2000.
11. See, for example, Ord, *The Precipice*, 217–41; and Baum, "The Ethics of Outer Space."
12. Originally paraphrased in Stan Lee and Steve Ditko, "Spider-Man," *Amazing Fantasy* 1, no. 15 (August 1962); exact phrase from Uncle Ben in J. Michael Straczynski, *Amazing Spider-Man* 2, no. 38 (February 2002).
13. Robert Wiblin, "Toby Ord on Why the Long-Term Future of Humanity Matters More Than Anything Else, and What We Should Do About It," *80,000 Hours Podcast*, September 6, 2017, https://80000hours.org/podcast/episodes/why-the-long-run-future-matters-more-than-anything-else-and-what-we-should-do-about-it/.
14. Questions like these form the premise of Kim Stanley Robinson's novel *The Ministry for the Future* (New York: Orbit Books, 2020).
15. Abney, "Ethics of Colonization," 60–63; Green, "Self-Preservation Should Be Humankind's First Ethical Priority and Therefore Rapid Space Settlement Is Necessary," 35–37; Jonas, *The Imperative of Responsibility*; and Smith, "*Homo Reductio*," 31–34.
16. Jared Diamond, "Ch. 14: Why Do Some Societies Make Disastrous Decisions?" in *Collapse: How Societies Choose to Fail or Succeed* (New York: Penguin Books, 2005), 419–40.
17. Paul Marks, "Satellite Collision 'More Powerful Than China's ASAT Test,'" *Space*, February 13, 2009), https://www.newscientist.com/article/dn16604-satellite-collision-more-powerful-than-chinas-asat-test/#ixzz6iEG1BLab.

*Chapter Six*

# Military, Dual-Use Activities, and International Relations in Space

**Figure 6.1. U.S. president Ronald Reagan's Strategic Defense Initiative (SDI, "Star Wars") graphic, showing satellites reflecting ground-based energy weapons at target satellites in Earth orbit**

## CHINA'S 2007 ANTISATELLITE WEAPON TEST

On January 11, 2007, the military of the People's Republic of China launched a missile with a kinetic kill vehicle that intentionally destroyed China's own Fengyun FY-1C weather satellite at an altitude of 865 kilometers and a closing speed of approximately 8 kilometers per second.[1] This successful antisatellite (ASAT) test demonstrated that they were capable of destroying satellites and potentially intercepting ballistic missiles as well. This demonstration of military power brought China into an elite "club" of nations that have proven they are capable of destroying satellites, including the USSR/Russia (1970),[2] United States (1985),[3] and India (2019).[4]

The demonstration of this new ASAT capability provoked a storm of controversy, not only for the military and political implications—China had accomplished a technically extremely difficult test and could now threaten the satellite assets of many nations—but also because of the enormous amount of long-term (decades to centuries) and dangerous space debris it created: thousands of trackable pieces (10 centimeters in size), tens of thousands of smaller but still very dangerous pieces (1 centimeter in size), and perhaps 150,000 tiny fragments, making the test "the largest debris-generating event in history."[5] Both aspects of the test, the political-military signal and the space debris legacy, point to the fact that space is not merely a place of peace but also one of contention, danger, and even war.

## RATIONALE AND SIGNIFICANCE: DEATH FROM ABOVE

According to Article IV of the Outer Space Treaty (OST), there can be no permanent weapons platforms and no nuclear weapons or other weapons of mass destruction in space, on the Moon, or on other celestial bodies. However, several times missiles have been used to shoot down orbiting satellites, and there are certainly many other military uses for space besides just keeping weapons there, such as space-based surveillance, communication, and navigational satellites. What is the present and future of military activity in space?

Additionally, dual-use activities in space (uses of space with both a benign and potentially more dangerous or sinister nature) are nearly impossible to avoid. Because the energies involved in getting into space, staying there, and returning from space are so extreme, nearly anything involved in space travel can also be used as a weapon. In fact, the U.S. military has proposed stationing kinetic weapons in space (for example, Project Thor, or "Rods from God"—long tungsten rods for orbital bombardment) because, being completely nonexplosive, they could be construed as not being "weapons" and therefore might not technically violate the OST (though they certainly violate the spirit of the treaty because their entire purpose is one of eventual use as a weapon).

These are significant questions for the future of humanity, not only on Earth but also in space. Will we bring our Earthly militarism into space? Or will we leave it behind in the hopes of making a fresh start, in the harshest of conditions, without the looming threat of war added to the natural dangers of space? And yet as long as there is conflict on Earth, there will always be the desire to control "the ultimate high ground" that is space.[6]

## ETHICAL QUESTIONS AND PROBLEMS

### What should be the relationship of Earth's militaries and space?

While Article IV of the OST bans weapons of mass destruction in space, as well as military installations, the treaty is also subject to revision or even withdrawal (as the United States did from the Anti-Ballistic Missile Treaty in 2002). Moreover, the treaty lacks a mechanism for international legal enforcement; thus, it represents at best an ideal for how space ought to be. In other words, while the OST is an international treaty with nominal legal force (when nations ratify international treaties they gain national legal force, but compliance, being at the national level, is voluntary), in fact it also provides a set of ethical norms for human conduct in space. The OST does serve as a hopeful beginning, albeit only a beginning.

That the OST provides this set of normative ideals should give us some consolation, as it is better than having no ideals at all. However, does the OST offer the *best* or *right* ethical guidance? The OST went into effect in 1967—does it still reflect what we think are our ideals today? Because what our current state is and *what it should be* may shift in and out of alignment with our ethical ideals, ethics requires a constant reevaluation of not only our present state—where we "are"—but also where we want to be in the future—our "destination."[7]

Finding the exact discrepancies between our reality and what we want our reality to be is difficult enough but is not in itself sufficient. Before we can find discrepancies between reality and ideals, we also need to ask ourselves what our ideals *ought to be*. This is an exploration of second-order desires—desire for desire, or what we should want to want—and it directly involves metaethics. After all, while many might argue that humans ought to be peaceful, others more militaristically inclined might argue that our capacity for war is more important and that therefore a future of peace is actually undesirable, as a notable episode of the American television show *The Twilight Zone* once suggested.[8] Various militaristic ideologies have existed throughout history among diverse peoples, religions, political movements, and nations, thus confirming that humankind seems naturally attracted to war, or at least finds war difficult to escape.

Therefore to answer the question of what the relationship of Earth militaries to space should be, we need to understand (1) the purpose of a military, (2) the benefits and risks of a military, and (3) the relationship of militarism, competition, and cooperation.

*1) The Purpose of a Military.* As an initial exploration, there are multiple reasons for the existences of militaries. One clear purpose of a military is to defend political entities, such as nation-states, against attack by outsiders or to deter such an attack before it happens.

A second purpose is to attack those outside of one's own political entity, whether for good or bad reasons. Poor reasons might include militarism, expansionism, imperialism, colonialism, and so on. But there also might be good, even altruistic, reasons to attack outsiders, such as defending the vulnerable in another country against their oppressive government (**humanitarian intervention**) or quickly moving to aid those in another nation after a natural disaster, as military agencies often do (though it should be noted that militaries have this role as a side effect, and other purpose-developed agencies might be better or more appropriate).

A third purpose would be to attack those inside a nation, whether as a form of oppression or as a way of defending against terrorists, rebels, or other dangerous social elements.

What do these three purposes have in common? They all "kill people and break things," that is, use force and/or violence (a *prima facie* moral bad) to pursue another end (only the most cynical militarists would say war is an end in itself). Militaries are then a means to both good and bad ends but use a bad means to get there, which therefore requires that the proportionality of the means and the end are crucial. Space weapons raise the stakes of a proportionate response, but perhaps not more than nuclear weapons already have, except, perhaps, in the most extreme cases (for example, redirecting an asteroid to hit the Earth, etc.).

*2) The Benefits and Risks of a Military.* What might be some of the benefits and costs of militarizing space as opposed to seeking peaceful uses of space?

The military use of space promotes most of the benefits of exploring and developing space mentioned in chapter 1. Particular benefits include technological development and spin-offs (such as satellite mapping and navigation) and general development of space launch infrastructure and preparedness toward space-based threats.

The costs of the military development of space, however, may be much worse and possibly not worth taking further (despite the benefits noted). Getting anything into space is terribly costly in terms of money and effort and despite having "the high ground" might still not be an economical expenditure given the existence of nuclear weapons, which already enable us with firepower millions of times greater than conventional explosives. Even orbital kinetic weapons would have great trouble matching this firepower (though it is theoretically possible, with enough effort, that is, using an asteroid as a weapon).

Space weapons also hardly need nuclear levels of energies for attacking other space-based assets, considering that satellites are often very fragile (built to be as light as possible, even bordering on flimsy), space habitats are extremely fragile and vulnerable to the slightest puncture, and in general everything in space is going extremely fast and therefore only need intersect a similar object to be smashed into a spreading cloud of debris.

Arms races are also extremely expensive, so any nation that begins the militarization of space will be committed to continuing that race with other nations. Just as the Cold War between the Soviet Union and the United States poured trillions of dollars into weapons, a space arms race could do the same, with no clear edge to the energies or expenditures. In space-based military matters, "the sky" is no longer "the limit."

There is one benefit that human militarization of space might achieve without resorting to arms races and endangering each other, however, and that would be a cooperative militarization of space for the sake of defending against intelligent extraterrestrial aggressors. Obviously no such aggressors are known to exist, so the entire premise is science-fiction fantasy with no empirical basis. But such a danger is a theoretical possibility, and if this likely very low risk scenario could have the benefit of uniting Earth's militaries, it could be beneficial—though at the same time putting humanity at risk for other such dangers such as rogue military leaders, coups, and so on. In the absence of any obvious external threat, there is no strong reason to militarize space in this manner, and many reasons not to.

*3) The Relationship of Militarism, Competition, and Cooperation.* There are good reasons to believe that militarism is not a preferable ethical value to the pursuit of peace—one reason being that with contemporary weapons, war always risks human extinction, which is a foolish risk in nearly all metaethical systems (because it undercuts the very existence of moral agency, it is hard to see how a moral imperative to undercut all moral imperative could make sense).[9] However, many thinkers, including philosophers and economists, have advocated for the value of competition,[10] including martial virtue.[11] It is likely that some combination of cooperation and competition yields the most beneficial outcome for humanity, however, exactly what that balance is, is not currently agreed upon. But surely militarism that risks human extinction is too far toward competition and ought to be avoided for the reasons mentioned earlier and many others.

Ultimately, what can this initial analysis of the ethics of peace and war tell us with regard to military and dual-use activities in space? Earth militaries and the nations they represent have legitimate interests in space, however, the progression of these military interests toward weapons seems unreasonable due to the arms race dynamic, increase of existential risk, and general opportunity costs of spending on military and not civilian ends.

## Should the OST be updated in order to better regulate military activities in space? What relevant moral issues might warrant revising the treaty?

The OST is decades old, and while it is still useful for providing guidance in the current world context, it needs to be updated, especially because technology has advanced so far. However, in its favor, the OST has worked so far and therefore should be built upon, not dramatically revised. As far as we know, with regard to military matters, the OST has not been broken by any nation, nor does any nation have explicit plans to break the treaty or withdraw from it in the near future. Even the U.S. Strategic Defense Initiative (SDI) of the 1980s, which would have (in many of its proposed forms) violated the treaty, was eventually scaled back to be treaty compliant, before being completely eliminated.

Given the overall success of the OST, it would make sense to revise the treaty rather than scrap it. There are at least three major issues related to military activities in space that should be addressed by a revised version of the OST: (1) voluntariness and lack of enforcement, (2) dual-use technologies, and (3) the proliferation of new actors in space. Each of these shall be discussed briefly.

### 1) How should activities in space be regulated and enforced given that international cooperation largely relies upon voluntary compliance?

In space, geopolitics becomes "cosmopolitics," and like everything else, its scale increases dramatically. While international relations typically rely on treaties that are voluntary and lack enforcement, when it comes to space, cooperation is absolutely necessary. Nevertheless, cooperation still comes in voluntary and involuntary forms, and both have their advantages and drawbacks. First, I will present an example, then the case for involuntary compliance, then the case for voluntary compliance.

Here is a scenario: a rogue nation decides to install one or more nuclear weapons in space in violation of the current OST (and presumably any future OST as well). These orbital weapons of mass destruction would not necessarily even have to be launched toward Earth in order to devastate an enemy: simply detonating a megaton-range nuclear weapon in place, a few hundred kilometers above an enemy (at orbital altitude), could generate an **electromagnetic pulse (EMP)** with energies in the range of 10,000 volts per meter: enough to destroy their electrical power grid, burn out most or all unshielded electronics, damage satellites thousands of kilometers away, severely damage their military capability, and cause nearly incomprehensible widespread and long-term harm.[12] Such a satellite's mere existence could itself be grounds for war and prompt an attack in order to destroy it. The risk presented by such a satellite is just too high. Therefore such satellites should be banned, as indeed the OST does.

However, the OST is voluntary, and a nation might decide to remain a party to the treaty and yet simply ignore it, secretly creating such weapons, or announce it is leaving. When the United States was planning the SDI, the second option would have had to happen at some point, unless the treaty was to be broken covertly (and covertly deploying such a system would nullify its deterrence value, as the 1964 movie *Dr. Strangelove* explored in a slightly different context[13]). If this had happened, how should the world have responded? Should the rest of the world have simply accepted U.S. weapons platforms in space, or should the rest of the world have taken actions to sanction the United States or even destroy the SDI satellites as they were placed in orbit (an act of war)?

These are unpleasant issues without good answers. If all humans had good hearts, such questions might never have to be confronted. But such a world does not exist. Therefore we should anticipate these scenarios and plan what sorts of enforcement mechanisms might be applied against those who ignore or defect from a new OST. One possibility might be to require all spacefaring nations to comply with the OST or else face increasing penalties in order to gain compliance, perhaps ranging from economic sanctions, to political sanctions, and even to military actions such as deorbiting noncompliant satellites or attacking a noncompliant nation's space infrastructure, such as launch facilities. Of course, any such enforcement would have to be from other nations, and then those nations could be labeled as aggressors by the treaty-noncompliant nation and justifiably attacked under most notions of sovereignty (even if the enforcing nations had the blessing of the United Nations). This obviously would be a very tense and dangerous situation, especially if the noncompliant nation were nuclear armed.

However, for agreements to mean something, they must be followed and enforced. For example, note that in the years preceding World War II, Nazi Germany's opponents tried both tactics, first following the "appeasement" strategy, allowing Germany to remilitarize the Rhineland and annex Austria (the Anschluss) and the Sudetenland (from Czechoslovakia) but then dropping appeasement and going to war when Germany invaded Poland. Sometimes both nonenforcement and enforcement "fail" depending on the definition of failure. If the goal was to avoid war, the measures first succeeded and then failed, but if the goal was to stop Nazi aggression, the strategy initially failed (during appeasement) then was eventually achieved by winning the war. Hopefully, taking extreme measures would never need to happen, but including such mechanisms in the treaty could help gain compliance—though it might deter nations from joining the treaty in the first place.

The alternative would be to keep the treaty voluntary and enforced by goodwill, which will make parties more likely to join the treaty but provide no deterrence to defection nor predictable response to such a defection. From an ethical perspective, it is worth noting that in consequentialist reasoning predictability is important, and therefore clear predictable consequences can be useful if widely known before a sanctionable action is taken. However, as a downside, if the sanctionable action is taken and no consequence results, the treaty is weakened because enforcement has failed. At the same time, if one has the choice to be a party to a treaty and thereby liable to its consequences for not participating, or not party to a treaty and thereby not liable, there would be a perverse incentive to simply not join the treaty, which then of course raises the next issue: Should nations not party to the treaty be forced to sign on, even involuntarily? This forcing compliance even on nonsignatory nations raises all the difficulties of enforcement discussed before. This conundrum is a subject of ongoing study in international law, to say the least.[14]

Once again, good alternatives are lacking: voluntary compliance leaves a system that is easy to set up but with no enforcement, and involuntary compliance leaves a system with enforcement but no one wanting to join it and risking war as the enforcement mechanism. It seems that for humans to truly avoid violence and war, we must first remove the desire for violence and war from our own hearts.

### 2) What should be done about dual-use issues in space?

In space, where high energies and speeds are necessary for simply being there, many dual-use issues arise. For example, devices that can get a person into orbit can also be used directly as kinetic energy weapons or can be used to deliver weapons. Even peaceful devices such as those constructed for the removal of space junk (that is, lasers, debris collectors, etc.) could also be used as weapons.

While the OST has stood the test of time with regard to space-based weapons platforms and weapons of mass destruction, it may not be standing the test of time with regard to dual-use concerns. Any satellite capable of steering is a potential weapon in Earth orbit, where velocities are near 8 kilometers per second. Any nation, or other actor, could fill Earth's orbit with numerous perfectly legitimate communication, navigation, observational, and other kinds of satellites yet always—if they still have fuel—potentially alter their use into weapons waiting to demolish another nation's satellites by merely intersecting their orbits. Indeed, such "derelict" satellites-waiting-to-become-weapons may already exist. This dual-use question will be considered in more detail later in this chapter.

### 3) What should be done about nongovernmental actors in space, such as commercial space exploration agencies, and what if there are someday paramilitary organizations in space, rogue actors, or "space pirates"?

Subnational entities are becoming bigger players in space as technology has progressed and democratized not only to other nations but to individual corporations. Currently, Article VI of the OST declares that states party to the treaty have responsibility "for national activities in outer space" whether governmental or not, but this standard may not be useful in the future if ideas such as **flag of convenience** (for example, registering a ship to a nation without much regulatory oversight) extend into space. Furthermore, unscrupulous groups may someday have the ability to threaten other groups in space, or nations, or threaten Earth. For example, a corporation could deploy a constellation of satellites that typically pursue normal communications activities but under certain circumstances might be redirected to impact other satellites as weapons. They could use this power to extort potential victims or even the entire planet by threatening to initiate a debris cascade. While the threatening party is likely to be vulnerable if based on Earth, if they were based beyond Earth, they could make themselves very difficult to control.

As another example, an unscrupulous asteroid mining company could set a valuable asteroid, worth one thousand times more than the cost to redirect it, on course for Earth. The company could at some point declare that the asteroid had gone out of control and it has run out of money, with the asteroid on a collision course. The people of Earth would certainly want to prevent this and would pay to redirect the valuable asteroid into a safe orbit. Who would then own the value of asteroid? Surely the cost of the Earth rescue operation would need to be recouped, but would the nations of Earth get more than that? Or would the unscrupulous corporation—or perhaps its investors, shielded by various laws—keep the value? After all, if the company (or its investors) were allowed to retain even 1/500 of the asteroid's value it would double its money while risking global catastrophe on Earth. And while the whole thing could be an accident, a truly unscrupulous company could do it on purpose, knowing that it would be bailed out by the people of Earth. Taken one step further, redirected asteroids could even be used to blackmail the Earth for payments, similar to "nuclear blackmail" scenarios that have been proposed (and perhaps even conducted, depending on the definition) in the past.

Paramilitary organizations, rogue actors, and pirates are currently science fiction, but they should be considered, as they might appear with little or no warning. For example, such an organization might appear as a mutiny or coup within a government's military or space operations. And while pirates are considered ***hostis humani generis*** ("enemies of humankind"), literal "outlaws" operating outside of national law and protection, and can therefore be killed with minimal due process, at least two problems appear with piracy in regard to space. First, pirates can operate under government cover, as did "privateers" in the past, and second, in space it might be possible for pirate parties to operate beyond the reach of law, in distant or well-hidden locations, perhaps operating by remote control or robotic intermediaries. This potential already exists on Earth with computer hackers, ransomware, and other sorts of manipulative attacks for financial gain. Space just makes the same situation more complex.

## ETHICAL CONCEPTS AND TOOLS

### International Cooperation and Treaties

From these explorations of needs for the revised OST, it is apparent that international treaty making is a complex affair to say the least. Various attempts at the UN to ban nuclear weapons and lethal autonomous weapon systems have failed to progress as major powers have refused to sign on, while the failures of the Anti-Ballistic Missile (ABM) Treaty and Intermediate-Range Nuclear Forces Treaty have demonstrated that treaties are only useful if people actually keep them.

Given this knowledge, we have to ask if it is worth it to attempt to make treaties at all. The answer to this is clearly yes: countless international treaties are extremely successful, for example, the Antarctic Treaty, Montreal Protocols (to control atmospheric CFCs), and United Nations Charter. But in their success we rarely hear about them. We only hear about treaties when they are experiencing difficulty, such as various violations of the Chemical Weapons Convention in recent decades.

As our world continues to grow in complexity, our treaties will need to grow in complexity as well, and the treaty-making process along with it. In the midst of a tense and uncertain world with many national and nonnational actors, this should encourage us to work harder to build strong international legal regimes, not discourage us from this crucial work.

### The Theory and Practice of Just War

Originating over history from ancient times until today, and in multiple civilizations,[15] **just war theory** seeks to limit the barbarity of war for the sake of protecting justice and preserving human dignity and the common good. Just war theory has three main parts, ***jus ad bellum***, ***jus in bello***, and ***jus post bellum***, and each part has several components. Sources will vary on the exact number and description of the components and here I give only a top-level overview.

*Jus ad bellum* refers to justice in the conditions necessary for going to war. To go to war a nation should

1. be governed by a **legitimate and competent authority**—one that has in mind the common good of the populace and typically one that is democratically elected and not one that is an authoritarian or military dictatorship;
2. have a **just cause**—a cause is just if, for example, it is attempting to prevent an outcome worse than war, such as genocide;
3. have a **right intention**—the intention should be to promote the common good, including what is good for one's opponent, not merely selfish political, military, or personal gain;
4. have a **probability of success**—the war needs to be winnable: if a war can only be lost, then such a futile action will merely cost lives rather than preserve the common good; and

5. have reached its **last resort**—all other options short of war should be utilized prior to entering war because war is such an extreme deterioration of the international order, humanitarianism, and respect for human dignity.

*Jus in bello* refers to justice in the conduct of the war itself. To wage a just war, a nation should

1. **discriminate between combatants and noncombatants**—also known as "noncombatant immunity," states that a just war should only intentionally target enemy combatants who pose a threat (for example, injured enemy combatants or prisoners should be cared for and not killed): civilians should not be intentionally targeted;
2. use **no means that are *malum in se***—intrinsically evil, such as sexual assault, torture, mutilation, mass starvation, razing cities, genocide, or otherwise killing in an intrinsically disproportionate or indiscriminate manner (such as using nuclear and biological weapons);
3. only attack targets of **military necessity**—military targets are legitimately targets and civilian targets such as power plants, bridges, and communications systems may only be targeted if the harm they do to civilians is not disproportionate; and
4. **exercise proportionality**—a military engaging in a just war should not use excessive force, for example, saturation bombing cities, using weapons of mass destruction, etc. Weapons or weapon effects that linger such as land mines, radiation, and toxins may violate not only proportionality but also combatant/noncombatant discrimination, the no *malum in se* criterion, and *jus post bellum* criteria.

*Jus post bellum* refers to justice in the conditions after a war has occurred. This aspect of just war theory remains under development relative to the first two. To complete a just war, in addition to all of the previously listed criteria that should extend through the restoration of peace, a nation should

1. **restore political peace**—an international peace treaty should be mutually agreed upon by legitimate authorities on all sides of the conflict, and this treaty may include reasonable reparations for war-related damages and injustices, but these reparations should be proportionate and swift or risk prolonging a conflict through feelings of injustice;
2. **restore social peace**—this may include war crimes trials, reconstruction of civil infrastructure, and reconciliation commissions, and prisoners should be returned; and
3. **restore environmental peace**—natural areas damaged or devastated during the war should be restored to a state similar to their state prior to the war.

In relation to space, the theory and practice of just war should help to inform future human activities in space. Space provides immense opportunities for helping humanity and huge risks for destroying it. Using space as a location for warfare—whether that warfare is directed back toward Earth, between Earth and other locations, or between other locations with Earth as a bystander—would be a terrible future.

Considering the justice in war in space, we can look at the *jus in bello* criterion of proportionality: space weaponry is almost intrinsically disproportionate in its destructive power. Velocities in space are almost always measured in kilometers per second, and this is often simply disproportionate. As mentioned in the last chapter, anything traveling at about 3 kilometers per second is roughly equivalent to its own weight in conventional explosives, and 3 kilometers per second is *slow* for space: velocities in low Earth orbit are about 7 to 8 kilometers per second, and Earth's orbital speed around the Sun is about 30 kilometers per second. Everything above 3 kilometers per second is just more violent energy (and the maximum speed limit is, of course, the speed of light in a vacuum: nearly 300,000 kilometers per second). This lack of proportionality leads to an inability to discriminate between civilians and combatants because civilians could often be harmed or killed as side effects of overly powerful weapons. In the case of war-spawned orbital debris, a debris cascade would destroy civilian as well as military assets in space, as well as potentially leaving long-term effects on whether humans can even reach space in the future, thus violating the rule to only target that which is militarily necessary. These highly probable *jus in bello* violations make *jus ad bellum* and *jus post bellum* just war criteria nearly impossible as well. For example, it is difficult to argue for a right intention when a war is likely to be exceedingly bad in its effects, and likewise, restoring peace after a conflict is likely to be difficult if disproportionate means have led to all sorts of grave events, even war crimes. Restoring a just society and environment could likewise be difficult if, for example, the Kessler Syndrome has cut Earth off from space and made the use of satellites impossible.

In short, it would be very difficult to justify a war in space. Because of this, humanity should try harder than ever before to avoid such a thing.

## The Ethics of Dual-Use Technologies, Part II

Dual-use technologies present a special threat with regard to space. There is an old Latin dictum: "*abusus non tollit usum*," which means "abuse does not remove use." In other words, just because a knife can be used for evil does not mean that it has no proper other use, such as cutting cloth or preparing food, and this proper use remains even if the improper use remains a risk. This phrase codifies in four words a standardized response to dual-use technologies. Dual-use technologies have been in existence ever since humankind has used any technology at all. But that does not mean that thinking about them has not changed; indeed, with some contemporary technologies abuse should take away use (or at least some kinds of use, requiring careful governance), particularly with technologies with potentially catastrophic misuses, such as nuclear weapons, **synthetic biology**, artificial intelligence (AI), or nanotechnology.[16]

There are differences between the "low" technologies of the past and the "high" technologies of today. Knives are very different from, for example, nuclear weapons or biological weapons. Partly that is because nuclear and biological weapons are precisely *weapons*, that is, their intended use is to cause harm to other humans, while a knife can be properly used for other purposes such as cutting vegetables. While weapons can have correct uses, such as defending the innocent, biological and nuclear weapons approach an ethical zone where technologies are *malum in se*—intrinsically evil—in other words, they may have no good, proper purpose, only an evil one. Nuclear weapons could potentially have a good use for, possibly, diverting asteroids, for example, and "biological weapons" can be defined differently, as **gain-of-function research**, and thereby have a good use in researching disease mutations.[17] But either technology, directed toward humans, would kill indiscriminately and therefore would clearly violate the norms of just war theory.

However, what of technologies that are not specifically designed to be weapons but can be used as such, or for other nefarious purposes, as their secondary "dual" use? We already looked at the example of a misdirected asteroid earlier—any space mining activities that involve moving dense cargo around at orbital speeds gain a secondary use as a weapon. Spacecraft can become kinetic weapons if misdirected; however, the clear, normal use of a spacecraft is to travel from place to place in space. And AI that can imitate a human voice can be used for conversations or to lie to people and manipulate them. (Indeed, AI is in some ways the ultimate dual-use technology because it can be directed toward anything at all, including war in space.[18])

Dual-use technologies should be regulated in proportion to their risk, that is, following the risk equation: their capacity for harm times the probability of that harm occurring. So, for example, an asteroid mining operation that sends chunks of rock or metal toward Earth at orbital speeds should be highly regulated because the harm is high and the probability could also be significant. Spacecraft, on the other hand, while constantly traveling between populated places and therefore at high risk for impacts, may burn up in atmospheres if their reentry geometry is imperfect and, as vehicles, tend to be valuable or carry valuable objects and/or humans who are probably not suicidal. The value of the craft and its contents lowers the risk of dual use simply because of self-interest (though the possibility of remote takeover or hijacking remains). Last, AI speech that may lie and manipulate is potentially high harm (potentially directing humans to evil actions), high probability (humans commonly lie, this is just a new means), and low cost (once produced it can be multiplied indefinitely).

Thus, we can see that there are at least four considerations related to dual-use technologies, which can nest under the "harm" and "probability" categories of the risk equation:

Capacity for Harm

- Does the dual use involve high energies (speed, fuel), dangerous environments (vacuum, toxicity), massive scale (region, planet, and larger), cascades or chain reactions (Kessler Syndrome, nuclear chain reaction), reproducing harms (biological), or very long-term or irreversible harms (death, extinction)?[19]

Probability of Use

- Does the dual use permit or enable an evil that humans already regularly do or threaten to do? In this case it could likely be used for this end.
- Does the dual use cost little or much to the perpetrator, for example, destroying something valuable to the perpetrator, such as life or property? The higher the cost, the less likely it might be used.
- Does the dual use put the perpetrator at risk, thus disincentivizing action, in terms of life, health, property, freedom, appearance, or other selfish interest? If so, it is less likely to be used.

Note that for all of these examples, space introduces complexities that might be absent in typical Earth-bound contexts. In space, velocities are almost always dangerously fast; energies are almost always high; environments are almost always fragile; cascades, chain reactions, and reproducing harms are often possible; and many courses of action are irreversible. The potential for harm is almost always high. Just to point out one example of the scales of harm involved: in Earth orbit, cascading disasters like the Kessler Syndrome are larger than the planet Earth itself, and their effects last longer than human civilization has existed.

On the issues of probability, there are plenty of evils we already do, such as killing each other and destroying things. Humans regularly threaten each other and sometimes carry out those threats, whether on the interpersonal, intergroup, or international level. Space technologies will not only enable previous manners of harming each other but also enable new ways. For example, just imagine the anti-American terrorist attacks of September 11, 2001, but with spacecraft. Nearly three thousand people died in those terrorist attacks, but with advanced spacecraft potentially many more could die if they slammed into a settlement (on Earth or elsewhere) or a space station.

Space technologies offer dual-use capacities that are potentially low cost in relation to destructive potential. Once again, due to the energies of orbital speed, even space junk can become a weapon. There would still be the matter of moving the object into the right trajectory to serve as a weapon, which would cost money, but costs are decreasing. There is even the potential to hack into satellite controls in order to hijack them as weapons, but typically the security on satellites is quite high.

Last, space allows perpetrators to be protected by distance and/or obscurity unlike any place on Earth. Earth is immense, the number of places to hide on Earth is incomprehensible, and yet Earth is just a speck in the solar system, which again is just a speck in the Milky Way Galaxy, which is again a speck in the universe. There are huge numbers of places to hide just in the solar system, and the distances involved make pursuit, attack, or capture extremely difficult. Space could dramatically lower the risks to perpetrators, thus incentivizing bad behavior.

In sum, dual-use space technologies do present serious threats and should therefore be carefully governed. These governing regulations should be preventive in focus, anticipating unethical uses of technology and keeping in mind the terrible destructive potential involved with space technology.

## The Ethics of Defense versus Deterrence

While defense resists and fights off an enemy who has already determined to attack, deterrence convinces a potential attacker not to attack in the first place. This has obvious benefits—wars and destruction can be completely avoided, saving countless lives and billions or trillions of dollars. Of note is that defense relies on power, while deterrence relies on psychology: the perception of power.

Much thought has gone into the value and ethics of deterrence due to the Cold War between the United States of America and the Union of Soviet Social Republics, in which thousands of nuclear weapons posed an ever-present threat to human life and civilization on Earth (and indeed, many thousands of these weapons still exist and still threaten). Integral to the theory of deterrence is that both sides know that if they attack, they themselves will be destroyed, thus making such an attack suicidal and therefore undesirable. This logic of destruction is horrible yet effective, at least under certain circumstances. While deterrence might seem to be a superior strategy because it may completely prevent "hot war" conflicts from happening, it does have some serious drawbacks.

First, deterrence can be incredibly costly. The United States spent about 5 *trillion* dollars on nuclear weapons during the Cold War, and the Soviet Union (despite having a very different economy, thus making direct monetary comparisons impossible) spent a similar amount of resources and effort on their nuclear weapons. Since then, both the United States and Russia have spent even more trillions of dollars on maintaining their deterrence capacity. Significantly, for deterrence to work it should be *showy*—new bombers, missiles, and submarines; nuclear tests to demonstrate efficacy; and so on. The money needs not only to be spent but also to result in material items of deterrence that can be shown off. Relatedly, these costly, showy weapons create a perverse incentive for arms manufacturers to promote international conflict and preparation for war because it advertises their products and they profit from it.

Second, deterrence relies on both sides having comparable levels of deterrent power. If one side feels that it is significantly stronger than the other, then deterrence may be lost, and likewise, if one side feels it is significantly weaker, then it will strive to hide that weakness with bluffing and blustery shows of strength or remedy that weakness with expensive and showy new weapons in order to maintain deterrence. Relatedly, if a new technology or technological product results in the balance of power rapidly changing, then deterrence is at risk. This is why defensive systems against nuclear war, for example, SDI or other defensive measures, were banned under the ABM treaty, because they were destabilizing to the balance of deterrence. Additionally, if one side believes that the other side is about to deploy a deterrence destabilizing system, then it might think it prudent to attack before that system is active, thus precipitating exactly what the deterrence system is seeking to avoid.

Third, deterrence relies on having accurate information about both sides. Therefore espionage, spy satellites, communications monitoring, and a multitude of other technologies become vital for surveillance. If one side believes that the other is weak—even if they are in fact not—the deterrence has been weakened or lost. This is why deterrence has to have an element of showmanship to it: both sides need to know that the other side is powerful enough to maintain the balance of power that keeps deterrence in place. Inaccurate information can lead to mistakes in judgment about exactly how irrational or suicidal an attack may prove to be.

Fourth, deterrence does not work against irrational or suicidal foes. If an enemy wants to die or believes that their life has no value or that their own death at the hands of their enemy will result in them achieving a goal such as proving an important point, entering Heaven, or fulfilling some purpose of the universe, then deterrence has failed. In fact not only has it failed, it has created a perverse incentive toward conflict, death, and destruction by giving the enemy an assured way to fulfill their suicidal goal. If that perverse incentive toward suicide did not exist—the entire point of deterrence, which should work on rational actors—then it could not be exploited by a suicidal foe. Deterrence relies on an enemy feeling they have something to lose, and if they feel they do not—or even feel they have something to gain—then deterrence fails.

Fifth, deterrence does not work if those paying the costs of the retaliatory strike are not the same as those making the decisions to engage in the first strike. For example, if the leaders of a country feel that they can hide in a bunker and be invulnerable to the effects of war, then they might not feel disincentivized from beginning that war because others will die and have their property destroyed, but not them. In fact, those leaders might feel that, relatively speaking, they would be quite well off after a conflict because there would be fewer people left, and those remaining alive would be relatively wealthier, own more land, etc.

Sixth, deterrence does not work if an attack cannot be attributed or the attackers are otherwise invulnerable to retaliation. If, suddenly, a nation is subject to a massive attack from submarine-launched missiles, it might not be clear who the attacker is, and therefore retaliation is impossible—deterrence has failed. Likewise, if a cyberattack destroys a nation's electrical system, water supplies, etc., and the attacker cannot be attributed, then retaliation is impossible and deterrence has failed.

Seventh, because deterrence requires constantly maintaining alertness and preparedness for destroying an enemy, it risks accidental war if the sensor systems or those in points of judgment make errors. There are many examples of this type of near mistake from the Cold War, and luckily none of them ever resulted in an attack actually being launched. Two famous examples include Soviet officers: Vasily Arkhipov prevented a submarine from firing a nuclear torpedo during the Cuban Missile Crisis in 1962 (the two other officers in charge wanted to attack, but Arkhipov refused), and Stanislav Petrov refused to escalate a sensor warning of an attack in 1983 (the sensor was later recognized as faulty).

Eighth, related to the previous point, the deterrence weapons must be survivable in order to maintain retaliatory ability and therefore deterrence capacity, otherwise deterrence relies on launch on warning (the tactic of launching nuclear missiles before the enemy's nuclear missiles have completed their flights and struck their targets). This is why nuclear missiles in silos are sometimes considered to be destabilizing weapons, because if an opponent strikes them first, then they are not capable of retaliation—thus, there is a strong incentive to fire the weapons before an enemy's attack destroys them, that is, while the enemy's weapons are still in flight. Once again, this can lead to egregious errors if a sensor or other system indicates an attack that is not real or launch officers otherwise become concerned that they need to fire before they are destroyed.

Ninth, deterrence relies on the firm knowledge that an opponent's chain of command will remain intact long enough in order to retaliate. For example, a "decapitation" attack might attempt to kill or isolate a nation's leadership, thus preventing them from ordering a response. This leads to the seeming rationality of launch on warning (attacking before you are hit), delegation of nuclear attack to lower-level authorities (bomber and submarine commanders), or even the creation of automated attack systems. The Soviet/Russian "Perimeter" or "Dead Hand" system is one such automated attack system, designed to ensure retaliation even if command systems are destroyed. It is a "fail-deadly" rather than a "fail-safe" system; once activated, it ensures destruction rather than safety. However, this automated attack system is itself a vulnerability for accidental attack if it is inadvertently or maliciously activated or makes a mistake.

Tenth, deterrence almost always involves threatening outrageously unethical acts, such as killing tens, hundreds of millions, or even billions of people, exterminating the human race, or destroying the entire biosphere. Because the logic of deterrence leads one to want to be the absolute worst nightmare of an enemy imaginable, it leads to the creation of the most nightmarish weapons imaginable, and then the flaunting of these weapons so that potential enemies recognize how dangerous, unstable, crazy, and/or evil one is. This perverse incentive toward evil is extremely unethical and corrupting of institutions and humanity.

Thus, while deterrence does have some benefits, it also has many very significant drawbacks. In relation to space, these benefits and drawbacks are merely escalated in their degree: everything becomes more extreme.

## Virtue Ethics, Part II

In a world of such complexity and conflict, trust becomes absolutely vital, as does coordination and collaboration. Virtue ethics is the approach to ethics that emphasizes good character. Aristotle conceived of virtues as lying between two vices, one of excess and one of deficiency. For example, courage deals with matters of fear and confidence and is the mean between foolhardiness (an excess of confidence and deficiency of fear) and cowardice (and excess of fear and deficiency of confidence). In the military, virtues such as loyalty and prudence are vital for force operations, and their vices of excess and deficiency would be, respectively, mindless loyalty and disloyalty and "paralysis of analysis" and impulsivity.

In international affairs, virtue ethics might especially emphasize virtues such as honesty, trust, and integrity, while the vices of excesses and deficiencies of these traits would be, respectively, brutal honesty and dishonesty, credulity and inappropriate wariness, and moral uptightness (not uprightness) and moral laxity. In situations in which the treaty-making parties are untrustworthy, wariness may not be a vice; indeed, skepticism might be quite appropriate, and thus measures for verification ought to be included in such a treaty. For example, the Russian proverb "trust, but verify" was picked up by U.S. president Ronald Reagan for use during arms control meetings with the USSR. While wariness might itself sometimes be a virtue, credulity, which allows one to be taken advantage of when one should have known better, is never a virtue, and so in the case of trusting others, the virtuous mean falls closer to wariness than to credulity.

With regard to space, virtue ethics can be an important ethical perspective because it relies more on the ethical agent (the decision maker) than on the ethical patient (subject or object in question). Military and international relations currently are built upon human-to-human interactions, and so humans are both the agents and the patients. There are instances in human history in which virtue ethics have become very important in war situations, for example, when actions may be "legal" by the rules of war but not ethical from the perspective of the agents involved. In these cases, the ethically right thing is done not because of rules but because the agents involved are acting from their own characters and to protect their own characters from harm. The ethical patients in these acts benefit from the virtuous choices of the agents they are subject to. Retrospectively, it is clear that the Western Allies in World War II, for example, gained some ethical legitimacy after the war based on their tendency to treat Axis civilians, prisoners, and so on in a decent manner (with some major exceptions such as the fire bombings and atomic bombings of Axis cities). In space humanity should remember that we are not only acting upon the universe but in all of our choices we are also acting upon our own character, making ourselves better and worse people. These considerations should, I hope, lead us to make a future better than our past and present.

## Common Good versus Partial Good

In general, ethics seeks to promote good at the largest scales. Because the universe is a whole, this holistic approach to ethics seeks to be unitary and comprehensive in the greatest sense. Of course, humanity only knows about its own small part of the universe and so cannot really make ethical decisions with the universe in mind but only based upon our experiences here on Earth and in the solar system. But based on that experience, there are important lessons to be learned from thinking about ethics at scale.

First, ethical prioritization needs to proceed from the largest scale first to smaller scales second. What this means is, for example, we should think about how to help all of humanity and not just a particular segment of humanity, such as those people who live in one nation-state. When partial groups assert their priority over the whole, not only are resources used inefficiently and unfairly but great injustices can also occur, for example, scapegoating of minorities or attacking neighbors. At the largest scales, and thinking idealistically, humanity should be considered as a whole, where all humans equally deserve justice, well-being, resources, etc.

Second, this is not to say that every nation needs to act perfectly altruistically, because in reality, perfect altruism can also lead to evils, namely harm to oneself. In other words, context matters. In a perfect world good is good and evil is evil, but in an imperfect world, sometimes doing what would be good in a perfect world actually turns out to be bad (for example, in a perfect world the Allies would have helped the people of Axis nations as much as the people of their own nations, but in the context of war against a vicious enemy that would be ridiculous). It would be nice to know that every other human would always be willing to help us, to share, to always do the right thing with regard to our own well-being, but we do not live in that world. Humans make bad choices and this will never be completely remedied, although we can certainly improve.

Additionally, with respect to ethical attention and resource allocation, sometimes justice does not mean equality but equity: giving more attention to some people than to others. For example, people who are poor, uneducated, or sick should have priority when it comes to welfare, education, and health care resources, respectively. The rich, educated, and healthy do not need those resources as much.

Sometimes the worst ethical outcomes occur because a part of a system mistakes itself for the whole. For example, cancer involves one part of an organism, the cancerous cells, growing at the expense of the organism. Likewise, in a war of aggression, one group or nation attacks another for the sake of its own benefit and to the detriment of the other. These "targeting errors" of resources and effort are disasters precisely because rather than benefitting humanity, they turn against it and mutilate humanity instead—and ultimately this evil can be self-destructive as well. Cancer grows well for a while, then kills itself when it kills its human host. Warmongering nations may succeed for a while, but often they reach too far, as the Axis did in World War II. Then these aggressor nations are heavily damaged, with millions of dead and their governments replaced with governments more able to cooperate in the contemporary world.

Militaries exist because they protect something important: human lives and the material resources necessary to support those lives. They protect these lives, in turn, by threatening to efficiently "kill people and break things" of potential enemies, so that those enemies either decide to never attack, or if they do attack, then get trounced. Either way, enormous resources are expended on deterrence and defense—on parts fighting other parts—rather than on the well-being of humanity as a whole. For now, in our imperfect world, sometimes lesser evils are the best that we can do.

## Justice in Resource Allocation: The Guns versus Butter Problem

How much should be spent on military as opposed to civilian life? Resources are finite, and, to quote former World War II supreme allied commander and U.S. president Dwight D. Eisenhower,

> Every gun that is made, every warship launched, every rocket fired signifies, in the final sense, a theft from those who hunger and are not fed, those who are cold and are not clothed. This world in arms is not spending money alone.
>
> It is spending the sweat of its laborers, the genius of its scientists, the hopes of its children. The cost of one modern heavy bomber is this: a modern brick school in more than 30 cities. It is two electric power plants, each serving a town of 60,000 population. It is two fine, fully equipped hospitals. It is some 50 miles of concrete highway . . .
>
> This, I repeat, is the best way of life to be found on the road the world has been taking. This is not a way of *life* at all, in any true sense.[20]

Military spending can ruin a nation and ultimately hurts human beings, including those in the nation those weapons are trying to protect. Sometimes military spending is a necessity: there are evils in the world that can only be resisted by force. But it should not be forgotten that military weapons could have been other things: schools, hospitals, civilian infrastructure, and so on. What, then, is a just balance between "guns" (military spending) and "butter" (civilian spending)?

There is a game that is sometimes played in university international relations classes. It is a multiround game in which students are seated around a table and are allowed to put one point each round into either "guns" or "butter." A point to butter makes your people happy. A point to guns makes your nation militarily stronger. After the first round, you can decide to attack your neighbor or not, and if you have more gun points they lose and are out of the game. If you have equal gun points then it is a tie. If you have fewer gun points than your opponent, then they drive you out of the game. This is reiterated through several rounds until all the peaceful butter-loving students are eliminated and only the most gun-loving students remain. The point of the game is that strength matters and more strength matters more.

But notice that the game doesn't have to demonstrate that. Everyone could agree to choose butter for every round and live in a perfectly peaceful and happy world—but that requires coordination and trust. In a world full of "gun" players, one is forced to also be a gun player. Justice requires it; defending one's own citizens, or the innocent residents of other nations, requires the ability to use force in order to stop evils from happening. Game theory, which investigates how cooperative and competitive interactions develop between agents under differing situations, has insights into these sorts of coordination problems and is well worth applying to ethical thinking.[21]

When it comes to justice on military spending, in the words of American pastor and civil rights activist Rev. Dr. Martin Luther King Jr., we have created a world of "guided missiles and misguided men."[22] We have excessive power and deficient ethics. We could make the world a better place where everyone does better together, but instead we have created a world with expensive weapons and starving children. But this past does not need to be our future and, when we consider the expansion of humanity into space, really should not be our future. We can solve these problems, but only if humanity works together. Until that time, there will also be a need for guns in addition to butter.

## Cooperation and Noncooperation with Evil

As a last concept for this chapter on war, there are sometimes points where one is not the ethical agent, but one must decide whether to cooperate with the evil actions of another ethical agent, or not cooperate with them. When it comes to space, that might mean agreeing to work for an aerospace company that also produces weapons or to develop technologies that could also have potentially disastrous dual uses. There is an ethical tool that helps to think about these specific cases of **cooperation and noncooperation with evil**.[23]

To decide whether to cooperate or not, the first question is whether one has the same intent as the agent that is doing the bad deed. Having the same intent as the evildoer is always wrong and is called **formal cooperation** with evil. For example, if someone robs a bank and you are driving the getaway car in order to get some of the loot, you are formally cooperating in evil. Your intent is for the bank to get robbed so that you can get some of the stolen cash.

Cooperating with the evildoer and yet having a different intent is called **material cooperation** with evil and may sometimes be justified but requires further investigation. There are two types of material cooperation: immediate and mediate. **Immediate material cooperation** means directly helping with the same evil act but for a different reason than the primary agent. An example of immediate material cooperation would be cooperating in a bank robbery because one is a bank teller handing bags of money to the robber at gunpoint because one does not want to get killed. The bank teller is literally helping to rob the bank by handing money to a robber, but this cooperation is under duress and for a completely different reason than the bank robber's reason.

**Mediate material cooperation** means not only is the intent different from the evildoer but also the actions taken are not the same as the evildoer. For example, a friend asks to borrow your car, you agree to it, and then, surprisingly to you, the friend uses your vehicle as a getaway car after robbing a bank. In this scenario you have no idea what your friend's evil plan is, and your cooperation has nothing to do with robbing the bank—but you did in fact materially help that evil act by supplying the getaway car. Maybe you should have known better. People may look at you suspiciously for a while. The circumstances around the incident help determine exactly how much culpability one might have, if any. Still there remains one more degree of cooperation to distinguish.

Mediate material cooperation can be proximate, such as car lending, or remote, such as being a mechanic who repaired the car last week that was later lent to be used in the bank robbery. While **proximate mediate material cooperation** can sometimes raise ethical questions (should you have known better than to lend your car to that no-good friend?), **remote mediate material cooperation** is not only permissible, and not only inescapable, it is actually typically required. If a mechanic looked at you and refused to repair your car, knowing that you had a suspicious friend, this would be unjust discrimination and one would be legitimately upset. Additionally, society would cease to function if everyone continually refused to help others on the slightest suspicion of their motives. Society requires trust, and it cannot be the duty of every person to pick through the theoretical ethical shortcomings of everyone else before a bad thing even happens. This is not a virtue; rather this is a moral disorder called scrupulosity: becoming excessively concerned with the ethics of one's actions. Instead of taking these duties upon oneself, one ought to wait to assist the proper authorities in solving these sorts of problems, should those problems actually ever arise. Figure 6.2 shows these distinctions schematically.

While remote mediate material cooperation can be required, the gradations between it and the more questionable forms of cooperation require some teasing out. Why are some forms of cooperation allowable when others are not? There are ten criteria that can help to settle this question.

1. *Gravity of the harm*—The more seriously bad or evil an action is, the less should one be willing to cooperate with it.
2. *Proportionality*—If cooperating with the act would cause more good than evil (for example, saving a life by cooperating with a robber), then it may be permissible to cooperate.
3. *Intention*—Is your intent the same as or different from primary, and if different, in what way and why? Just because one has a different intention from an evildoer does not

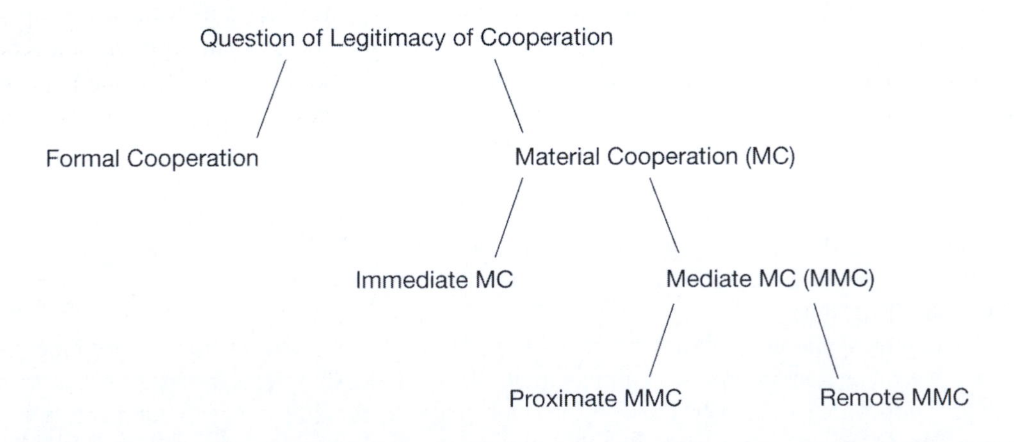


**Figure 6.2. Cooperation is classified based on intent, action, and moral distance. Formal cooperation shares the evil intent. Material cooperation does not share the intent, but if it participates in the same act, it is immediate cooperation. If it is a different intent and a different action, then it is mediate cooperation, and can run the spectrum from proximate to remote, depending on circumstances.**

automatically mean an intention is good: people can do the right thing for the wrong reason.

4. *Proximity*—Are you actually involved, and if so, how close are you, physically and ethically? Can you do something? Can you alter the way the decision maker acts in order to mitigate their evil activities?
5. *Duress/necessity*—If there is danger or force involved, one's culpability is diminished due to that duress or threat.
6. *Are you replaceable?*—If you are replaceable, then your refusal to cooperate may ultimately have little effect on the outcome of the situation. Alternatively, if your skills are unusual or unique, your refusal might completely stop the evil action, and therefore you are more responsible.
7. *Legitimate authority*—Do you have a right to act? If you are a public safety officer and you see a criminal, not only are you authorized to act to stop them, it is also expected of you, unlike with a regular citizen, who instead might be expected to run away from a criminal.
8. *Probability/risk of outcomes*—Is the act you may be (non)cooperating in really likely to have an effect? Is the evil outcome 50/50? Is it 90 percent certain? The more certain the evil outcome is, the more one should try to avoid cooperating.
9. *Confidence of moral judgment*—Sometimes you may be uncertain of your own moral judgment. The more certain you are in your judgment that an act is evil, the less you should cooperate. On the other hand, the more certain that you are that an act is not evil, the more you should cooperate. And if one is very uncertain either way, then one should think more, seek further counsel, and work to increase one's certainty one way or another.
10. *Scandal*—Last, choosing to remotely cooperate with evil may seem reasonable from one's own perspective, but from the outside it could cause a scandal, which, in a technical sense, means confusing or mixing good and evil in a public way: making evil look good or good look evil. Scandal is a serious problem because it damages ethical clarity and authority in a public way and ultimately can erode public confidence in ethics and discourage public ethical behavior.

Questions of legitimate cooperation with respect to space will be as inescapable as all of the dual-use powers that space exploration develops. Every dual-use technology has a good use that might legitimately be cooperated with and a bad use that perhaps should not be cooperated with. Exactly how to deal with these situations will require a deep analysis of the legitimacy of that cooperation.

## BACK IN THE BOX

What can foreseeable future problems involving the military use of space enlighten in our current world? All three of the inextricably connected topics of this chapter—the military, dual-use technology, and international relations—are also problems for us on Earth right now. Perhaps the lesson we can learn from the foreseeable problems of space exploration is that we ought to begin solving these problems now rather than later. International cooperation could be a higher priority. Governance of dual-use technologies could be a higher priority. Cooperative reductions in military spending could be a higher priority. All of these long-term, civilization-level problems tend to be overlooked in the day-to-day worries of global civilization. But they should not be; indeed, they are some of the most important problems of all, and we will not have forever to solve them.

## CLOSING CASE

The United States Space Force (USSF) was established on December 20, 2019. The United States had previously conducted space-related military activities under the Department of the Air Force as the Air Force Space Command, and even now, the Space Force is still under the Department of the Air Force, though now as its own uniformed service.[24] Of course, the existence of a "Space Force" raises the question of ethics. Is having a Space Force good?

There are certainly many things for the military to do that involve space, for example, keeping track of friendly and unfriendly satellites, thinking about power projection, communications and information awareness, logistics, and so on. In fact, conducting these activities within the Air Force had grated upon Air Force Space Command leadership for decades, as they were always of lower priority than core Air Force interests such as aircraft. Giving the Space Force more autonomy should free them to more fully concentrate on space. But will this make the United States—and the world—safer or less safe? Certainly the concerns discussed in this chapter highlight much of the complexity of that question. The USSF could spark an arms race, for example. It could provoke the leadership of the United States or other nations to violate the OST. It could deter enemies from attacking the United States, knowing that the "Guardians of the high frontier" (the old motto of the Air Force Space Command) are *Semper Supra*—"Always Above" (the new Space Force motto).[25] For now, the consequences of creating this branch of military service remain obscure. Hopefully good ethical judgment will lead the Space Force to be a stabilizing influence that enhances international cooperation and not something that causes risk or provokes destabilization.

## DISCUSSION AND STUDY QUESTIONS

1. What are your first reactions to the idea of military involvement in space? Why are those your reactions?
2. Do you think there should be military involvement in space exploration, use, or even settlement? Why or why not?
3. How do you think the OST ought to be revised in terms of military involvement in space? Do you think the OST is too strict on military involvement with space, not strict enough, or about right?
4. Of the ethical tools and concepts discussed in this chapter, which do you find to be of most interest and why? Which do you find to be of least interest and why?

## FURTHER READINGS

Dan Deudney, *Dark Skies: Space Expansionism, Planetary Geopolitics, and the Ends of Humanity* (New York: Oxford University Press, 2020).

United States Space Force, https://www.spaceforce.mil/.

United States Space Force, "Space Power: Doctrine for Space Forces," *Space Capstone Publication*, June 2020, https://www.spaceforce.mil/Portals/1/Space%20Capstone%20Publication_10%20Aug%202020.pdf.

## NOTES

1. Covault, "Chinese Test Anti-Satellite Weapon"; Geoffrey Forden (introduction by Noah Shachtman), "How China Loses the Coming Space War (Pt. 1)," *Wired*, January 10, 2008, https://www.wired.com/2008/01/inside-the-chin/.
2. "Historical Information 'Satellite Fighter'—Program," Army.lv, October 9, 2016 (in Russian), https://web.archive.org/web/20161009144831/http://www.army.lv/ru/istrebitel-sputnikov/istorija/894/342.
3. Peter Grier, "The Flying Tomato Can," *Air Force Magazine*, February 2009, 66–68.
4. Chethan Kumar, "India Shows Off Tech to 'Kill' Satellites, Will Also Help Tackle High-Altitude Missiles," *The Times of India*, March 27, 2019, https://timesofindia.indiatimes.com/india/india-shows-off-tech-to-kill-satellites-will-also-help-tackle-high-altitude-missiles/articleshow/68602482.cms.
5. "Chinese ASAT Test," *Center for Space Standards & Innovation*, December 5, 2007, http://www.centerforspace.com/asat/ and "ISS Crew Take to Escape Capsules in Space Junk Alert," *BBC News*, March 24, 2012, https://www.bbc.com/news/science-environment-17497766.
6. Douglas S. Anderson, "A Military Look into Space: The Ultimate High Ground," *Army Lawyer* (November 1995), quoting Donald B. Rice, "The Air Force and U.S. National Security: Global Reach—Global Power" (Washington, DC: U.S. Air Force Department, 1990); the phrase has developed and gained prominence over time through Thomas S. Moorman Jr., "Space: A New Strategic Frontier," *Airpower Journal* (Spring 1992): 14, 17; Steven J. Bruger, "Not Ready for the First Space War: What about the Second?" *Naval War College Review* 48, no. 1 (Winter 1995), 82; and Benjamin S. Lambeth, *Mastering the Ultimate High Ground: Next Steps in the Military Uses of Space* (Santa Monica, CA: RAND, 2003).
7. MacIntyre, *After Virtue*, 51–56.
8. Carter Scholz and Alan Brennert, "A Small Talent for War," *The Twilight Zone*, January 24, 1986, https://www.youtube.com/watch?v=fbT1fCHOjfI.
9. See, for example, Abney, "Ethics of Colonization"; Green, "Self-Preservation Should Be Humankind's First Ethical Priority and Therefore Rapid Space Settlement Is Necessary"; and Jonas, *The Imperative of Responsibility*, 43.
10. For example, David D. Friedman, *The Machinery of Freedom: Guide to a Radical Capitalism*, third edition (Chicago: Open Court, 2014).
11. For example, Aristotle, *Nicomachean Ethics*, 41 [III.6].
12. Paul P. Craig and John A. Jungerman, *Nuclear Arms Race: Technology and Society*, second edition (New York: McGraw-Hill, 1990), 319–34.
13. Stanley Kubrick, Terry Southern, and Peter George, *Dr. Strangelove or: How I Learned to Stop Worrying and Love the Bomb* (Culver City, CA: Columbia Pictures, 1964).
14. See, for example, Beth Simmons, "Treaty Compliance and Violation," *Annual Review of Political Science* 13 (2010): 273–96; Jonas Tallberg, "Paths to Compliance: Enforcement, Management, and the European Union," *International Organization* 56, no. 3 (Summer 2002): 609–43; Abram Chayes, Antonia Handler Chayes, and Ronald B. Mitchell, "Managing Compliance: A Comparative Perspective," in *Engaging Countries: Strengthening Compliance with International Environmental Accords*, ed. Edith Brown Weiss and Harold Jacobson (Cambridge, MA: MIT Press, 1998), 39–62; Abram Chayes and Antonia Handler Chayes, "On Compliance," *International Organization* 47, no. 2 (Spring 1993): 175–205; and Abram Chayes and Antonia Handler Chayes, "Compliance Without Enforcement: State Behavior Under Regulatory Treaties," *Negotiation Journal* 7 (1991): 311–30.
15. Paul Robinson, ed., *Just War in Comparative Perspective* (London: Routledge, 2003).
16. Green, "Ethics Is More Important than Technology"; and Green, "Emerging Technologies, Catastrophic Risks, and Ethics."
17. Arturo Casadevall and Michael J. Imperiale, "Risks and Benefits of Gain-of-Function Experiments with Pathogens of Pandemic Potential, such as Influenza Virus: A Call for a Science-Based Discussion," *American Society for Microbiology* 5, no. 4 (July/August 2014): e01730–14.
18. Keith A. Abney, "Space War and AI," *Artificial Intelligence and Global Security* (July 15, 2020).
19. See, for example, Jonas, *Imperative of Responsibility*; and Turchin, "Processes with Positive Feedback and Perspectives on Global Catastrophes."
20. Emphasis added; Dwight D. Eisenhower, "Chance for Peace," April 16, 1953, https://millercenter.org/the-presidency/presidential-speeches/april-16-1953-chance-peace.
21. See, for example, Bruno Verbeek and Christopher Morris, "Game Theory and Ethics," in *The Stanford Encyclopedia of Philosophy*, ed. Edward N. Zalta, winter 2020 edition, accessed March 5, 2020, https://plato.stanford.edu/archives/win2020/entries/game-ethics/; John R. Chamberlin, "Ethics and Game Theory," *Ethics and International Affairs* 3 (1989): 261–76; Thomas C. Schelling, "Game Theory and the Study of Ethical Systems," *Journal of Conflict Resolution* 12, no. 1 (1968): 34–44; John C. Harsanyi, "Morality and the Theory of Rational Behavior," *Social Research* 44, no. 4 (Winter 1977): 623–56; and Richard Bevan Braithwaite, *Theory of Games as a Tool for the Moral Philosopher* (Cambridge: Cambridge University Press, 1955).
22. Martin Luther King Jr., "The Man Who Was a Fool," in *Strength to Love* (Minneapolis, MN: Fortress Press, 2010 [1963]).

23. This section is derived from John Berkman and Brian Green, "Lecture 19: Cooperation with and Appropriation of Evil," Fundamental Moral Theology, Dominican School of Philosophy and Theology, Spring 2008. For further resources, see Kevin L. Flannery, SJ, *Cooperation with Evil: Thomistic Tools of Analysis* (Washington, DC: Catholic University of America Press, 2019).

24. United States Space Force, "About the Space Force," accessed December 31, 2020, https://www.spaceforce.mil/About-Us/About-Space-Force/.

25. James Richardson, "The U.S. Space Force Logo and Motto," U.S. Space Force Public Affairs, United States Space Force, July 22, 2020, https://www.spaceforce.mil/News/Article/2282948/the-us-space-force-logo-and-motto/.

*Chapter Seven*

# Protecting Earth from Hazardous Asteroids and Other Extraterrestrial Dangers

**Figure 7.1. The Tunguska blast area, Russia, taken by the Leonid Kulik expedition in 1929**

## THE TUNGUSKA BLAST AND CHELYABINSK METEOR

Sometime after 7:00 a.m. on June 30, 1908, witnesses in the sparsely settled and remote region within a few hundred kilometers of the Tunguska River, Russia, reported a bright light crossing the Siberian sky, followed by a blinding flash, concussion waves, and ongoing rumbling. The "Tunguska blast," also known as the "Tunguska event," was seen for hundreds of kilometers around and heard and felt across Europe and Asia. For several days afterward the night skies glowed as vaporized meteoric debris shrouded the upper atmosphere.[1] It is variously estimated that the Tunguska explosion was the equivalent of a nuclear weapon between 3 and 50 megatons.[2] At the time, the phenomenon was not well understood, and there are still mysteries related to the blast even today (such as whether it was an asteroid or a comet).

Fast forward nearly 105 years to 9:20 a.m. on February 15, 2013, in Chelyabinsk, Russia, west of Tunguska. Again, a bright light appeared in the sky, but this time it was recorded by myriad video cameras and scientific instruments both on Earth and in space. At a height of 27 kilometers the object burst in an explosion equivalent to more than 500 kilotons of TNT. Witnesses reported the light was blinding, some even receiving radiant sunburn-like effects to their skin. Within minutes after the brightest light of the explosion, a shockwave hit the city of Chelyabinsk, injuring over one thousand people.[3]

These meteors are just two examples of many in the kiloton to megaton range that have impacted the Earth in (relatively) recent history, mostly over the oceans.[4] Reaching farther back into time, the size of impactors grows, even causing global devastation, cooling the planet into an "impact winter" and contributing to mass extinctions, including that which killed the dinosaurs. Humanity has the historical record and astronomical data to make clear the dangers we are exposed to. We also have the scientific knowledge and technical power to perhaps do something about these dangers. Now the question is: What should we do about it?

## RATIONALE AND SIGNIFICANCE: PREPARING FOR CATASTROPHE

Astronomers and scientists recognized only several decades ago that **Earth-crossing objects**, including **near Earth objects** (**NEOs**, asteroids and comets that approach within 1.3 **astronomical units** of the Sun),[5] have been important throughout geologic time for delivering important elements and compounds to our developing planet. They also have been responsible for catastrophic impacts, including correlating to at least one global mass extinction at the Cretaceous-Paleogene boundary sixty-six million years ago. Some key issues with asteroid defense are detection, deflection, and decision making in an international context.

Following the previous chapter on the military, dual-use technologies in space, and international relations, the issue of dealing with dangerous asteroids is perhaps the ultimate extension of these questions. Detection, deflection, and decision making require international sharing of knowledge, the application of enormous amounts of very risky power, and the international cooperation necessary to make it happen. Thanks to concerted international effort, most—but not all—of the riskiest objects in the most obviously threatening orbits have been identified. Deflection may be possible, but we are not yet capable of rapidly implementing any of the possible methods. And recent simulation exercises have shown that international cooperation on decision making may not be as simple as we might hope.[6]

Additionally, while impactors should be one major focus of our extraterrestrial risk mitigation and adaptation efforts, there are other dangers this chapter will consider to a lesser extent, such as solar flares and **coronal mass ejections (CMEs)**, geomagnetic storms, supernovas, gamma ray bursts, and the eventual expansion of the Sun.

ETHICAL QUESTIONS AND PROBLEMS

### What can humanity do to protect itself against asteroids, comets, and other space dangers?

There are actually numerous proposed strategies for dealing with asteroids and other space-based dangers that may threaten the Earth. This section will not be an exhaustive list but merely an overview to show that many dangers exist and yet many options are available.

First, before any defensive measures can be taken against asteroids, before we can even set up systems to detect them, we must overcome the social and political resistance to addressing infrequent but catastrophic disasters. The good news is that a lot has been done in order to set up these systems already, though more can be done still. Perhaps the hardest challenges are the political ones focused on actually taking the problem seriously and devoting funds to preparation. The next level of political difficulty would be if a threat were detected, how should Earth respond? But while such things can be simulated in advance, they can never be fully planned before the specifics of the situation become real.

The next step is risk detection, which we are well on the way toward with large Earth-crossing asteroids. Thousands of these objects have been detected by various projects and missions, and while not all have been discovered, the majority have been and none seem to present a near-term threat. In addition to asteroids we should also be concerned about comets, which are harder to detect due to their relative rarity in near Earth space because they have longer and more distant orbits. Comets require another level of research and preparedness, but luckily, the difficulties with detecting comets (their rarity in Earth's locale and distance) also lower their probability of impact.

Asteroids and comets can both be defended against in four main ways. First, and perhaps most obviously, nuclear weapons can be used upon asteroids and comets. While the first thought might suggest blowing asteroids to bits with bombs, in reality, creating a cloud of debris is not a good solution to the problem. A better solution is to use the energy of the nuclear weapon, exploded at a distance from the asteroid, to induced heating in its surface and the vaporization of that surface material, which then acts like a diffuse rocket engine, nudging the asteroid onto a slightly different path. Of course, the Outer Space Treaty bans nuclear weapons in space, but for the purposes of protecting the Earth, presumably the majority of humanity would prefer to ignore the treaty or reclassify the nuclear weapon as something other than a weapon, for example, as a "device" or "tool" (redefinition being one tactic for avoiding moral and legal rules). A second approach is to use high-speed and heavy **kinetic impactors** to nudge the asteroid or comet into another orbit. This technique redirects the asteroid by transferring momentum from the impactor to the asteroid, thus slightly altering its orbital vector. A third option is to use lasers, **ion beams**, or **concentrated solar** energy to try to induce ablation on one side of the asteroid and thereby nudge it, similar to a nuclear weapon but from a potentially much greater distance and over a longer period of time. Unlike a nuclear bomb, which delivers all of its energy in one massive pulse, these methods would deliver energy over a longer time span, thus lowering the forces on the asteroid. Fourth, and more complicated, is the possible use of a **gravity tractor** in which one or more satellites go into orbit around the asteroid but consistently apply just a bit of engine power to move their orbit in a manner that pulls the asteroid along with them. This very subtle technique is the least violent and is less risky than the other alternatives; however, it would likely take more time to work than the other more violent approaches.

Moving onward from impactors, solar flares and CMEs are powerful waves of radiation and particles from the surface of the Sun caused by the activity of the Sun's magnetic field.

When these waves hit the Earth's magnetic field, they cause geomagnetic storms, which can damage satellites and destroy electrical systems and electronics. Flares and CMEs can be detected by telescopes watching the Sun and other **space weather** monitoring systems, and thus Earth can have several days' notice of bad space weather. During this time, vulnerable systems on Earth could be prepared for the impact of the geomagnetic storm, though it is likely that even with significant protective measures there could still be serious negative effects. This is at least partially due to the fact that wires are effective antennas for geomagnetic storm energy and would therefore collect the energy and direct it toward both ends—if electrical transmission wires, then both power plants and power consumers. Most crucially, the geomagnetic storm energy would be forced through the electrical transformers that step voltages up and down in power systems, which could cause the transformers to burn out, thus causing long-term, potentially months-long damage to the electrical grid. Because it takes electrical power to run the factories that make transformers, if there are insufficient transformers in reserve, then the disaster could be even longer and worse.

The simple defense against this, then, is to have reserve equipment available for quick replacement into power systems. Similarly, it would be highly beneficial to have the ability to disconnect large segments of the electrical grid so that geomagnetic energy does not propagate unimpeded until it destroys vital equipment. Satellites are another matter; however, there are ways to "harden" satellites against radiation and electromagnetic energy (indeed, militaries often do this with their satellites and other equipment in order to prepare for a nuclear-induced electromagnetic pulse). These shielding techniques can also be used on Earth to protect important pieces of electrical and electronic infrastructure.

Defenses against supernovae and gamma ray bursts would be again different from the defenses against the previously discussed risks. Supernovae and gamma ray bursts primarily cause their danger through the release of high-energy X-rays and gamma rays, which, while potentially quite lethal to unshielded life in space, would not propagate far through Earth's atmosphere. These rays would, however, have a potentially harmful effect on the atmosphere, damaging the ozone layer and creating oxides of nitrogen that would act like a global layer of smog, potentially darkening the sky, as well as combining with water to form nitric acid and acid rain.[7] While this effect could be terrible, it would not likely be catastrophic, unless it created something like a **winter** scenario analogous to **volcanic or nuclear winters**, which would disrupt agriculture. These winter scenarios are quite dangerous; however, there are solutions to make sure that everyone stays fed even despite such a disaster, for example, by rapidly scaling up mushroom farming and other forms of agriculture that do not require sunlight (of which there are surprisingly many).[8] For humans and other Earth life forms in space or on the surfaces of space objects that lack atmospheres, on the other hand, this wave of high-energy radiation could be quite deadly, so as we continue into space, this potential disaster ought to be kept in mind, and radiation shelters (which in any case are necessary for protection from solar flares and CMEs) should be constructed and available whenever possible.

## Should humanity be responsible for protecting the Earth from asteroid impacts?

The overarching question behind this chapter, indeed behind this entire book, is one of human power. Before the past few decades, humans have never been able to stop an asteroid impact—to completely avert a catastrophic natural disaster. Now, with some time and effort, we could. If such an event occurred without our being prepared for it, would we be guilty of **ethical omission** or a "sin of omission" —failing to do what we ought to have done? Has what was formerly excusable weakness become **culpable inaction**, one for which we should credibly be held "guilty" if we allowed it to happen?

It is unquestionable that humans have become a very powerful species, capable of taking very forceful actions over nature and each other (though certainly, only a few of us actually wield such power; as noted by C. S. Lewis, it is not that humans have power over nature but that some humans have power over other humans by using nature as an implement).[9] We have gathered the power of the stars in our hydrogen bombs, begun to change the Earth's climate, and landed men on the Moon. Every time a new power comes into our grasp, we become liable to the responsibilities that go along with that power. With nuclear weapons, we are responsible for nuclear war. With the advent of anthropogenic climate change, we are now responsible for maintaining Earth's climate. With space travel, we are responsible for human life and health, forward and backward contamination, as well as myriad other concerns.

The relevance of ethics for human power lies not only in how we apply the new powers themselves but also in how we choose not to apply them, whether intentionally or not. For example, we have the power to move objects in space—that is what spacecraft do. With this power and the understanding that asteroids can impact the Earth and that there are more asteroids out there that potentially endanger the Earth, we have inadvertently become responsible for preventing possible asteroid impacts upon the Earth. Because "ought implies can," meaning that we can only be ethically responsible for that which is within our power, "can" also potentially implies "ought," meaning that if there are ethical implications of our powers, then we are responsible for those implications and ought to meet those ethical expectations.[10] If one *has* power, then one *is* ethically responsible for it, whether one uses it or not.[11] Choosing not to act may make one liable for failures of duty: moral omissions.

In general, as we grow in power, how might we be able to identify when these types of situations of increased culpability occur? For certainly humankind now collectively knows many things and is capable of tremendous works, and yet, in general, we do not do many of the good things that we are capable of and instead use our powers to do many bad things instead. For example, it is easily within our power to feed everyone in the world; give them adequate water supply, elementary health care, and sanitation; and give everyone a primary education. Yet humanity, collectively, does not do this. But nowhere in history did we collectively *decide* not to do these good deeds, we have simply stumbled along this course of action as though we lacked these powers: because in the past we did. The world has changed, but our minds and cultures have not.

Thus, something of a change of mind should be in order. With each new technological discovery and advance in power we should consider what good we might achieve through it. This should mean a comprehensive search for good uses of our power: not just thinking of traditional good activities such as giving to the poor, healing the sick, and educating the young but more sophisticated tasks such as creating better systems for global economic enrichment, facilitating ways to develop cooperation and trust, sharing sophisticated medical technology with doctors in impoverished areas, using technology to improve education, and even literally outlandish ideas such as settling other planets and moons or diverting dangerous asteroids.

Interestingly, the artificial intelligence (AI) community is engaged in something like this now, with various efforts to support "AI for good" or "AI for social benefit."[12] Perhaps because in some ways AI is a dream technology that offers us the possibility of achieving so many different ends, perhaps that possibility has attracted dreamers who have seen both its promise and its perils and therefore have sought to get ahead of them. Whether they will succeed waits to be seen. In any case, other communities of technologists, including the space community, would do well to learn from the example of the AI community. Humankind has the power to make the world better, and therefore we have the responsibility, whether we recognize it or not.

### Beyond identification, how can we also gain the motivation to act correctly based upon this new responsibility?

Of course, merely identifying a complex and difficult problem for which humankind has newfound responsibility does not solve it. The next step beyond identification, with any problem in general and with asteroid dangers in particular, is to motivate ourselves to develop plans and specific means to solve the problems. It is one thing to have a spacecraft capable of reaching an asteroid; it is another to develop one specifically capable of affecting the asteroid's course, should that need arise.

While it might seem like avoiding the risk of global catastrophe ought to be motivation enough, perhaps strangely, this has not been an adequate motivation in nearly any case that humans have encountered. Nuclear war, climate change, and dangerous asteroids are just three areas where the human survival instinct has failed to solve the problem, partly because our interests are construed too narrowly both spatially (for example, at the national level) and temporally (for example, just the immediate future: the next few days, weeks, months, years, not decades or centuries). Without large-scale thinking in terms of space and time, problems that are spatially and temporally large will never be solved.

The motivation to do great things can be achieved, but it takes leadership as well as the proper political structures within which that leadership can operate. In the United States, President John F. Kennedy gave a speech that motivated America to go to the Moon, saying, "We choose to go to the Moon . . . and do . . . other things, not because they are easy, but because they are hard."[13] Despite his declaration that such a project would be difficult, Kennedy motivated his nation toward his vision of the future.

Certainly other leaders have motivated other great projects around the world, through various means, both good and evil. It will be challenging to find the right ways to use our increasingly powerful technologies to build a better future, but it is a challenge that we should be able to fulfill. Already leaders as different as Greta Thunberg and Pope Francis have provided inspiration and motivation on issues such as climate change, and political leaders in some nations have been quite effective in some ways. But the concerted effort of international cooperation and leadership remains.

### What of other extraterrestrial dangers?

There are other significant natural dangers in outer space besides just asteroids. For example, the Sun is quite capable of producing flares and CMEs that can induce geomagnetic storms that will wreak havoc upon human satellites, electrical systems, and communication systems. Solar outbursts of this kind have occurred in the past, one of the most remarkable being the "Carrington event" or "September 1859 geomagnetic storm," in which an extremely powerful CME hit Earth. Very strong auroras were seen in both hemispheres, and even into the tropics. Despite the world at the time being significantly less reliant on electromagnetic technologies, this storm still managed to interfere with telegraph systems, causing unusual effects such as large sparks, electric shocks, and the transmission of messages on auroral energy alone.[14] If such a magnitude event as the storm of 1859 were to occur in today's world, it would have the potential to cause years' worth of damage: destroying satellites, burning out components in the electrical grid, and ruining en masse electronics and computers. This would be disastrous and, depending on the magnitude, potentially even catastrophic for civilization, if the electrical grid were to be disabled for months or years and global communications collapsed.

Still pertaining to the Sun but on the opposite end of the risk spectrum is that of the gradual expansion of the Sun over stellar time scales. This expansion is absolutely certain but will not lethally affect the Earth's biosphere for hundreds of millions of years, perhaps even a billion years. However, eventually the Earth will become too hot, and a runaway greenhouse effect will lead to the oceans boiling and the death of all life on Earth. Billions of years after this, the Sun will grow into a red giant star that engulfs the Earth and sends the planet's remains plunging into the core of the Sun. Thankfully, even the near-term effect of this expansion—Earth overheating—is still extremely far into the future, much farther away from us living today than the age of the dinosaurs is away from us in the past. By this time humanity will either be long dead or we will engineer a way to maintain the Earth's biosphere (for example, with solar shades or other techniques of terraformation), and/or our descendants will be spread to other stars, carrying Earth life and civilization with it. Thus, while the expansion of the Sun is inevitable, it need not be an apocalyptic disaster for the human race if the people of the future take measures to protect the Earth and/or settle farther into space.

Shifting on the risk spectrum again, there are a few very low-probability and high-harm events that should be acknowledged, even if they are not worth serious worry. These less probable risks include nearby supernovas (nearby exploding stars that shower Earth in radiation) and precisely aimed gamma ray bursts (enormous explosions that shoot beams of gamma rays across the universe). In his book *The Precipice*, the philosopher Toby Ord describes the probabilities of these risks as "very small," with the probability of supernovae-induced catastrophe being something 1 in 50,000,000 in the next century, and the probability of gamma ray bursts being something like 1 in 2,500,000. Compared to other natural risks such as impactors or supervolcanoes, and anthropogenic risks such as nuclear war or misaligned AI, these risks are nearly insignificant.[15] Risk is always relative, and therefore very improbable risks are of much lower priority than other risks, typically anthropogenic, which might be a thousand to a million times more likely to cause catastrophe within the next century.

## ETHICAL CONCEPTS AND TOOLS

### International Trust, Coordination, and Collaboration with Dual-Use Technologies for Asteroid and Comet Deflection and Redirection

Words such as trust, coordination, and collaboration need some defining before this issue can be parsed. **Trust** means believing that another entity will attempt to work with you in good faith, to carry out duties with you as agreed upon. **Coordination** means working in a cooperative fashion, even if not directly intentional or trusting. **Collaboration** means working together in a cooperative fashion, intentionally, with some level of trust.

International trust, coordination, and collaboration are hard enough on Earth; moving these difficulties into a new frontier will only complicate matters. However, there are clear reasons why coordination ought to occur when dealing with issues such as asteroid redirection because not only could the scale of destruction be planetary but the effort to divert an asteroid could require the resources of several countries. Additionally, there are other clear benefits to international coordination in space. Because the environment is so hazardous and human life so fragile, perhaps it will be clearer that coordination is necessary because our main "competitor" in space is not primarily other humans but the space environment itself.

The issue of trust, however, is a different matter from that of coordination. Coordination can occur even in situations without trust, given the right circumstances (for example, some forms of market economy allow for coordination without trust). Indeed, with respect to space currently, there is fairly good international coordination when it comes to launches, satellites, the International Space Station, etc., yet many of the nations involved cannot be said to fully trust each other, for example, the United States and Russia. Trust is something that is earned over time by being *trustworthy*, which includes fulfilling one's duties in a competent and faithful manner. Developing a trustworthy reputation is crucial for individuals, organizations, and nations, and failure to develop a trustworthy relationship can greatly impede coordination between parties and prevent collaboration.

Collaboration involves more intention and trust than coordination. Two or more parties operate toward a common goal with assurance that the other parties will also fulfill their obligations, thus leading to the desired outcome. When it comes to space operations, coordination is necessary, but collaboration is desirable. The risks inherent in space require coordination or disastrous consequences could occur at nearly any time. Given this situation, without trust, coordination could all too quickly collapse due to second guessing and the deliberation required whenever it is necessary to question the motives of others. With trust, coordination is more likely to succeed due to faith in the other parties and the accelerated speed that can be had when it is not necessary to question the motives of others. Obviously, in the fast-paced environment of space, where orbital velocities are in kilometers per second and automated responses may be necessary in many cases, it is not only important to trust other nations and the individuals from those nations but to also be able to trust their equipment and computer systems, whether made by close collaborators or otherwise.

When it comes specifically to trust, coordination, and cooperation surrounding the issues of asteroid and comet detection, deflection, and decision making, various scenario exercises have demonstrated the difficulty in achieving good outcomes. NASA's Jet Propulsion Laboratory's simulation exercises are conducted on a regular basis, with groups within the simulations playing different roles, including playing representatives from various nations on Earth.[16] At times the representatives disagree, and at other times they agree, but at all times the coordination is unsteady and vulnerable. Cooperation cannot be taken for granted, and neither can even basic coordination. As humankind moves further into space we need to become better at coordination and cooperation, and that means humanity needs to become better at trusting each other, which first requires that more of humanity—in the form of the national governments in which we live—be trustworthy.

## Cognitive Bias in Ethics and the Human Inability to Properly Assess Risk, Especially of Rare Yet Catastrophic Events

**Cognitive biases** are biases in the way that people think; they represent trends in thought that can be either good or bad, depending on the context. Humans tend to be cognitively biased toward familiar hazards, such as criminal activity or local disaster conditions, and think little of rare but much larger disasters. However, we know from the history of the Earth, as well as scientific observations within our solar system, that rare catastrophic events, which are beyond recorded human history, do occur and can be devastating.

These events include not only extraterrestrial risks such as asteroid strikes and the others noted earlier but also terrestrial risks such as natural climate change, oceanic anoxia, pandemics, and massive volcanic eruptions (including supervolcanoes and flood basalts).[17] The SARS-CoV-2 coronavirus pandemic (causing the disease COVID-19), which began in 2019 and expanded worldwide in 2020, provided a prime example of this bias toward the familiar, with nations of the world that had recently dealt with other pandemics responding in a swift and orderly manner, while other nations that had perhaps not had to respond to a pandemic since the 1918 influenza tending to respond in a less effective manner (though certainly more variables were involved).

Humans are biased so that we are much less likely to properly consider the risk of rare, high-impact events relative to the risk of common, low-impact events. Anthropologist Brian Fagan has asserted that humans are also predisposed toward trading frequent small disasters for less frequent but larger disasters, for example, trading decade-frequency small floods for century-frequency massive floods.[18] This is at least partially due to human memory: we remember things that happen frequently and forget those things that do not. But it also has to do with several clear cognitive weaknesses:

1. A bias toward optimism: assuming things will turn out fine, that one is special or lucky, forgetting how bad things can be, etc.
2. Biases that make us want to socially "fit in": peer pressure, desiring to act the same as others and not seem different (for example, by caring about infrequent risks).
3. A bias toward the familiar: discounting unusual ideas, assuming the future will be the same as the present.
4. Fatalism biases that manifest as **TINA**, "there is no alternative," and **IIDDISEW**, "if I don't do it, someone else will."
5. Biases that excuse us from responsibility and/or hope someone else will do the work, sometimes called "**ISEP**": "it's someone else's problem."

There are several more of these biases, as noted in Eliezer Yudkowsky's instructive chapter on the topic: "Cognitive Biases Potentially Affecting Judgement of Global Risks."[19] Geographer Jared Diamond's book *Collapse* also considers some of these biases, with respect to societal collapse, classifying them as **rational biases** (ones that make sense by benefitting at least one person) and **irrational biases** (ones that do not make sense because they benefit no one).[20]

What should we do in the face of strong biases that endanger our civilization? The response should be to directly fight these irrational impulses by institutionalizing systems and organizations that both combat the biases directly (for example, schooling, media, etc.) and that directly combat the risks themselves (for example, focus on responding to infrequent yet high-impact disasters). These efforts do exist but have met with slow progress.[21] More efforts are needed.

## Consequentialism and Virtue Ethics as Approaches to Catastrophic Risk

Consequentialism highlights some important points in asteroid scenarios, partially because it works in a way that runs counter to our cognitive biases toward the familiar. Because consequentialism takes seriously the impacts of events, consequentialist ethics takes seriously the possibility of even rare events such as asteroid strikes. For example, the effective altruism movement, inspired by utilitarian thinkers such as Peter Singer and others, is one of the major movements thinking about existential risks, including asteroid impacts.[22]

However, just as consequentialism succeeds in overcoming some human biases with respect to certain types of disasters, it fails in the same way: very few people are drawn to thinking about the world in this way. This is something of an intrinsic difficulty, unless consequentialist movements can determine ways to bypass common opinion and institutionalize their ideas. These institutionalized consequentialist preparations are already seen to a limited extent when governments prepare for war, prepare for pandemics, search for NEOs that could impact the Earth, and so on, but very few nations do most of this work. There is a great incentive to be a **free rider** if others are doing the work, some such work is often more for show than for real effect (for example, military **deterrence**), and the best laid plans can often result in less than effective outcomes.

Virtue ethics, rather than focusing on the consequences of decisions and actions, instead focuses on helping people become good decisions makers and promoting overall virtue in a population. Philosopher Shannon Vallor makes the case for emphasizing virtue with respect to future ethical decisions by noting that we currently live in a state of "**acute technosocial opacity**" in which it is increasingly difficult to predict even the near-term future.[23] Given this situation, consequentialism becomes an increasingly difficult and complex endeavor (perhaps in some cases verging on useless), as unpredictability and volatility gain sway over events. However, there is a solution to this problem: we may not know the contents of what the future may bring, but we do know that it will be *human decision makers* who will be there making decisions. Even if those human decision makers delegate their powers to AI agents or other automated systems, it will still be humans who made those decisions to delegate, and because of that, these humans ought to be virtuous, ethically focused people who have good judgment.[24]

## Responsibility for the Common Good

The common good is a moral framework with ancient origins in the communal life of humanity. The common good should not be thought of as merely the good of individuals writ large, nor as the aggregate of a utilitarian calculus, but rather as the conditions for the flourishing of society, including relationships, social institutions, communities, etc.[25]

A crucial aspect of the common good is that because all of the people in a society can rise and fall together (as frequently happens in disasters such as wars, hurricanes, earthquakes, etc.), all of the people in a society need to cooperate in order to prevent or overcome hardships together. However, not all people in a society have the same resources available in order to promote these efforts for common good. The poor and powerless may have little ability to help their own situation, much less that of others (though certainly heroism of this sort does exist). On the other hand, the rich and powerful are in a position not only to help themselves but to help others as well. While governments tax the public in order to (hopefully) promote the common good, societies that are more unequal lay more of this task on individual wealthy people. If these resource-rich people step up to the task, then society improves, as with numerous philanthropic initiatives in many countries around the world. However, if these individuals do not step up and instead allow society to continue to degrade while they live above the fray, it becomes the duty of the government to promote the common good by seeking the resources that it needs in order to do so; that is, taxes.

Of course, taxes should always be at the lowest level possible that also allows for the promotion of the common good. Exactly how low this can be is often subject to long debates and varies greatly depending on circumstances. In a war or other disaster more resources are necessary; and even if wars and other disasters have not happened but may happen, it makes sense to prepare. And when it comes to taxation, just as it makes no sense to drill for oil where there is no oil, it makes no sense to tax society where there is no money, for example, taxing the poor. Rather, it makes sense to tax the wealthy, in a progressive taxation scheme, not a regressive one. How does this apply to space?

Specifically with respect to space-originating disasters, preparation can be expensive. Detection of NEOs requires telescopes and humans to build and use them. Deflection of potential impactors requires research, technological development, testing, and powerful launch capabilities. And decision making requires international coordination, cooperation, and preferably trust, which in itself can be quite expensive to promote. The rarity of these disasters then raises the question: Is the cost worth the benefit?

While cautionary tales abound, the calculation of "worth it" with regard to safety and risk will always be partially subjective. If the cost is too high and the benefit too low, the wealthy tax base may flee to cheaper nations where they can live as free riders, possibly benefitting from efforts that reduce risk while not paying for that benefit. If the cost is too low, the systems created to protect the common good may be inadequate to the task, and all may pay a high price in the form of a disaster. In any case, defending civilization is a costly endeavor and quite underfunded considering the dangers, so hopefully more resources will be devoted to it in the future, whether through philanthropy or through government tax dollars.

## The Importance of Truth for Ethical Decision Making

Having truthful information about the situation one is in and the options one might have in the context of that situation is absolutely crucial for ethics. Making decisions without adequate information can be a foolhardy endeavor. In any courtroom or other place that is pursuing justice, the first thing to do is to establish the facts of the case. In political regimes in which lies replace truth, the social situation tends to rapidly degrade as political reality diverges from social and natural reality. And in science and technology, when the full causes or effects of actions and decisions are unknown, great damage can occur relatively rapidly, as with environmental toxins, climate change, and so on (note that this relates to the precautionary principle: if there is a plausible but as yet uncertain risk, one can justifiably attempt to prevent it if the damages are beyond the acceptable level of risk).

In space ethics, this need for accurate information can be a very difficult matter. For example, in the early days of the space program, leaving space junk in orbit might not have seemed like a serious ethical problem. However, a little bit of thought, as Kessler and Cour-Palais did in 1978, would have made the problem apparent.[26] Because humans prior to this either lacked the facts or had the facts but did not care, early space missions began to cause the accumulation of junk in Earth orbit, thus leading to the situation that we have today.

With regard to asteroids, we have also learned some morally relevant facts in the past few decades. We know that asteroids do sometimes hit the Earth; in fact minor hits are rather frequent. We know that major hits occur at a lower rate but that when they do hit, they can cause massive devastation, such as the extinction of the dinosaurs. We also know that we could do something about these asteroids if we put in the effort. With these three facts, we should realize that we have enough information to make a moral judgment that we are culpable and therefore ought to act.

If we lacked this information, we would be in a completely different situation, ignorant of the threats of space and ignorant of our ability to respond to it. Likewise, if some people had discovered this threat and then lied about it, either never telling the rest of us or actively denying it, we—humanity overall—would similarly be oblivious and unprepared. And while there is a saying that "ignorance is bliss," ignorance is not bliss when it is about a growing problem that one has the ability to solve, if only one knew about it.

Ethics requires accurate information. We need not only the sensors and systems to compile the data that we collect but also the theory to understand it and the ethical-social-political know-how to know what to do with that significant information. **Ignorance** can be **culpable** (ignorance that is one's fault) or **nonculpable** (ignorance beyond anyone's fault), but in a world where humankind knows more and more, our ignorance is increasingly culpable if we fail to draw inferences from the data before us.

## The Parts of an Ethical Action, Commission versus Omission, and Second-Order Intentions

Just as ignorance (lack of knowledge) can be culpable or nonculpable, lack of action can be ethically blameworthy or nonblameworthy as well. Justice and injustice are not only the products of actions but can also be the products of inaction. Indeed, perhaps much of the injustice in the world is due to inaction rather than action, as we blithely go about our lives and fail to give justice, in all its forms and requirements, its due regard.

The choice not to choose is a choice in itself. Simply not acting does not excuse any person or organization from culpability. But what exactly are some of the relevant moral differences between **commission** and omission?

Commission implies an action taken that could have been not taken. It may not have a direct and/or intentional relationship to the desired outcome, for example (recalling double-effect reasoning, see chapter 4), space debris exists as a side effect of actions that were committed intentionally. The creation of the debris was foreseeable in that other choices could have been made that might not have created the same level of problem. However, the effects of that orbital debris were not intentional, and more indirectly, the potential for a Kessler-like debris cascade was not intentional, *and yet* through intentional decisions such a situation was foreseeable and may be in process, in a slow sense, even now. Acts were committed that yielded this outcome, and therefore the current situation with orbital debris is an act of commission.

Commission of an action involves three parts.[27] First is the action itself. With respect to orbital debris, the action is simply dumping debris in Earth orbit in order to be done with it. Second are the layers of intention behind the action. With respect to orbital debris, the intention was not to create a debris field around the Earth but rather other simpler intentions such as getting rid of unwanted rocket parts, abandoning satellites without the cost of deorbiting them, testing antisatellite weapons to demonstrate military deterrence, etc. However, these simple intentions had the implicit further effect, wanted or not, of polluting Earth orbit with dangerous debris. This double effect was foreseeable and not intended but nevertheless is now having a disproportionately negative effect on the Earth's orbital environment. Third are the circumstances surrounding the action. In the case of orbital debris, the problem in the past was minor, but this minor problem is now becoming a major one as orbital conditions have changed and entire orbits have become unusable due to debris. Other circumstances include improving knowledge of the situation, improving technology for preventing orbital debris and potentially to clean up orbital debris, and the growth in numbers of actors involved in space and the overall growing military and economic importance of space for our civilization on Earth.

With the choice to redirect an asteroid for the purposes of avoiding impact with the Earth, the action itself is to alter, via various means (which need specification), the orbit of an asteroid. The first intention would be to cause the asteroid to miss impacting the Earth. A further intention might be to make the asteroid permanently nonthreatening to Earth or to capture the asteroid so that it may be mined or otherwise made use of. The circumstances around the case would involve the specifics of the means of redirection, the specifics of the results of the act, any anticipated side effects of such a redirection (for example, will the asteroid need to be redirected again at some point in the future, or will it impact another solar system body?), etc. There is a saying that "circumstances make the case" because circumstances really can completely alter the nature of moral actions. Redirecting an asteroid to avoid the Earth really is quite different from redirecting an asteroid to impact the Earth. Changing the circumstances, for example, which side of an asteroid a nuclear device is detonated on to alter its course toward or away from Earth, can make the case for the action being good or bad (as well as the intention). There are many other relevant circumstances as well, in fact, sometimes approaching infinite, and therefore not only is fact gathering very important when it comes to ethics, but so is knowing when to stop gathering facts because further data has become too cumbersome, uncertain, or irrelevant.

Due to the potentially massive complexity of circumstances, ethics is always bounded by expectations of what is reasonable, including limitations on our abilities to predict the future and take into account details that don't seem relevant. These assumptions can be wrong, even terribly wrong, but they can be justified due to expectations of what are reasonable limits on what a human being or group of humans are reasonably capable of foreseeing. There is no way to make ethics perfect; ethics is always the best we can do, given what we know, because our knowledge is always limited. The **reasonability standard** is not always a clear standard, but given the complexity of the universe and circumstances it is often the best that we can do.

Returning to the topic of omissions, omitted acts take the three previously discussed categories and invert them in some ways. No action is taken in the first place, which is, of course, a major difference and might seem at first to be morally exculpatory. However, there are many cases in which actions are not done that ought to be done, for example, if a child is in one's care, one ought to care for it, and omitting that care would be potentially criminal. Likewise, if we see someone injured and in need of help yet do nothing, this can sometimes be morally or even criminally blameworthy, depending on our role in the situation (as a prime example, a medical doctor who refuses to help someone who is seriously injured in his or her place of work could be legally culpable for negligence).

The role of intention in omission is an interesting one. Often there is no specific intention to harm by failure to act, but sometimes there is, as perhaps in the previous case of the medical doctor. More often, however, omissions are due to higher prioritization of other actions. No one intends for an asteroid to hit the Earth and harm or even end human civilization; however, it might happen simply because we are prioritizing other actions, such as developing the economy, competing in geopolitical struggles, paying attention to entertainment, etc. This then raises the matter of **second-order intentions**, goals, and desires, or *what we should intend to intend*. We should *desire to desire* the good, for example, and our goals should be good goals. This is not mere tautology; instead it reveals a deeper truth that sometimes, while we may think we are spending our effort on good things, they are not in fact the best good things but lower ones. In other words, we could be intending better if we were being more intentional.

Last, the circumstances again make the case. If an asteroid is bound toward the Earth and humankind has the power to act, but no one acts to do anything about it, then a grave evil will occur that might have been prevented. This is extremely ethically blameworthy. However, if we back up millions of years into the past, could we say the same thing about the impactor that hit at the Cretaceous-Paleogene boundary? Of course, creatures of the time had no technological civilization and no ability to protect themselves. They were hapless victims of a terrible event. (Yet by severely damaging the environment, causing a mass extinction, that impact correlates with the subsequent rise of mammals, and therefore humankind, thus demonstrating the potential vagaries of judgment of ethical consequences: something good came of something bad.)

Of course, we do not need to back up so far in history to see how circumstances change the case. If we back up only one hundred years into the early twentieth century, failing to prevent an asteroid impact would not have been morally blameworthy for many reasons, including our lack of knowledge of such things, our lack of understanding of the literal impacts of such events, and our lack of ability to do anything about such a disaster. Humans at that time were completely incapable of any action in response to an asteroid impact and therefore had no moral responsibility. They were mere **moral patients**, not **moral agents**. However, we no longer live in that time, even though we may feel as though we still do. The scale of our feeling of moral responsibility has not kept pace with our actual moral responsibility. In the case of humanity, where we highly value ourselves and our environment (which of course would also be devastated by an asteroid or comet impact), and we have the knowledge, understanding, and power to act, we now have a moral imperative to act responsibly and not merely be acted upon.

## The Ethical Difference between Natural and Human-Made Disasters

While obviously humans are responsible for human-made disasters, it would seem that no one is responsible for natural disasters. However, as explored in the previous section, this may, in some cases, no longer be true. Asteroid impacts are a type of natural disaster that is now potentially preventable. Other natural disasters such as earthquakes, volcanoes, cyclonic storms, tornadoes, floods, and so on are perhaps not yet as avoidable, however, there are proposals (of varying levels of practicality) for ways to potentially limit or control some such disasters.[28]

Importantly, it should be noted that the categories of "human-made" and "natural" disasters are becoming blurred as human power increases. For example, the 9.1 magnitude Tohoku earthquake in Japan in 2011 triggered a tsunami (together killing over fifteen thousand people), and both in turn contributed to triggering three nuclear reactor meltdowns at the Fukushima Daiichi nuclear power plant. The power plant could have been built farther uphill, out of the reach of tsunamis, or with longer-lasting battery backups or myriad other safety improvements, or not built at all, but because of choices made—choices with severe ethical ramifications—one of the world's worst recorded earthquakes also became one of the world's worst nuclear disasters. Of note is that despite the devastation, much of Japan's infrastructure survived the earthquake and tsunami with minimal damage, thus demonstrating that even such a terrible disaster as this can be withstood if there is proper preparation.

While the Tohoku earthquake triggered a human-made disaster, human activities can also trigger what we might typically have considered to be natural disasters. For example, **induced seismicity** is a well-known phenomenon in which human activities trigger earthquakes. Causes of induced seismicity can include geothermal energy generation (typically associated with underground water injection), hydraulic fracturing ("fracking") of underground rocks (for hydrocarbon exploration), underground wastewater injection, mining, and reservoir impoundment. This last phenomenon—common enough to have a name: reservoir-induced seismicity—correlates to, for example, the 2008 Sichuan earthquake, an 8.0 magnitude temblor that killed over sixty-nine thousand people, which was associated with the nearby Zipingpu reservoir (though a causal connection cannot be conclusively proven).[29] Human activities have even caused the eruption of a mud volcano in Indonesia in 2006,[30] though some dispute that nearby hydrocarbon exploration caused the volcano.[31] In both of these cases, if human activities triggered these events it was due to providing the activation energy needed to release naturally occurring stresses. But in all of these cases, the line between a purely human and purely natural disaster has become unclear indeed.

Now for the ethical question: *Should* humans have a responsibility to develop responses to, or even controls for, natural disasters? Certainly it is prudent to prepare for known disasters that are common to an area of the planet, whether they be earthquakes, volcanoes, storms, etc. Structures ought to be strong enough to withstand the forces to which they will be exposed, whether the forces are seismic, fire, wind, or water. With adequate preparation, even incredibly powerful disasters such as the Tohoku earthquake and tsunami can be much less terrible than they would otherwise be (for example, compare the Tohoku earthquake with the 1923 Kanto earthquake and fire, when building standards were lower, which killed over 105,000 people). Much of the destruction of natural disasters is actually destruction caused by poorly designed and prepared human technology.

The most effective ways to deal with disasters will vary depending on the disaster type. For example, it would not be reasonable to design Earth structures to withstand an asteroid impact (perhaps leading to the movement of human civilization underground) when a much more cost-effective solution (detection and deflection planning) could be pursued. Instead, for each disaster type, the most appropriate means of response need to be determined, and then those pursued. There are certainly many more wrong ways to solve a problem than there are right ways, and efficacy depends a lot on the circumstances of the situation.

While we do have a responsibility to prevent what evils we can—for example, wildland fires can be partially limited by active land management, and asteroids impacts potentially prevented by diligent detection and deflection measures—many natural disasters are simply beyond our ability to prevent, and there do not seem to be realistic ways to completely prevent them in the foreseeable future: for example, volcanoes, cyclonic storms, tornadoes, floods, etc.[32]

Given this difficulty, we should only try to do what we reasonably can do, given the circumstances. For example, human-caused disasters are completely under our control and therefore ought to attract much of our attention for disaster prevention. Nuclear war is completely preventable, though it is obviously a terribly difficult situation to completely resolve. Climate change, in contrast, is already happening and needs to not only be stopped but reversed, and the environment restored to an earlier state, which will be exceedingly costly (though much less costly than the property losses incurred by flooding every coastal city in the world). Natural disasters are different and each will require their own approach. Impact prevention ought to attract our attention because it is not only exceedingly beneficial to avoid the horrors of an impact but also relatively easy compared to dealing with an impact if we prepare ourselves for it now.

## The Ethics of Risk Mitigation versus Adaptation

When it comes to risks associated with climate change, "mitigation" means "avoiding the unmanageable" and "adaptation" means "managing the unavoidable."[33] These terms work for other areas of risk reduction as well,[34] including space ethics, if we recognize the general rule as well as the shift in context: the role of mitigation is to reduce the risks we are faced with in the first place, while the role of adaptation is to prepare for the risks that cannot be reduced. If we completely avoided human space travel, we would completely mitigate the risks of space travel to human beings. But if we do choose to send humans into space, then we should mitigate the other risks we are exposed to, such as avoiding traveling into planetary radiation belts such as those around Earth and Jupiter, etc. We could prepare to send humans into radiation belts, but this could easily become unmanageable, not to mention unnecessary. Adaptation, on the other hand, seeks to manage the unavoidable, for example the "normal" risks of space exploration discussed in chapters 3 and 4. Dealing with the unavoidable risks of human spaceflight requires thinking about radiation, microgravity, and so on, all areas that have been the subject of detailed research ever since human spaceflight began.

Another way to think of what mitigation and adaptation mean is that mitigation addresses the probability side of the risk equation (risk = harm × probability), while adaptation addresses the harm side of the equation. In other words, mitigation attempts to reduce the probability of disasters and/or decrease their scale and intensity, while adaptation seeks to endure and survive the disaster when it happens. For example, through the mitigating efforts of international cooperation we might be able to reduce the probability of war or decrease the magnitude of a war from nuclear to conventional. Through adaptation, the harm of war could be reduced by speeding victory, reducing suffering, or, in the worst-case scenario of nuclear war, surviving a nuclear attack by, respectively, preparing a strong military defense, stockpiles of emergency resources, and nuclear bomb shelters.

However we analogize mitigation and adaptation, the point is that mitigation attempts to avoid the worst of disasters, while adaptation lives with the remaining component of a threat that cannot be mitigated. For example, because earthquakes cannot be prevented, instead earthquake-prone areas typically develop construction standards that adapt to the risk by either decreasing the likelihood of building collapse (for example, building out of flexible materials) or reducing the harm if a building does collapse (making them out of light materials or automatically shutting off gas in case of leaks, etc.).

Returning to space, asteroid impacts can be mitigated through comprehensive efforts at detection and development of the necessary deflection technologies. Through these means the asteroid-induced disaster might be completely avoided or greatly reduced (if, for example, the asteroid merely grazed the upper atmosphere and returned to space, without an explosive impact). Adaptation to possible impacts might include some sort of shelter preparation or development of food supplies to help people survive an "impact winter" scenario, but it is worth realizing that impacts on Earth are so rare—as well as potentially being so severe—that these adaptations might not be worth it. (For example, there is a narrow window in which particular adaptations are helpful. If an impactor is small, less than 1 kilometer, then an impact event is more probable but less harmful, perhaps not worth making serious global-scale adaptations against. If an impactor is larger than 10 kilometers, then the event is extremely rare and at the same time so harmful that no adaptation is really possible. Therefore only mid-level threats from 1 to 10 kilometers should be adapted toward, being both common enough and harmful enough to adapt toward.[35]) Surely it might make sense to take these two proposals seriously and create shelters for a certain number of people as well as preparing food supplies, but it might not be realistic to do much more than that, though certainly some countries have prepared extensive civil defenses for nuclear war (for example, Switzerland and China) that might also be usable for impact scenarios, so there are ways to utilize adaptations that some nations have already taken or repurpose adaptations to disasters that are more probable, such as nuclear war.

When it comes to the ethics of mitigation and adaptation, the first thing to consider is that mitigation, under certain circumstances and with adequate preparedness, can sometimes completely prevent disasters from occurring. Asteroid impacts are now at least partially in this category: with adequate preparedness and the right circumstances, an impact might be prevented. However, no amount of mitigation can completely reduce all risks to zero. Consider, for example, that in 2020, comet NEOWISE (designated C/2020 F3) was first detected in March 2020 and passed closest to the Earth in July of that year. While comet NEOWISE was never near enough to the Earth to pose a risk for impact, if it had been heading toward Earth it is unlikely that much could have been done: comets are too fast and the time to prepare was too short. Comet NEOWISE was approximately 5 kilometers in diameter, more than enough to cause a global catastrophe and even pose an existential risk to human survival.[36]

While mitigation and adaptation toward disaster are noble goals, and a reasonable level of preparedness is ethically necessary, we must also remember that complete safety is an impossible goal. Mitigation and adaptation can be tremendously beneficial, as, for example, various national responses to the SARS-CoV-2/COVID-19 pandemic have demonstrated: some nations were more prepared than others, those that were more prepared are ethically commendable, and those that were less prepared are ethically culpable—that is, blameworthy—for that lack of preparedness. But even while some nations responded very well to the pandemic, those same nations might still be harmed by flooding, earthquakes, volcanoes, and so on. Preparedness is sector specific, though many sectors do have overlap in preparedness.

Perfection is impossible, but better and worse are very real. We can do better, not only with respect to asteroids and other extraterrestrial source of risk but also with the myriad sources of risk we engage here on Earth, whether natural or anthropogenic, right now.

## BACK IN THE BOX

While space holds numerous threats toward humanity, we also face many threats right here on Earth without looking to space. There are many potentially catastrophic situations, both natural and anthropogenic, that require using powerful technologies in the midst of international cooperation. The natural challenges include natural climate shifts, ocean anoxia, pandemics, and large volcanic eruptions. Anthropogenic challenges include nuclear war, bioweapons, attacks on cyber-physical infrastructure, malicious or misaligned AI, nanotechnology, environmental toxicity, climate change, climate engineering accidents and climate warfare, ocean acidification, and uses of redirected asteroids as weapons.[37] What can we do to prepare for such risks? What institutional structures should exist in order to deal with these challenges? Thinking about space can help us recognize that our situation here on Earth is not only so much better than anything we will find in space any time soon but also so much more precarious than we appreciate, and the risks we face are in need of much greater attention. Policies should be adopted to help mitigate and adapt to these risks and there should be an overarching international strategy to maximize efficacy. But who should enact such preparations and pay for them? As always, action is owed by the powerful more than the weak, and costs can only be paid by those who have money, not those who do not. Simple logic dictates that the more powerful nations lead the way and the richer nations pay the cost. But hopefully as our world develops together toward the future, all the people of the world will become more powerful and wealthy—as well as wise and good—and the burden of care for the Earth and humanity can be shared by all.

## CLOSING CASE

On March 13, 1989, two CMEs hit the Earth's magnetic field in quick succession. In something of a "perfect storm," the CMEs burst forth from the Sun over a day apart but arrived within seven hours of each other, due to the first CME clearing the way for the second. Upon reaching the Earth, electrical systems in several nations began to experience large fluctuations due to geomagnetic energy, and in Quebec, Canada, the effects were particularly pronounced. Satellite images showed an intense auroral band extending around the north pole from west to east, spanning from near Hudson's Bay to nearly Chicago, which began to wreak havoc on electrical stability. Geomagnetically induced currents were driven through the Hydro-Québec power system, causing a variety of effects upon transformers and other equipment until eventually safety systems shut everything down. Similar effects occurred in the United States, United Kingdom, and Sweden.[38] While the primary initial effect of this event was interrupted electrical service, one beneficial secondary effect was the recognition by power system operators that they were vulnerable to geomagnetic storms and their subsequent decision to better harden their systems against them.

## DISCUSSION AND STUDY QUESTIONS

1. What should be the justifiable limits of humankind's responsibility? Why? What should be the justifiable limits of an individual's responsibility? Why?
2. How much effort should humankind put into searching for asteroids and comets that might threaten Earth? Why that amount of effort and not more or less?
3. Should humankind try, ultimately, to eliminate all disasters, both natural and human made? Why or why not?
4. If some types of disasters should be eliminated, why those, and if some types of disasters should not be eliminated, why not those? What is the reason for drawing the line between the two categories where you think it should be?
5. Of all the types of disaster discussed in this chapter, which do you find to be the most neglected and in need of more work? Do you think any have received too much attention?
6. Of the ethical tools and concepts discussed in this chapter, which do you find to be the most or least useful and why?

## FURTHER READINGS

Nick Bostrom and Milan M. Circovic, eds., *Global Catastrophic Risks* (Oxford: Oxford University Press, 2008).
Toby Ord, *The Precipice: Existential Risk and the Future of Humanity* (New York: Hachette, 2020).

## NOTES

1. Giuseppe Longo, "Chapter 18: The Tunguska Event," in *Comet/Asteroid Impacts and Human Society, An Interdisciplinary Approach*, ed. Peter T. Bobrowsky and Hans Rickman (Berlin: Springer-Verlag, 2007), 303–30.
2. Olga P. Popova et al., "Chelyabinsk Airburst, Damage Assessment, Meteorite Recovery and Characterization," *Science* 342 (2013).
3. Ibid., and Leonard David, "Russian Fireball Explosion Shows Meteor Risk Greater Than Thought," *Space.com*, November 1, 2013, https://www.space.com/23423-russian-fireball-meteor-airburst-risk.html.
4. NASA, "New Map Shows Frequency of Small Asteroid Impacts, Provides Clues on Larger Asteroid Population: Bolide Events 1994–2013 (Small Asteroids that Disintegrated in Earth's Atmosphere)," *NASA Jet Propulsion Laboratory News*, November 14, 2014, https://www.jpl.nasa.gov/news/news.php?release=2014-397.
5. NASA, Jet Propulsion Laboratory, Center for Near Earth Object Studies, "NEO Basics," accessed March 14, 2021, https://cneos.jpl.nasa.gov/about/neo_groups.html.
6. Paul Chodas, "Overview of the 2019 Planetary Defense Conference Asteroid Impact Exercise," *EPSC Abstracts* 13, EPSC-DPS Joint Meeting, Geneva, Switzerland, September 15–20, 2019; Paul Chodas, "Planetary Defense Conference Exercise—2019," Center for Near Earth Object Studies (CNEOS), NASA Jet Propulsion Laboratory, https://cneos.jpl.nasa.gov/pd/cs/pdc19/; and more scenarios here: https://cneos.jpl.nasa.gov/pd/cs/.
7. Osmel Martin, Rolando Cardenas, Mayrene Guimarais, et al., "Effects of Gamma Ray Bursts in Earth's Biosphere," *Astrophysics and Space Science* 326, nos. 61–67 (2010), https://doi.org/10.1007/s10509-009-0211-7; Brian C. Thomas, "Gamma-Ray Bursts as a Threat to Life on Earth," *International Journal of Astrobiology* 8, no. 3 (July 2009): 183–86.
8. David Denkenberger and Joshua M. Pearce, *Feeding Everyone No Matter What: Managing Food Security after Global Catastrophe* (London: Elsevier, 2015).
9. C. S. Lewis, *The Abolition of Man* (New York: Harper Collins, 1944), 55.
10. Stern, "Does 'Ought' Imply 'Can'?" citing various passages in Kant, including *Religion Within the Boundaries of Mere Reason*, 92, and *Critique of Pure Reason*, 637.
11. Jonas, *The Imperative of Responsibility*, 129–30.
12. For example, Michael Chui, Martin Harryson, James Manyika, Roger Roberts, Rita Chung, Ashley van Heteren, Pieter Nel, "Notes from the AI Frontier: Applying AI for Social Good," McKinsey and Company, December 2018, https://www.mckinsey.com/~/media/mckinsey/featured%20insights/artificial%20intelligence/applying%20artificial%20intelligence%20for%20social%20good/mgi-applying-ai-for-social-good-discussion-paper-dec-2018.pdf.
13. Kennedy, "We Choose to Go to the Moon."

14. Elias Loomis, "The Great Auroral Exhibition of August 28 to September 4, 1859—2nd Article," *The American Journal of Science* 29 (January 1860): 92–97, https://babel.hathitrust.org/cgi/pt?id=uva.x001679511&view=1up&seq=112.
15. Ord, *The Precipice*, 77–79.
16. Paul Chodas, "Hypothetic Impact Scenarios," Center for Near Earth Object Studies (CNEOS), NASA Jet Propulsion Laboratory, https://cneos.jpl.nasa.gov/pd/cs/.
17. Green, "Emerging Technologies, Catastrophic Risks, and Ethics," 1.
18. Brian Fagan, *The Long Summer: How Climate Changed Civilization* (New York: Basic Books, 2004).
19. Eliezer Yudkowsky, "Cognitive Biases Potentially Affecting Judgement of Global Risks," in *Global Catastrophic Risks*, ed. Bostrom and Circovic, 91–119.
20. Diamond, *Collapse*, 419–40.
21. For example, various initiatives to promote rationality, such as the Center for Applied Rationality (CFAR), https://www.rationality.org/, and prevent existential risks, such as the Centre for the Study of Existential Risk (CSER), https://www.cser.ac.uk/; the Future of Humanity Institute (FHI), https://www.fhi.ox.ac.uk/; the Future of Life Institute (FLI), https://futureoflife.org/; and the Global Catastrophic Risk Institute (GCRI), https://gcrinstitute.org/.
22. For example, Effective Altruism, https://www.effectivealtruism.org/; and 80,000 Hours, https://80000hours.org/
23. Vallor, *Technology and the Virtues*, 6.
24. Ibid.
25. See, for example, Manuel Velasquez, Dennis Moberg, Michael J. Meyer, Thomas Shanks, Margaret R. McLean, David DeCosse, Claire André, and Kirk O. Hanson, "A Framework for Ethical Decision Making," Markkula Center for Applied Ethics, 2009, https://www.scu.edu/ethics/ethics-resources/ethical-decision-making/a-framework-for-ethical-decision-making/.
26. Kessler and Cour-Palais, "Collision Frequency of Artificial Satellites."
27. In more archaic Western terms these would be the object, intention, and circumstances of the act.
28. David C. Denkenberger and Robert W. Blair Jr., "Interventions That May Prevent or Mollify Supervolcanic Eruptions," *Futures* 102 (September 2018): 51–62; Ross N. Hoffman, "Controlling Hurricanes," *Scientific American* 291, no. 4 (October 2004): 68–75.
29. In favor of the hypothesis: Richard A. Kerr and Richard Stone, "A Human Trigger for the Great Quake of Sichuan?" *Science* 323 (January 16, 2009), 322. And opposed: Shiyong Zhou, Kai Deng, Cuiping Zhao, and Wanzheng Cheng, "Discussion on 'Was the 2008 Wenchuan Earthquake Triggered by Zipingpu Reservoir?'" *Earthquake Science* 23 (2010): 577−81.
30. Mark Tingay, Oliver Heidbach, Richard Davies, Richard Swarbrick, "Triggering of the Lusi Mud Eruption: Earthquake versus Drilling Initiation," *Geology* 36, no. 8 (August 2008), 639–42; Richard J. Davies, Maria Brumm, Michael Manga, Rudi Rubiandini, Richard Swarbrick, Mark Tingay, "The East Java Mud Volcano (2006 to Present): An Earthquake or Drilling Trigger?" *Earth and Planetary Science Letters* 272, nos. 3–4 (August 15, 2008): 627–38.
31. A. Mazzini, H. Svensen, G. G. Akhmanov, G. Aloisi, S. Planke, A. Malthe-Sørenssen, and B. Istadi, "Triggering and Dynamic Evolution of the LUSI Mud Volcano, Indonesia," *Earth and Planetary Science Letters* 261 (2007): 375–88.
32. For example, while perhaps physically possible, interventions such as those proposed by Denkenberger and Blair, "Interventions That May Prevent or Mollify Supervolcanic Eruptions"; and Hoffman, "Controlling Hurricanes," are still far from being implemented.
33. Mearns and Norton, "Equity and Vulnerability in a Warming World," 8, citing the Scientific Expert Group on Climate Change, *Confronting Climate Change*.
34. Green, "Emerging Technologies, Catastrophic Risks, and Ethics."
35. Alexey Turchin and Brian Patrick Green, "Aquatic Refuges for Surviving a Global Catastrophe," *Futures* 89 (May 2017): 33.
36. Mikayla Mace, "Comet NEOWISE Sizzles as It Slides by the Sun, Providing a Treat for Observers," Near Earth Object Wide-field Infrared Survey Explorer (NEOWISE)/Infrared Processing and Analysis Center, California Institute of Technology, July 8, 2020, https://neowise.ipac.caltech.edu/news/neowise20200708/.
37. Green, "Emerging Technologies, Catastrophic Risks, and Ethics."
38. D. H. Boteler, "A 21st Century View of the March 1989 Magnetic Storm," *Space Weather* 17, no. 10 (2019): 1427–41, https://agupubs.onlinelibrary.wiley.com/doi/full/10.1029/2019SW002278.

*Chapter Eight*

# Astrobiology and the Search for Extraterrestrial Life

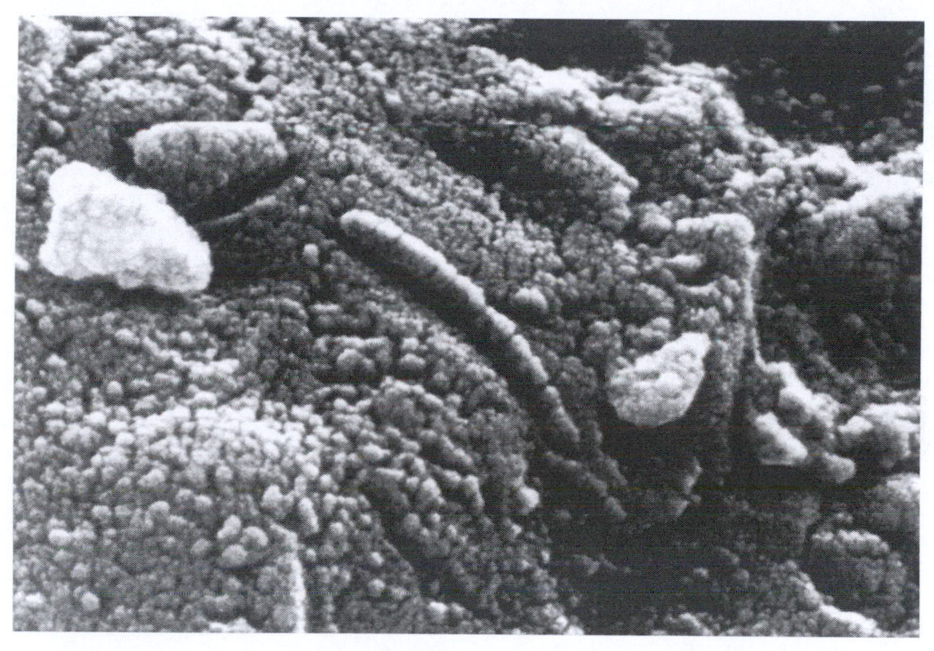

**Figure 8.1. Electron micrograph of the Martian meteorite ALH84001, showing possible signs of fossil microorganisms**

## A MARTIAN METEORITE

Roberta Score discovered the meteorite Allan Hills 84001 (commonly referred to as ALH84001) in Antarctica in 1984.[1] After study, scientists concluded that ALH84001 was a Martian meteorite; it was blasted off of Mars billions of years ago and eventually came to rest in Antarctica, where it was found. After even further study, in 1996 scientists announced it might contain fossilized microbial life forms from Mars's wet past, when water flowed on its surface. The story was a media sensation and became so important that the president of the United States, Bill Clinton, even gave a brief speech about the meteorite.[2]

The rest of the scientific community was not as impressed, however. There were other possible causes for the shapes and minerals in ALH84001, and jumping to the conclusion that these phenomena were caused by ancient life forms was unwarranted, they said. Eventually most scientists decided that the "fossils" in the meteorite were not fossils at all but rather just shapes in the rock.[3] While this conclusion was anticlimactic, it proved to be a historical turning point in the public understanding of the possibility of life on Mars, one that shifted the conversation on extraterrestrial life (ETL) from science fiction into the possibility of science fact.

## RATIONALE AND SIGNIFICANCE: IS THERE LIFE OUT THERE?

Astrobiology is the study of the origin, extent, and future of life in the universe. It asks such fundamental questions as "Does life exist beyond the Earth?" and "Do we have adequate definitions of life?" and "What is the destiny of life in the universe?" The first question is a scientific one, the second is a philosophical one, and the third is both scientific and philosophical, but from the ethical perspective there come additional sets of questions such as "How should we value life?" and "What ethical consideration should we give to ETL that is microbial, multicellular, animal-like, or intelligent?" and "How hard should we look for ETL?" Included in these evaluations must be considerations of the unknown and the role of uncertainty in decision making. For example, coming from our limited Earth-based knowledge, we may not really know what is of ethical value in the universe or if we have, in our searches, really encountered ethically valuable entities.

## ETHICAL QUESTIONS AND PROBLEMS

### Where might we find life?

When looking for ETL we might first think to look for life in the sorts of places that look "lifely" on Earth, for example, in a forest or a pond. And if we carried this assumption into space, we would quickly discover that there are no forests in the solar system except on Earth and not much liquid water, at least on planetary surfaces, either.

However, there are many places in the solar system that might harbor life that is not found in places like those on Earth. For example, Mars is a cold desert but might harbor life in protected places such as geothermal areas or even deep underground. Europa and other large ice moons have large subsurface oceans that could harbor life, though it could be very difficult to access it to determine for sure. And there might be other life-bearing locations in the solar system as well, for instance, a recent announcement proposed that Venus might have life in its high-altitude clouds.[4] We really don't know what the extent of life in the solar system might be, much less in the entire universe. This makes it all the more important to look diligently for life, even in places that we might not expect. Once we gain more information about our solar system and the universe, then we might be able to say more conclusively where we are or are not likely to find life in our further explorations. Tools such as "The Life Detection Ladder" already exist for aiding this search,[5] and much other work has been done in this area.[6]

### Some scientists believe we have already found fossils of ancient life from Mars. What should we think based on this, and what of further discoveries?

Getting the facts right is crucial for forming ethical judgments: bad information can lead to bad judgments. And being right or wrong on the presence of Indigenous life on another planet is a huge matter to be right or wrong about. Uncertainty—whether epistemological or moral—makes ethics much more difficult, and the more important the matter, the more important the wrong or right judgment is. And the question of the existence and ethical status of ETL is highly uncertain both epistemically and morally.

If it turns out that investigations for life on Mars indicate that there is no life there, then that simplifies further ethical judgments. And if investigations discover that there is life on Mars, then that also simplifies further ethical judgments, though in a different direction than the no-life discovery. Either discovery simplifies further ethical judgments because in either case uncertainty is removed and replaced by certainty—even if it is a certainty that complicates this in its own way (as the discovery of life would do). Last, if investigations continue to be inconclusive, then ethical uncertainty remains and judgment remains exceedingly difficult.

It is worth noting that discovering life is conclusive, but the failure to discover life is inconclusive. This is because one positive discovery of a native Martian life form demonstrates clearly that there is life on Mars. However, the failure to discover life adds inductive strength to the hypothesis that life may be absent but never conclusive proof. As an old saying goes, "absence of evidence is not evidence of absence." No search can be thorough enough to look for life everywhere, and even if we could eventually look everywhere, motile life forms could move between searches (though it would be unlikely that all life on a planet would be motile in this way and take all its evidence with it upon moving). Searching without discovery will decrease uncertainty, but never to zero. Within this uncertainty, humanity will likely determine thresholds at which some activities become more or less acceptable, such as destructive activities like mining or releasing chemical or biological pollution.

Fossils are a special case of uncertainty because they demonstrate that life previously existed and could still survive. If all we find are fossils and nothing alive, then uncertainty remains at a higher level, despite the nondiscovery of life, which leads to the next question.

### What should our policies be during searches for ETL or upon its discovery, and how should we come to these decisions in an international context?

During the search for ETL the highest levels of planetary protection should be followed, not only for lowering contamination risk but also for the sake of maintaining the best scientific standards, which are themselves a necessary prerequisite for the discovery of life. During this time of exploration, the various environments of a planet that might harbor life ought to be thoroughly searched, including at multiple depths, starting with the most hospitable areas. There should be general policies in place preventing environmental contamination and pollution, and reversibility ought to be at least a theoretical possibility should life be discovered.[7]

Several worst-case scenarios for Martian astrobiology are foreseeable. In one scenario, in which uncertainty is the dominating factor, the discovery of fossil life forms and then a fruitless search for living organisms would stretch on indefinitely, for decades or even centuries. In a second scenario, in which harm is the dominating factor, a large amount of infrastructure is set up on the planet only to discover later that a deadly organism exists on Mars that necessitates the evacuation of the planet. However, none of the evacuees could return to Earth due to their potential exposure to the harmful life, thus leaving large numbers of people with no infrastructure and nowhere to go.

Regarding the uncertainty scenario, at some point, a fruitless search for life should be considered sufficiently conclusive. As to what that threshold might be, that is a problem requiring scientific, technological, political, ethical, social, and many other forms of input. But there are at least three considerations that ought to inform this threshold.

First, the spatial extent of the search is vital. If humanity explored and published reports on ten places on Mars and, not finding life in any of these places, therefore declared that Mars was lifeless, there would be a justifiable outcry at the incompleteness of the search and recklessness of the conclusions. An entire planet cannot be reasonably characterized by ten samples. But how many samples *would* be enough? A thousand? A million? A billion? A general rule can emerge: the higher the number of samples, over a wider distribution, with greater thoroughness, will gradually build inductive strength toward some level of confidence in a tentatively negative conclusion. Exactly what the right confidence should be is a difficult matter relating to risk and harm, as we have already noted. All resources are finite, so a *complete* search of a planet is not a realistic proposal (not to mention it would itself be damaging to the environment). However, given the risks involved, the more complete a search is the better.

Second, the temporal extent of the search. This considers not only the thoroughness of the search from a scientific sense but also the limits of human psychology and, in particular, human patience. While patience is a virtue (classified by Thomas Aquinas as a form of courage because one is enduring suffering for the sake of a higher good[8]), every virtue can become a vice in its excess or deficiency. Excessive patience is no longer a virtue but can become something like cowardice, an irrational fear of action that paralyzes and halts progress—this should be avoided. Once again, there is a balance to pursue between reasonable action and inappropriate risk taking. Returning to the scientific sense, the more time allotted to a search, the more thorough it will be. This is not only due to the likely correlation of spatial, scientific, and technological thoroughness with the passage of time (this includes assumptions about the sharing and accumulation of data, analysis, etc., that may be incorrect if information is lost, etc.) but also due to the likely progress of technology over time, which will probably produce improved tools for finding life.

Third, thoroughness in a scientific and technological sense is absolutely crucial. Even if significant space and time are covered, if the investigators merely used their naked eyes, never turn over stones, never dig into the ground or use sophisticated instrumentation, then the investigation cannot be considered to be thorough. Any search for life ought to include the best instrumentation to find not only direct evidence of life but also its indirect evidence, such as metabolic by-products, fossil and traces, mineral interactions, etc. Additionally, it should not be ignored that all searches for life can only search for life as we know it, and with our very slight experience of the cosmos, as well as theoretical problems with defining life, etc., that experience and theory are likely to be inadequate.

Regarding the harm scenario, it would make sense to avoid any major commitment to a settlement effort before sufficient certainty about the presence of life can be obtained from investigation. Note that this relies on information collected not only in the form of lack of life discovered but also, if life is discovered, on the type of life itself. Any life at all is much riskier than a complete lack of life. If no life is discovered, this may be an ideal situation (at least from the perspective of biological contamination—in other respects it means that the environment is especially harsh and therefore dangerous to Earth life, including human life). If any life is discovered, then the danger is much higher, even if the first life forms discovered may seem completely harmless. Benign life forms can become threatening under changed circumstances, and benign life forms may also indicate the presence of dangerous life forms as yet undiscovered.

Standards for these sorts of considerations ought to be pursued in an international context and enforced through either an update to the Outer Space Treaty or a completely new treaty. This will require the coordination of many nations and components of society, including scientists, engineers, lawyers, politicians, ethicists, government officials, and so on. In sum, a thorough search for life is vital for space exploration, use, and settlement, and the risks of such exploration, use, and settlement—both forward risks to the local environment and backward risks to humans and Earth—should not be underestimated.

## How should we anticipate any potential long-term interaction with ETL? What dangers are presented by the search for ETL, if any?

As discussed in the next chapter, long-term interactions with ETL will involve ongoing planetary protection measures until interactions with the ET life forms can be considered safe. But of course there are many more considerations as well. Here are some scenarios.

First scenario: ETL may be useful—scientifically, technologically, economically, or otherwise. If this is the case, then protections should be created in order to keep the ETL from being exploited, damaged, exterminated, or otherwise harmed by humans. These measures will likely be legal in nature, and legal enforcement in space may be problematic, as has already been discussed, for not only political reasons but also for technological and physical reasons; for example, it can be difficult to enforce laws at great distances because laws rely on the projection of force.

Second scenario: ETL might be not useful. In this case it might be a mere curiosity, perhaps utilizing resources that do not at all compete with people or Earth life. This would be especially clear with resources in distant star systems that humankind cannot possibly get to for many years, but it also might occur closer to home if the ETLs are microbes living deep underneath the surface of Mars or other objects in the solar system, where humans have no interest in going.

Third scenario: ETL could be in the way of something else that is useful. This is an unfortunate and common occurrence on Earth: humans want a resource, but there is a forest on top of it, so we destroy the forest to get the resource. Note that if the forest itself was the resource, this would be the same as the first scenario. This scenario has appeared in science-fiction movies such as *Avatar*[9] and just about every other space movie that involves ETL and mining.

Fourth scenario: there are dangers just from searching for ETL. Searching for ETL will use resources on Earth, and to some this might seem to be a waste of resources better spent on other things. Such a search for ETL might also alert hostile ETIs to our presence and thereby be dangerous.

Last, there are dangers in the finding of ETL and attempting to coexist along with it. The next chapter on planetary protection deals with this in some detail so I will not belabor it here, but almost needless to say, ETL could be very dangerous, whether physically, chemically, biologically, or otherwise, and therefore we ought to be very careful when dealing with it.

## ETHICAL CONCEPTS AND TOOLS

### Extraterrestrial Claims: Ownership, Territoriality, and Respect

In his *Cosmos* television series Carl Sagan opined that "if there is life on Mars, then I believe we should do nothing to disturb that life. Mars then belongs to the Martians, even if they are microbes."[10] Similarly, Charles Cockell and Gerda Horneck have argued for a "planetary park system" wherein extraterrestrial environments would be kept untrammeled by human contact, or if not completely untrammeled then at least human behavior there would be controlled in very specific ways (with much of their interest being to protect from disturbance human space exploration sites).[11] Richard York picked up on this "planetary park" idea in his discussion of a "Martian land ethics."[12] And likewise, Lupisella and Logsdon argued for the care of vulnerable ETL in their parable of the "overseeing aliens,"[13] and Lupisella considers what "rights" Martians might have.[14] But how sturdy are these arguments really?

First, the concept of property ownership, as in private ownership of objects like automobiles and cell phones, is (as far as we know) fairly exclusive to humans on Earth, and even among human groups can be quite variable. This form of ownership is likely related to our human necessity for possessing tools to use, and even more so complex technology, etc., hence its tendency to become a more developed concept in more technologically developed societies. The concept of property ownership, as in "real estate," belongs to an even narrower set of human cultures, and so envisioning microbes on another planet with property rights seems rather odd. So neither of these concepts of property are particularly helpful for our case—they seem too specific to certain human groups.

While rights claims such as property "ownership" by nonrational life forms may not really make sense, "territoriality" does make more sense. Many creatures are territorial, and certainly territorial integrity may be necessary to fulfill such goals as nonaggression and noninterference between groups, especially if one of them is intelligent enough to be able to respect ethical rules. Certainly, the concept of "territory" goes deeply not just into the human mind but even into that of nonhuman animals, and both humans and many kinds of animals (singly or in groups) will fight to defend their territory. And this certainly makes sense from an evolutionary standpoint: if the resources of a certain area are necessary for the survival of one or several animals, then those animals have a clear incentive—linked to their survival and reproduction—to defend those necessary resources.

From the perspective of humankind, we also need resources, and due to economic growth we need more resources all the time, but if there are other organisms already relying upon those resources, then, ethically speaking, humankind ought to think twice before simply deciding we can take them. Humanity has the ability to develop technology, and this allows us to create new resources where once there were none, and ultimately to do more with less. Rather than destroying other life forms for the sake of our own growth, there might well be ways to expand economically while not at the same time depriving other life forms of the resources they need to survive. Whether that means that humans can have no contact at all with an extraterrestrial territory occupied by life forms of nonterrestrial origin or whether we must merely be mindful of it is the next question, and it relies on numerous further factors such as what exactly the organisms in question are, including their complexity, their dangerousness, their rarity, their communication or ability to express interests, whether they are endemic, etc. In general, the more like ethical agents that organisms are, the more ethical consideration they ought to be given, but that is not to say that organisms very unlike agents therefore lack any consideration; indeed, perhaps even completely nonliving environments still retain demands upon our consideration.

Various ethical traditions give different considerations to nonrational life forms. Based on their ability to sense and suffer, utilitarianism would give microbial life forms very little consideration. Likewise, with deontology, because nonrational creatures lack dignity they would also be excluded from moral consideration. However, a virtue ethics perspective actually can support giving consideration to nonrational life forms.[15] The reason for this is that virtue ethics not only thinks about what might be the *telos* of the organism in question (their *telos* is their purpose, goal, and reason for existing: at least to survive and reproduce, preferably without outside interference), it also asks what the *telos* is of the ethical agent, and that *telos* is virtue, not vice. Virtuous agents give reasonable consideration for other beings, including nonaggression, noninterference, and nonmaleficence. Only with good reason would an agent cause harm to another being, and even then, the ethical claims of the wronged life form are not extinguished; they remain in consideration, and perhaps gain even more consideration, the more wrong they have suffered.

So from a virtue ethical perspective, territorial integrity can be argued purely from the side of human agents. As humans, we understand the concept of being outside of our native territory and the respect that we should have for territory that is not our own. Based on this awareness, we ought to recognize that everything beyond the Earth is not our territory and that good human beings (whom we are called to imitate) would treat this space and everything in it with respect, as if we were guests entering "someone else's land" or even "someone else's house." At the very least, if entering "someone else's" territory, even if that "else" is not a person or is even a microbe, there should be some measure of respect shown, and one's actions and deportment should reflect that respect. Moral principles such as nonmaleficence, nonaggression, and noninterference, then, ought to be on the minds of humans who are put into such a situation. Just as one should not walk into a foreign country on Earth and start damaging tourist attractions, nor go into a forest or other ecosystem and wantonly wreck it, neither should humans go to other worlds and act in such a manner.

## Comparative Ethical Approaches in Astrobiology: Deontology, Utilitarianism, and Virtue Ethics

The encounter with ETL of various kinds is not only a fascinating subject for science but also a great opportunity for ethics and for stress testing the theoretical foundations of ethics: its ethical approaches.

One of the great things about philosophy is that it is and always has been open to "thought experiments" in which ideas are proposed via a brief story, scenario, or fictional case, and then others react to them. Typically, these thought experiments are for the sake of making some point; for example, Philippa Foot's classic Trolley Problem dilemma is meant to highlight the differences between deontology and consequentialism through the hypothetical case of a trolley hurtling down a track toward several people, with an alternative track with only one person on it, and the decision maker standing at the switch between tracks.[16] What *should* they do?

With this in mind, we can then view the potential discovery of ETL as a thought experiment and can use it to test various approaches to ethics, such as Kantian deontology, Millian utilitarianism, and Aristotelian virtue ethics. With each of these three approaches we can propose three hypothetical encounters: nonlife, nonintelligent life, and intelligent life. How does each approach handle each encounter?[17]

Kant's deontology gives unquestioning ethical value to other intelligent forms of life. In fact, when writing the *Foundations of the Metaphysics of Morals* Kant specifically said that his ethical system applies to "all rational creatures," not to humans exclusively.[18] However, when encountering nonintelligent life and nonlife, Kant's approach renders both to be of merely instrumental value.[19] They are not persons with dignity but things subject to outside judgment of their value.

However, Kant has an interesting caveat on this point. In his *Universal Natural History and Theory of the Heavens*, Kant states his belief that planets lacking life and/or intelligent life would someday develop life and intelligent life, some of which would likely be more intelligent than humankind.[20] This opens up the interesting question of whether the intelligences that may someday appear on these planets have some sort of "anticipatory dignity" that applies to the planets themselves. Kant himself does not go into this question, but the tension is there and it is interesting, especially now that humanity could soon be faced with such questions as we travel forth to other planets. Should we try to anticipatorily respect life and intelligence that might evolve on another planet? In some way, probably, yes. This comes as a ramification of having respect for rationality not only spatially but also temporally, at very large scales, as well as the Kantian context wherein rationality, ethics, and teleology are seen as the ultimate purpose of the universe.[21]

The utilitarianism of John Stuart Mill reacts quite differently to the space environment compared to Kant's deontology. Mill focuses not on rationality but on the ability to sense pleasure and feel suffering.[22] He does distinguish between intellectual and other kinds of pleasures and ranks them higher, so this does give intelligent species some priority over nonintelligent ones.[23] With respect to intelligent ET life, then it is clear that ETIs ought to be included in the utilitarian calculus. However, because utilitarianism is consequentialist, just being included in the calculus does not necessarily mean that one is protected or has any "rights" (indeed, rights are a foreign concept to utilitarianism). The utilitarian calculus has to weigh the pleasure and suffering of the ETI versus that of humanity, and it is possible that either side could lose out and be swamped in the calculation. Obviously, the ideal would be for all involved parties to become friendly and make the universe an overall happier place, but what if that ideal is remote from reality? For example, if there are a billion times more ETIs than humans, and each one can feel pleasure and suffering more acutely than any human can, then that would seem to indicate that they deserve more consideration in the utilitarian calculus both individually and collectively. If, moreover, the ETIs found humans to be abhorrent and disgusting, and our mere existence caused them much suffering, this could be grounds for raising the overall happiness level of the universe by eliminating humanity. Note that the reverse might occur as well: if there were vastly more humans and very few ETIs, or who had little in the way of sensation or feeling, or whom humanity found quite worrisome or disturbing, this logic could work in reverse as well.[24]

In this case we might pause to ask if utilitarianism might be leading us into unethical territory.[25] These results are counterintuitive and, to be frank, evil: bigger groups do not get to justifiably exterminate smaller groups just because they hate them—so much of human history has already taught us the horror of this path. And so here we may have found a place where an ethical theory has broken under the stress test of space ethics. Utilitarianism has a weakness when it comes to this topic. Other oddities that are difficult for utilitarianism would be "insensate" aliens—ones that feel no pleasure or suffering, such as an **ET artificial intelligence (ETAI)** or **ET artificial general intelligence (AGI/ETAGI)** might be, and ETIs or humans that derive immense pleasure from inflicting suffering on one or the other, as a Nozickian "utility monster" would, thus throwing the utilitarian calculus hopelessly out of balance.[26]

On the issue of nonintelligent life (never mind the difficulty of defining intelligence in this case), utilitarianism is more nuanced than Kantianism, though its conclusions can still be quite odd. For example, if ETL pleasure and suffering are on the same quantitative scale as humanity, then strange situations might be possible, even with nonintelligent life (resembling in some ways philosopher Derek Parfit's "repugnant solution" in which sheer volume of sensation outweighs other considerations[27]). If ETL pleasure and suffering are qualitatively inferior to human pleasure and suffering, then they will tend to be overruled by human interests, perhaps even to the extent of eradication, if doing so would make humans happier (for example, terraforming a world to suit humans and killing the native life as a side effect). Once again, utilitarianism causes an outcome that seems intuitively wrong to many people.[28]

Last, utilitarianism and nonlife is an easy calculation: nonlife cannot experience pleasure or suffering and so it has no ethical value or consideration. Its value is purely instrumental, in what it can do for creatures that can experience pleasure and suffering.

Aristotelian virtue ethics comes to space ethics from a different perspective, that of agents rather than actions. With respect to living things, all living things have an inbuilt **entelechy** that they are trying to fulfill. For most life forms it is simple: survive and reproduce. But for humans we have higher virtues that we are called toward, such as living in a political organization and seeking truth. If humans encountered intelligent aliens, the ideal ethical outcome would be for both humanity and the ETIs to seek their *telei* of flourishing in the way most appropriate to them. Hopefully human and ETI *telei* are nonconflicting, that is, the ETIs will not rely on harming humans to achieve their intrinsic ends or vice versa.[29] In this nonconflicting case, humankind and ETI could work together toward a better future. However, if conflict were unavoidable, either side would be justified in self-defense because their flourishing would rely on not being harmed. Relatedly, ethical agents are prioritized by their capacity for virtue, so those less capable of virtue would have lower ethical priority. In Aristotle's time this led to such ideas as some humans being "brutes" and "natural slaves" only capable of taking orders, and this is a danger that people using Aristotelian ethics need to now avoid.[30] However, the idea of reducing the freedom of those less capable of virtue is actually something that we do practice, the most basic example being imprisonment of those convicted of crimes. Therefore if ETIs attempted to commit unethical actions toward humans we might be justified in attempting to constrain their behavior—and likewise, if humans misbehaved badly, ETIs might feel justified in restricting our behaviors. This is a sticky balancing problem that in practice unfortunately all too often reduces to rule by the most powerful—who also think they are the most ethical (see Earth's history of imperialism, colonialism, etc.). But it can make intuitive sense under certain circumstances, especially when we remember that the flourishing of both groups is the paramount goal.[31]

Nonintelligent life forms also have the capacities for excellence and virtue that are specific to them, once again centered on survival and flourishing. These entelechies are worthy of respect, to varying degrees as suits their natures (that is, animals more than plants, plants more than microbes, etc.),[32] even if it is not as much respect as would be accorded to an intelligent being or even more so a virtuous intelligent being. But because all life seeks its own good by nature, that natural inclination should be respected as much as is possible in any particular case. Parasites, pathogens, and predators are naturally at odds with the flourishing of other creatures, but even they probably ought not be controlled unless they threaten intelligent lives or interests lest the delicate balance of nature become disrupted.[33]

For Aristotle even nonlife is worthy of some respect, as all things are on their own trajectories toward the future and attempt to maintain their being and seek their own ends (in a nonliving way).[34] Gravity pulls things down for example: the innate tendency of a rock, then, is to follow its nature and go down. Obviously the interests of nonliving things are lowest on the rankings of ethical value. But just the same, in this view, humans ought not to destroy rocks—whether on Earth or in space—for no good reason.[35] From an Aristotelian perspective, such a thing would be a wanton violation of nature.

In virtue ethics, the character of the agent is also in question, not just the entelechy of the moral patient ETI, ETL, or nonlife.[36] A good person does not kill without reason or destroy without reason. Accustoming ourselves to such destructive behavior is a vice, not a virtue. There are good reasons to cause environmental degradation, for example, mining for necessary materials, farming land, or sustainably hunting for food. Each of these activities destroys or repurposes nature in its own way, the first most destructively and the last least destructively. A wise decision maker will know when it is prudent to make which decisions, given the circumstances, and becoming one of these wise and discerning decision makers should be a goal of every human being who seeks virtue (and we all should seek virtue).[37]

It is clear then that which ethical approach one takes might dramatically affect how one makes decisions, especially in the extreme environment of space. Our Earthly theories of ethics can help illuminate the salient features of ethical decision making but can also themselves bend and break under the strain of such extremes.[38] There is something to be said for **ethical intuitionism** in these cases in which the practical results of theory seem wrong: if an ethical theory seems to guide us astray, then the theory perhaps ought to be discarded. And if someone has such terrible judgment that they cannot tell that a theory is wrong, then they should not be making ethical decisions and ought to consult a practically wiser person.

## BACK IN THE BOX

How should thinking about life on other worlds affect our thinking about life on Earth? Surely, the possibility of life on other planets seems exotic. But what we have on Earth seems like an unbelievable plenitude of life compared to the barest possibilities of microbes on Mars.[39] While Mars might have some sort of life hiding underground or in geothermal pockets, and that life might well be very interesting, it would be nothing like the diversity of life on Earth. Jupiter's moon Europa and other icy moons might also have thriving ecosystems miles under their icy crusts, and these might rival Earth in their living diversity, but it will be years before we can know such things. Until then, we probably ought to better appreciate the life that we are certain about right here on Earth.

Of course, we cannot perfectly appreciate all life on Earth: a fair amount of Earth life might like to eat us, after all, or otherwise want to fight us off if we tried to eat it. Parasites, pathogens, and predators abound, and we have to defend ourselves against them with various biological means (our immune systems), chemical means (medicines, cleaners), and physical means (weapons, barriers, sanitation, cooking, etc.). Life can be both beautiful and dangerous, and as we go into space, if we find living things, we will do well to keep that in mind. Luckily, the many scientists and other experts already working on these issues are already doing that.

## CLOSING CASE

While this chapter mainly focused on the possibility of life in the solar system, recent surveys of the galaxy have confirmed the existence of over four thousand exoplanets, many of which are in the habitable zones of their stars, as well as five thousand more candidate exoplanets.[40] While at this point we cannot tell much more about these planets than their mass and how close they are to their stars, in the future, missions such as the James Webb Space Telescope could potentially determine, with just a few hours of study, if these planets have atmospheres.[41] The more pieces that fall into place, the more we will know regarding how common habitable planets are in the universe, and if at some point the gas contents of the atmosphere can be spectrographically determined, we might be able to indirectly determine if a planet has life. An oxygen atmosphere, for example, could indicate that a planet has life forms actively producing it. There is much more to learn.

## DISCUSSION QUESTIONS

1. Do you think it is likely that there is life in other places in the solar system? If so, which places and why there?
2. How should we treat ETL if we discover that it is from the same "genesis" as Earth life—in other words, the ETL is genetically similar to us, shares our biochemistry, that we are on the same "tree of life"?
3. How should we treat ETL if we discover that it is from an independent "second genesis" and is unrelated to Earth life? If there is a difference from your answer to question 2, why is that?
4. Do you find that some ethical systems and concepts (for example, deontology, utilitarianism, or virtue ethics) are more useful for thinking about ETL and some less so? If so, why? Does this "stretching" of our Earthly ethical systems indicate that some of our ethical systems are weaker or stronger than we might have previously thought?

## FURTHER READINGS

Gustaf Arrhenius, "Astrobiology at NASA: Life in the Universe," Astrobiology at NASA, 2021, https://astrobiology.nasa.gov/.

Simon Conway Morris, *Life's Solution: Inevitable Humans in a Lonely Universe* (Cambridge: Cambridge University Press, 2003).

Richard O. Randolph, Margaret S. Race, and Christopher P. McKay, "Reconsidering the Theological and Ethical Implications of Extraterrestrial Life," *CTNS Bulletin* 17, no. 3 (Summer 1997): 1–8.

Kelly C. Smith and Carlos Mariscal, eds., *Social and Conceptual Issues in Astrobiology* (New York: Oxford University Press, 2020).

Peter D. Ward and Donald Brownlee, *Rare Earth: Why Complex Life Is Uncommon in the Universe* (New York: Copernicus, 2000).

## NOTES

1. William A. Cassidy, *Meteorites, Ice, and Antarctica: A Personal Account* (Cambridge: Cambridge University Press, 2003), 138.
2. William J. Clinton, "President Clinton Statement Regarding Mars Meteorite Discovery," The White House: Office of the Press Secretary, August 7, 1996, https://www2.jpl.nasa.gov/snc/clinton.html.
3. Matt Crenson, "After 10 Years, Few Believe Life on Mars," *USA Today*, August 6, 2006, https://usatoday30.usatoday.com/tech/science/space/2006-08-06-mars-life_x.htm.
4. Jonathan O'Callaghan, "Life on Venus? Scientists Hunt for the Truth," *Nature* 586 (October 2, 2020), 182–83, https://www.nature.com/articles/d41586-020-02785-5.
5. NASA, "Life Detection Ladder," Astrobiology at NASA, accessed January 1, 2021, https://astrobiology.nasa.gov/research/life-detection/ladder/, citing Marc Neveu, Lindsay E. Hays, Mary A. Voytek, Michael H. New, and Mitchell D. Schulte, "The Ladder of Life Detection," *Astrobiology* 18, no. 11 (November 13, 2018), https://www.liebertpub.com/doi/full/10.1089/ast.2017.1773.
6. A. Allwood, D. Beaty, D. Bass, C. Conley, G. Kminek, M. Race, S. Vance, and F. Westall, "Conference Summary: Life Detection in Extraterrestrial Samples," *Astrobiology* 13, no. 2 (2013): 203–16; I. Almar and M. S. Race, "Discovery of Extraterrestrial Life: Development of Scales Indicative of Scientific Importance & Associated Risks," *Philosophical Transactions of the Royal Society A* 369 (2011): 679–92, http://rsta.royalsocietypublishing.org/content/369/1936/679.full.pdf+htm.
7. Christopher P. McKay, "Biologically Reversible Exploration," *Science* 323, no. 5915 (February 6, 2009), 718.

8. Aquinas, *Summa Theologiae*, II–II, 123.6, 1705.
9. James Cameron, *Avatar* (Century City, Los Angeles, CA: 20th Century Fox, 2009).
10. Carl Sagan, Ann Druyan, and Steven Soter, "Episode 5: Blues for a Red Planet," *Cosmos: Collector's Edition* (Studio City, CA: Cosmos Studios, Inc., 2000 [1980]), minute 52:36–48.
11. Charles Cockell and Gerda Horneck, "A Planetary Park System for Mars," *Space Policy* 20, no. 4 (November 2004): 291–95; and Charles S. Cockell and Gerda Horneck, "Planetary Parks—Formulating a Wilderness Policy for Planetary Bodies," *Space Policy* 22, no. 4 (November 2006): 256–61.
12. Richard York, "Toward a Martian Land Ethic," *Human Ecology Review* 12, no. 1 (2005): 72–73.
13. Lupisella and Logsdon, "Do We Need a Cosmocentric Ethic?"
14. Mark Lupisella, "The Rights of Martians," *Space Policy* 13 (1997): 89–94.
15. Brian Patrick Green, "Ethical Approaches to Astrobiology and Space Exploration: Comparing Kant, Mill, and Aristotle," special issue "Space Exploration and ET: Who Goes There?" *Ethics: Contemporary Issues* 2, no. 1 (2014): 29–44.
16. Philippa Foot, "The Problem of Abortion and the Doctrine of the Double Effect," *Virtues and Vices and Other Essays in Moral Philosophy* 19 (1978); and popularized by Judith Jarvis Thomson, "The Trolley Problem," *The Yale Law Journal* 94, no. 6 (May 1985): 1395–1415.
17. Much of this section is related to Green, "Ethical Approaches to Astrobiology and Space Exploration," 29–44.
18. Immanuel Kant, *Foundations of the Metaphysics of Morals and What Is Enlightenment?* 32, 45.
19. Ibid., 45.
20. Immanuel Kant, *Universal Natural History and Theory of the Heavens*, trans. Ian Johnston (Arlington, VA: Richer Resources Publications, 2008), 131, 133, 137–40.
21. Immanuel Kant, *Critique of Judgement*, trans. J. H. Bernard (New York: Hafner Press, 1951), 284–86.
22. Mill, Utilitarianism *and* On Liberty, 185.
23. Ibid., 186–89.
24. Green, "Ethical Approaches to Astrobiology and Space Exploration," 35–36.
25. For a more sympathetic interpretation of consequentialism, see Seth D. Baum, "The Ethics of Outer Space," 118–22, and Seth D. Baum, Jacob D. Haqq-Misra, and Shawn D. Domagal-Goldman, "Would Contact with Extraterrestrials Benefit or Harm Humanity? A Scenario Analysis," *Acta Astronautica* 68, nos. 11–12 (2011): 2114–29.
26. Robert Nozick, *Anarchy, State, and Utopia* (New York: Basic Books, 1974), 41, as cited in Green, "Ethical Approaches to Astrobiology and Space Exploration," 36.
27. Derek Parfit, "Overpopulation and the Quality of Life," in *The Repugnant Conclusion: Essays on Population Ethics*, ed. J. Ryberg and T. Tännsjö (Dordrecht: Kluwer Academic Publishers, 2004), 7–22; see also Derek Parfit, *Reasons and Persons* (Oxford: Oxford University Press, 1986).
28. Green, "Ethical Approaches to Astrobiology and Space Exploration," 36–37.
29. Green, "Convergences in the Ethics of Space Exploration."
30. Aristotle, *Politics*, trans. Benjamin Jowett, in *The Basic Works of Aristotle*, ed. McKeon, 1131–35 (I.3–6); and Aristotle, *Nicomachean Ethics*, 99–100 (VII.1) and 106–7 (VII.5).
31. Green, "Ethical Approaches to Astrobiology and Space Exploration," 39–40.
32. Aristotle, *History of Animals*, trans. d'A. W. Thompson, in *The Complete Works of Aristotle: The Revised Oxford Translation*, ed. Jonathan Barnes (Princeton, NJ: Princeton University Press, 1984), 774–993.
33. Green, "Ethical Approaches to Astrobiology and Space Exploration," 39–40.
34. Monte Ransome Johnson, *Aristotle on Teleology* (Oxford: Oxford University Press, 2005), 293.
35. James, "For the Sake of a Stone?"
36. Saara Reiman has made this point in "Sustainability in Space Exploration: An Ethical Perspective," in *MarsPapers*, Mars Society, 2011, accessed August 29, 2014, http://www.marspapers.org/papers/Reiman_2011_paper.pdf.
37. Aristotle, *Nicomachean Ethics*, 90–91 (VI.7) and 163–67 (X.7–8), and Aristotle, *Metaphysics*, trans. W. D. Ross, in *The Basic Works of Aristotle*, ed. McKeon, 689–93 (I.1–2).
38. Green, "Ethical Approaches to Astrobiology and Space Exploration," 41.
39. Charles Cockell has considered this problem in "Ethics and Extraterrestrial Life," in *Humans in Outer Space—Interdisciplinary Perspectives*, ed. Ulrike Landfester, Nina-Louisa Remuss, Kai-Uwe Schrogl, Jean-Claude Worms (New York: SpringerWienNewYork, 2011), 80–101.
40. Pat Brennan, "Exoplanet Exploration: Planets Beyond Our Solar System," Exoplanet Exploration Program and the Jet Propulsion Laboratory for NASA's Astrophysics Division, January 2, 2021, https://exoplanets.nasa.gov/.
41. Christine Pulliam and Laura Betz, "Astronomers Propose a Novel Method of Finding Atmospheres on Rocky Worlds," NASA, December 2, 2019, https://www.nasa.gov/feature/goddard/2019/astronomers-propose-a-novel-method-of-finding-atmospheres-on-rocky-worlds.

*Chapter Nine*

# Contamination, Planetary Protection, and Responsible Exploration

**Figure 9.1. After returning from the Moon in 1969, Apollo 11 astronauts in the quarantine module visit with U.S. president Nixon aboard the aircraft carrier USS *Hornet*.**

## EXPLORATION OR CONTAMINATION?

On November 27, 1971, the robotic lander of the Soviet Mars 2 mission crashed into the southern hemisphere of Mars, near Hellas Planitia. For the purposes of this chapter, unlike other chapters on safety, etc., the details of what went wrong are not important. What is important is that Mars was touched at all because this, for the first time in human history, brought the possibility of a human-made vehicle introducing Earth life to Mars. In **planetary protection** parlance, this presented a risk for **forward contamination**. Construction protocols for the lander instructed that it be made completely sterile, and strong measures were taken to ensure this.[1] But this contact raises a question—was it really sterile? Certainly Earth microbes are known for their abilities to get into all sorts of places that humans do not want them. Were we merely "contacting" Mars or "contaminating" it? After Mars 2's unceremonious impact, the Soviet Mars 3 lander arrived on Mars a few days later, on December 2, and while this time the landing was successful, radio contact was lost within a few seconds.[2] Including these two landers, as of 2020, some fifteen missions have contacted Mars (some controlled and some not controlled) and not all have been sterilized; some have merely had their microbial load reduced, not eliminated.

## RATIONALE AND SIGNIFICANCE: RISKS TO AND FROM TERRESTRIAL AND EXTRATERRESTRIAL LIFE

This chapter will focus on planetary protection—the practice of controlling forward and **backward contamination** between objects in space. Like the chapter on space debris, at first glance this might seem mundane compared to the ethical issues associated with warfare and human health, but as with all things ethical in space, first looks can be deceiving. Planetary protection, in its very name, strikes at the heart of safety on Earth and for all other objects humans—or our robotic emissaries—may ever touch on other worlds.

At the bare, legal level, the Outer Space Treaty (OST) mandates, in Article IX, that

> States Party to the Treaty shall pursue studies of outer space, including the Moon and other celestial bodies, and conduct exploration of them so as to avoid their harmful contamination and also adverse changes in the environment of the Earth resulting from the introduction of extraterrestrial matter and, where necessary, shall adopt appropriate measures for this purpose.

While a reason is not given in the text, we can surmise from the use of the word "harmful" that forward contamination could be potentially, at least, harmful to the scientific value of such places, if not worse, harming any potential native life there.[3] Ethically speaking, the OST embodies implicit ethical values, and other groups, such as the Committee on Space Research (COSPAR) that preceded the OST and influenced Article IX, have continued to this day working on these ideas with respect to protecting space science, sometimes explicitly addressing ethics.[4]

The phrase "adverse changes in the environment of the Earth," contrarily, is highlighting a different concern, that backward contamination could introduce extraterrestrial "matter" (perhaps life forms, perhaps otherwise) that might damage the Earth's environment. Controlling forward, backward, and **lateral contamination** (once a network of interconnected locations begins to form in space) is a significant ethical issue not only in space but on Earth as well because contamination presents a huge risk to environments, living organisms, and science. Invasive species have wreaked havoc on many Earth environments, and the risks of any potential space life doing this to Earth—or of Earth life doing this in space—are major risks that should be mitigated against and adapted toward during space exploration, use, and settlement.

## ETHICAL QUESTIONS AND PROBLEMS

### COSPAR's Planetary Protection Policy

COSPAR's planetary protection policy specifically concerns the protection of scientific research and lists five categories for missions to various locations in space. I will here provide some quotes because the terminology of the policy is highly relevant as background for the ethical analysis of planetary protection.

Category I: includes missions that are "flyby, orbiter, or lander" to "**undifferentiated, metamorphosed asteroids**,"[5] Jupiter's moon Io, and "others TBD." These bodies are "not of direct interest for understanding the process of chemical evolution or the origin of life" and no planetary protection is warranted or required.

Category II: includes all types of missions to Venus, the Moon, comets, **carbonaceous chondrite asteroids**, Jupiter, Saturn, Uranus, Neptune, Ganymede, Callisto, Titan, Triton, Pluto/Charon, Ceres, Kuiper-Belt objects, and "others TBD." These are bodies "where there is significant interest relative to the process of chemical evolution and the origin of life, but . . . only a remote chance that contamination carried by a spacecraft could compromise future investigations. The requirements are for simple documentation only . . . a short planetary protection plan . . . brief Pre- and Post-launch analyses . . . and a Post-encounter and End-of-Mission Report which will provide the location of impact if such an event occurs."

Category III: includes missions that are flyby or orbiters to Mars, Europa, Enceladus, and "others TBD." These missions are to places of "chemical evolution and/or origin of life interest and for which scientific opinion provides a significant chance of contamination which could compromise future investigations. Requirements will consist of documentation . . . and some implementing procedures, including . . . the use of cleanrooms . . . and possibly bioburden reduction. Although no impact is intended for Category III missions, an inventory of bulk constituent organics is required if the probability of impact is significant."

Category IV: includes lander missions to Mars, Europa, Enceladus, and "others TBD." The target body has "chemical evolution and/or origin of life interest and for which scientific opinion provides a significant chance of contamination which could compromise future investigations. Requirements imposed include rather detailed documentation . . . including a bioassay to enumerate the bioburden, a probability of contamination analysis, an inventory of the bulk constituent organics and an increased number of implementing procedures . . . [which] may include . . . cleanrooms, bioburden reduction, possible partial sterilization of the direct contact hardware and a bioshield for that hardware. Generally, the requirements and compliance are similar to *Viking*, with the exception of complete lander/probe sterilization."

Category V: include any Earth return missions. "The concern for these missions is the protection of the terrestrial system, the Earth and the Moon. (The Moon must be protected from back contamination to retain freedom from planetary protection requirements on Earth-Moon travel.) For solar system bodies deemed by scientific opinion to have no indigenous life forms, a subcategory 'unrestricted Earth return [Venus, Moon; others TBD]' is defined. Missions in this subcategory have planetary protection re-

> quirements on the outbound phase only, corresponding to the category of that phase (typically Category I or II). For all other Category V missions, in a subcategory defined as 'restricted Earth return [Mars; Europa; others TBD]' the highest degree of concern is expressed by the absolute prohibition of destructive impact upon return, the need for containment throughout the return phase of all returned hardware which directly contacted the target body or unsterilised material from the body, and the need for containment of any unsterilised sample collected and returned to Earth. Post-mission, there is a need to conduct timely analyses of any unsterilised sample collected and returned to Earth, under strict containment, and using the most sensitive techniques. If any sign of the existence of a nonterrestrial replicating entity is found, the returned sample must remain contained unless treated by an effective sterilizing procedure. Category V concerns are reflected in requirements that encompass those of Category IV plus a continuing monitoring of project activities, studies and research (i.e., in sterilization procedures and containment techniques)."[6]

There are further extensive details in the policy, reflecting the depth of thought that has gone into planetary protection thus far. With respect to ethics, this is laudable as it sets a standard for seriousness and consideration not only for science, but also for the common good that science presents, as well as considerations for safety on Earth and elsewhere.

## Planetary protection is protection from and for both the known and the unknown. But by definition the unknown is unknown. How do we protect ourselves—and the unknown—from what we do not know?

The unknown can be a frightening thing. In the absence of information, we might imagine risks to be much less than they really are, much worse than they really are, or, perhaps, much worse than we can even imagine. This last option—the "failure of the imagination" option—is a real threat when considering the exploration, use, and settlement of space. Things might be a lot better than what we are preparing for; perhaps the risk is zero. Or things could be a lot worse than what we are preparing for, and the long tail on the risk chart goes far beyond what we are expecting. In situations such as this, it is good to remember that the opportunity for a finite gain is not worth the risk of an infinite loss.[7] Yet at the same time, continuing in the vein of the chapter on space-based risks, the inevitable destruction of the Earth and extinction of humankind due to the expansion of the Sun (assuming we last that long) is also assured if humankind does not explore space. Therefore the less risky scenario—in the very long term—is to explore, but with great care. In the immediate, short term the waters are muddier, but given the known risks that we face on Earth, prudent exploration seems like a reasonable risk, though certainly not without danger both to humankind and the extraterrestrial environments that we enter. Between the risks of exploring and not exploring, there is room for reasonable people to disagree.

Returning to the immediate needs of now, robotic space exploration may present contamination issues for extraterrestrial environments. For typical robotic missions that are not searching for life, there are reduced stringencies for biological load. However, if a probe is searching for life, then the parts of the probe involved in the search must be sterile or risk contaminating their own scientific experiments and generating false positive readings.

With regard to contamination-related safety, while we may not know what we will encounter elsewhere, we do know ourselves and our own vulnerabilities. We know that we can be harmed by various forms of contamination, and therefore we must be ready to protect ourselves and our environments against such threats. However, in space where environments are so diverse and so unknown, it may be very difficult to know whether a situation is safe or not, even after quite a bit of testing.

Likewise, we know what environmental harms will likely accompany us as we go into space. Not only may our robotic probes carry microbes, but as humans travel to other planets or space objects we will certainly carry our entire microbiome and likely other associated life forms as well. There is no doubt that when American astronauts left the Moon, they left behind uncountable numbers of microbes, not only on such iconic objects as the lunar landers and rovers but also in such unseemly items as the diaper-like devices required when wearing spacesuits for long periods of time.

The Moon, after some exploration, appears otherwise dead, though it is still in Category II for planetary protection. There is little risk that potential Moon microbes might present to the Earth, and conversely, little risk that Earth microbes present to any living environment on the Moon because there apparently is none. But other objects in space may not be the same. Mars was warmer and wetter for hundreds of millions of years, and life may have evolved there and may still exist there. Europa and other icy moons may harbor life miles beneath their icy surfaces, in the high-pressure watery environments of their under-ice oceans. And other space objects may harbor life forms that we cannot yet imagine. In the midst of such uncertainty, precaution is advisable.

## How carefully should we avoid the exchange of potential contaminants and/or organisms between outer space and Earth?

The simple answer to this question is that we should be very careful. COSPAR's policy highlights the importance of care when it comes to responsible exploration. Contamination, especially biological contamination, can be massively destructive and irreversible, as numerous examples from human history attest. We need only consider the introduction of any new disease to a vulnerable population to see the dangers possible, perhaps the foremost example being the multiple pandemics created by the human contact of the Western Hemisphere by the Eastern, which killed so many Indigenous Americans that we will never truly know how many people, and entire civilizations, were lost.[8]

Less dramatically, but still extremely serious, are any of the innumerable invasive species making their way around the Earth right now, thanks to humankind's global travel network. From the disease-bearing mosquito *Aedes aegypti* to the voracious agricultural pest *Zachrysia provisoria* (a snail), invasive species are a constant threat to people, animals, plants, and all living things. Even the nonliving environment can be changed by invasive species, for example, some types of invasive plants (for example, *Phragmites australis*, *Tamarix ramosissima*) can dry up or otherwise alter the rivers and marshes they invade, thus affecting the entire habitat and even geology of the area.[9] Given these historical examples, the utmost care should be taken when protecting planetary environments against forward and backward contamination.

Even travel to places that seem devoid of life and incapable of supporting Earth still require precautions because it might be damaging to scientific exploration to find Earth microbes, tardigrades, and so on, even dead ones, strewn across a research environment. Each new and unknown environment should be explored with great care, and as its risks are evaluated, only then might uncertainty become more certain, whether warranting less caution or more. If we discover an environment that requires very high levels of protection, we might even need to abandon it and never venture there again.

## What risks should we take not only for ourselves but for "others" in space, even if they are only microbes?

Ethically speaking, risks need to be "worth it." There should be some benefit from taking a risk that makes the downside a fair trade-off, given the circumstances, and acceptable risk tolerances. Recalling the risk equation, where risk = harm × probability, when exploring extraterrestrial environments, it is a high-risk situation because both the probability and the harm are high. The probability of some level of forward contamination is near 100 percent in many cases, the probability of backward contamination is often unknown, and the harm is also of an unknown level (ranging from benign to extreme) that cannot be fully grasped until it is too late.

Risks that affect entire planets have the potential for the gravest possible outcomes, up to and including multiple extinctions and ecosystemic collapse. And yet, for example, on Mars where we don't even know if there are any native life forms to endanger or be endangered by, the likelihood of the worst outcome might seem rather low. But in fact this is not the correct assumption to make: the probability is not low, the probability is unknown—we do not know what the true probability is. Uncertainty and ignorance are not the same thing. And existing in a state of ignorance does not allow us to propose a probability of our choice; instead it demands that we seek more information so that we can make a truly informed decision rather than simply blundering into the unknown. Fact gathering may take time; decades have already elapsed with no clear results in the case of searching for life on Mars (the only clear result is that Mars still might harbor life). But if we blundered ahead assuming things will turn out well, it could actually go quite badly, with native Martian life being driven extinct by Earth life forms, or conversely, Martian life forms harming settlers or returning to Earth to wreak havoc on the Earth environment.

It is one thing to put ourselves at risk if the risk is only to us, the decision maker. It is an entirely higher level of responsibility to be making decisions for other human beings. And it is again an entirely higher level of responsibility to be making decisions for all life on planet Earth. Yet with space explorations we are not only making decisions about risks to the Earth but to other planets as well. This responsibility is beyond anything humans have previously had, and in other cases when humans have had moral obligations with serious responsibility, they have often not fulfilled those obligations. Space ethics requires us to mature into the best ethical decision makers that we can be—or else potentially face unprecedented disasters.

## Robotic missions and human missions will necessarily have different standards for planetary protection (because humans carry an entire microbiome, and if we bring plants and animals even more so), but what standards should be set and why?

There is no way that we could make human travel to Mars sterile because humans are not a sterile environment. In other words, forward contamination would become unavoidable. We carry more microbes that we do cells in our own bodies (because microbes are so small compared to human cells, many times more of them can fit inside us). It would also be difficult or impossible not to bring many other contaminants and organisms aboard the spacecraft, such as bacteria and fungi, on human skin, in our clothing, in our hair, etc. Not to mention we cannot leave our microbiome behind—we exist symbiotically with much of it, so we really *ought* to bring it, for the sake of human health. And in that case, wherever we go will become contaminated because the biosafety measures simply cannot be adequate, in practice, for indefinite periods of time. Much has been written about this.[10]

There still might be ways for humans to go to other planets and yet not contaminate them, however, if for briefer periods of time. Biosafety level 4 facilities on Earth, for example, take the most dangerous pathogens and manage to do research on them in relatively safe conditions. If humans went to another planet, we could attempt to live within biosafety level 4 conditions relative to the outside, thus protecting the planet from us. However, we might also have to live in biosafety level 4 conditions to protect ourselves from the planet, thus perhaps doubling the safety infrastructure. These conditions would be quite cumbersome and, as people grew tired of them or just sloppy, would be unlikely to last for long. They still might be reasonable while we are exploring for life on Mars. If no life is discovered, then the safety levels could be reduced. However, if life were to be discovered, then not only would this unwieldy and unsustainable system of safety remain necessary, it would potentially become permanent. Then the most reasonable action from a planetary protection angle, for the sake of both Mars and Earth, would be to leave the planet entirely.

Of course, people could choose to ignore this safest course of action and instead take risks that will lead not only to the Mars environment's exposure to Earth life but also to human exposure to Mars life. Conducting this unethical experiment could help to demonstrate whether, in fact, all of this planetary protection infrastructure is actually necessary or not. After all, even if there are Mars life forms directly interacting with Earth life forms and humans, it does not necessarily mean that anything bad will happen. Indeed, discovering this could be ideal for those who want to settle other worlds. But there is incredible risk associated with this experiment. Not only are human lives put at risk, but ultimately the ecosystem of Mars (if there is any) and potentially the ecosystem of Earth as well.

These risks, ranging from nearly nothing to nearly everything, are what make planetary protection such a difficult ethical topic. Given these terrible uncertainties, an approach to risk grounded in precaution—the precautionary principle—is advised. Missions ought to begin with the highest level of precaution possible that still allows for achieving their goals. If the mission is merely scientific exploration, robotic rovers might well suffice. If the mission is to start a self-sustaining civilization on Mars (as is the dream of Elon Musk[11]), then humans must be present and the risk situation is greatly changed. Between these goals are many other possible courses of action, and they are not necessarily mutually exclusive. But we need to recognize the risks at play when we make world-changing decisions.

### What of potential Mars sample return missions, or what of the return of astronauts from Mars? Should contaminated astronauts be allowed to return to Earth? What are some issues with human quarantine, both on Earth and off?

As noted in the COSPAR policy, different celestial bodies will have different needs for planetary protection. When astronauts first returned from the Moon, they were put into quarantine. It now appears that the Moon is lifeless, though the quarantine precautions were sensible at the time given the unknowns of the situation, and the Moon is still categorized as a Category II celestial body. However, for many other space objects we cannot be so sure, and in fact for Mars we should be quite careful (it is Category IV) and sample return missions from Mars are "restricted" Category V. Standards already require biosafety level 4 conditions for sample return missions. Perhaps after careful inspection these samples could then be placed in lower safety levels or even treated normally. However, if samples do contain life, then they would need to remain in biosafety level 4 indefinitely, experimented upon for safety relative to Earth life, sterilized, or removed from Earth.[12]

When astronauts agree to explore extraterrestrial environments, they should be prepared to encounter situations with much uncertainty, including uncertainty about whether they will be allowed to rejoin the rest of human society. While objects and primitive life forms might be reasonably sterilized or permanently excluded from contact with human society, astronauts cannot be treated so simply. Astronauts, if contaminated with something potentially harmful, could not be allowed to return to Earth; instead they would need to stay in quarantine. With this in mind, the situations that astronauts are placed in should not be so risky that their destiny is ever in doubt. If Mars has contaminants or organisms, then perhaps humans ought not set foot on it, or, if there are not contaminants or organisms dangerous to humans but instead dangerous to other Earth life, and Earth life does not harm Mars life (itself potentially a difficult or impossible question to assess), then the journey to Mars ought to be one way, permanently.

## How should we manage the risk for events with very low probability but potentially highly harmful consequences?

This question, in general, is a highly contested one in ethics of risk management. As mentioned in chapter 3, philosophers have come up with various approaches to ethical uncertainty, for example, the precautionary principle, prudent vigilance, and the gambler's principle.[13] In the case of high-harm, low-probability events, the gambler's principle, for example, gives fairly safe advice: do not bet more than you can afford to lose. However, risk tolerances tend to be very context specific and can differ significantly between cultures and in different times and places. For example, in the Western world many types of risk aversion have increased dramatically in the past one hundred years, while others have decreased, and in other cultures risk taking behaviors have not necessarily changed in the same way.

Certainly in the past, many governments, authoritarian and democratic, have made disastrous choices when it came to starting wars, preparing for pandemics, and so on. Perhaps if their populations had been more directly consulted on their choices things might have turned out better—or perhaps they would have turned out worse. But in any case, the more people who agree to a decision, the more legitimacy it has, and therefore also the more accountability—both for good and for bad—can be assigned to those people. While a representative democratic government might well do things that are contrary to the public will (again, for better or for worse) and the public might well want things that it should not want, the broader the support something has, the more legitimately assigned is blame or honor as a consequence of that decision.

Some notion of societal informed consent would be useful at this point, as well as risk tolerances around levels of consent. For example, if 50 percent of all people were willing to take or tolerate a risk, would that be sufficient to provide societal informed consent? What if 40 percent or 60 percent? In all of these scenarios *billions* of people are still not in agreement with the risk being taken on. Is that just "too bad" for them?

In addition to the matter of proportion is the matter of whether or not society is actually informed. What if, for example, those who knew the most about a subject agreed that it was either too risky or not risky at all, while the less informed people though the opposite? In this case, whose opinion would matter more, or would all opinions still matter equally? These are hard questions that various world cultures have resolved in various and quite different ways. In some places, the will of the elites, whether informed or not, is always obeyed, while in others, the public, again, whether informed or not, has a greater say in public policy. However, because humans tend to specialize tasks into particular groups of people through division of labor, and the world is becoming more complex (both technologically and socially) every year, there may be good reasons why more specialization of decision-making capacity is warranted.[14] Democratically elected representatives who are capable of making nuanced discernments (not a guarantee, by far) would seem to be a reasonable choice for making these sorts of decisions.

### What are the risks of failure of containment?

With any system there is always a certain probability of failure; however, with safeguards, the probability of failure can be substantially reduced. The nuclear energy and aviation industries offer two examples of industries that have taken risk very seriously and have done very well at reducing the risk of disasters. However, despite their remarkable safety records, airline crashes and nuclear accidents still have happened. Biocontainment likewise generally has a fairly good safety record, but not a perfect one. For example, in 2014 in the U.S. State of Maryland, vials of the smallpox virus were found in a Food and Drug Administration freezer at the National Institutes of Health campus in Bethesda, where they were not supposed to be. Apparently, the samples had been lost track of and accidentally stored under inadequate biosafety conditions for decades.[15]

Under much worse circumstances, in September 2001, anthrax stolen from a U.S. government biodefense laboratory at Fort Detrick, Maryland, was deliberately used as a weapon by being sent through the U.S. postal system, resulting in five deaths, seventeen injuries, and a massive decontamination operation and investigation.[16]

From this it should be obvious that if humanity is relying on containment, at some point containment will almost certainly fail, whether due to intention or accident. Therefore if there is no 100 percent safe solution to planetary protection, we (that is, all the inhabitants of Earth or some other outer space body) must either be tolerant of some amount of risk or else not transfer objects between planets at all.

### What should we do if we discover life on a planet after we have already started exploring it robotically?

Robotic exploration for life is one of the better ways to potentially find life on other planets. However, again, these robotic probes may be harboring Earth life that could disrupt the local planetary ecosystem, if there is one. So how, then, can we search for life (in order to know whether we need to take biosafety precautions) if our robotic probes are themselves possible sources of contamination?

Christopher McKay has argued for the idea of "reversible" exploration, which is that we should explore extraterrestrial bodies in such a way so that if we do later find native life forms we can "Ctrl-z" ("undo") the exploration and decontaminate the planet of introduced Earth life.[17] While this is a worthy idea, it may not actually be that practical. As mentioned before, most of the robotic probes that humanity has sent to Mars are not sterile; only a few are sterile or a few parts of them are sterile. While theoretically people could remove the probes, it is unlikely that we could remove all of the biological contamination that they have left behind, if, say, biological contaminants had fallen off onto the Martian surface, or were blown off and carried away in the wind. These biocontaminants could be quite unlikely to survive on the Martian surface, given its extremes of aridity, temperature, chemistry, and radiation, but it is not certain that they would die; indeed, even such relatively complex Earth life such as lichens are capable of living in a Mars-like environment at least for several weeks.[18] We might be able to do a lot to reduce the load of biocontaminants carried by our landers, but it is unlikely that we could truly, completely "undo" that contamination. There are simply no ways to explore a planet and keep the planet perfectly pristine and risk free at the same time. It is a trade-off that we need to consider carefully, but we can also do much to make it as low risk a trade-off as possible.

## ETHICAL CONCEPTS AND TOOLS

### Risk Ethics, Part II: Risk Assessment under Conditions of Uncertainty

The ethics of risk assessment under conditions of major uncertainty and potentially great harm is a serious consideration with planetary protection. As has been noted, these are perhaps the most difficult conditions under which ethical decisions can be made, and the level of protection desired depends greatly upon the risk tolerance of the decision makers in question (and, more legitimately, of the constituents they represent). The highest level of risk aversion in the case of planetary protection includes the precautionary principle and the gambler's principle, and both are helpful for thinking about risk in this case.

Freedom of movement and freedom of knowledge (exploration) are traditionally held in high regard by some societies on Earth, and for good reason. These two values have led to incredible growth and dynamism in society. And yet these freedoms need to be balanced with the ethical mandate to reduce harm (or even the risk of harm) to extraterrestrial environments, which could be extremely fragile and vulnerable to human interference, not to mention the possibility of backward contamination putting Earth at risk. Given these trade-offs, even traditionally valuable freedoms could be legitimately curtailed because they potentially put at risk even the most fundamental freedoms of many other people—perhaps even everyone and the entire Earth ecosystem. In these sorts of situations other ethical values should predominate.

The **principle of proportionate care**, for example, states that the riskier a situation is, the more care ought to be taken in it. Low-risk situations require low amounts of care, but extreme risk situations require extreme amounts of care. Especially with potentially irreversible actions such as self-propagating disasters and invasive species, proportionate care should be extremely high.

Critically, the uncertainty of planetary protection should not remain so uncertain forever. Instead planetary bodies should be inspected for life in great detail and the uncertainty reduced as much as is possible. If a place is reasonably determined to be free of contaminants and organisms, then it can be treated with some certainty. However, if any form of life is found on a space object, then certainty has been found, but in the opposite direction: the risk is greatly increased and even more precaution—as well as even more investigation—must take place. If life is found it would be one of the most momentous discoveries in history, not just because of the scientific opportunity but also because of the risk it presents and that is presented to it by humanity. Microbial life would need to be studied and exposed to Earth life (and Earth life exposed to it) in the most controlled and safest possible conditions. And these exposures would need to be exhaustive: everything that it can interact with would need to be studied before the true safety of the situation could be assessed. This would likely require decades of painstaking research, which raises the next issue.

It is worth noticing that for some people with expansionist motivations, there might be a perverse incentive to *not* discover life when exploring other worlds: the discovery of life could ruin their expansionist aspirations. And yet they could be the ones with the ability and resources to conduct such explorations because they would be on or near the site and intent on getting closer to it. This presents a **conflict of interest**. Scientists with conflicts of interest, or biases, either in favor of finding life or against it, should either not be involved in these sorts of investigations, should be mixed with or checked by other scientists with opposite inclinations, or should attempt to overcome their own biases. However, overcoming one's own bias is a difficult request: anyone interested enough in the subject matter is also likely to have an opinion. When possible, searching for life in new places should be done by neutral third parties who are genuinely interested in science as an objective pursuit of fact. This will require high moral standards of those conducting these studies, and fortunately many scientific fields do have these high ethical expectations. It could be beneficial to codify them and give them some legal force, perhaps through a professional society, before such searches for life begin to run the risk of motivated reasoning.

As one last point on risk, it is worth considering the use of risky remedies in case of out-of-control contamination. If we biologically contaminate another environment, would it be right to attempt to reverse the contamination via synthetic biological or robotic countermeasures to target invasive species? Certainly there are many invasive species on Earth that humans feel free to destroy, but this "free-ness" of destruction lends itself to questions of means: How destructive can we be toward these invasive species? (And note that our tools for biological control are limited against microbes.) If total eradication (within the invaded area) is the goal, then might we use any means to pursue it or only some? The answer is (once again) that means should always be proportionate to the problem itself. If an invasive species is apparently harmless, then few means should be applied toward controlling it. On the other hand, if an invasive species is causing economic harm, damaging ecosystems, driving other species extinct, or harming or even killing people (as with the invasive and disease-bearing *Aedes aegypti* mosquito), then more powerful means of control should be applied. To eradicate invasive mosquitos, rodents, and so on, researchers are considering using synthetic biological tools such as "**gene drives**," which force a gene through a population in order to control them or drive them extinct.[19] To kill crown-of-thorns sea stars that are responsible for severely damaging the Great Barrier Reef in Australia, automated submersible drones (named COTS-bots) are being used to seek and destroy these pests.[20] Both of these tools of eradication are highly focused on proper targeting; drones and synthetic biological vectors are designed to be very, very specific. However, there is a danger of off-target effects on nearby species or of inadvertently artificially selecting species in a way that makes them no longer a target of the drones or vectors. Obviously, if these technologies get out of control, they could cause even worse problems (this is more the case for synthetic biology due to its ability to self-propagate, at least until drones become more common). For now, these tools are being debated both for their technical efficacy (will they work safely or go awry?) and their ethicality (is it right to use these powers to drive entire species locally extinct?).

With respect to space, these problems, of course, get bigger. Should a biocontamination problem appear, people might need to assign enormous numbers of drone hours to seeking out and eradicating an invasive life form[21] or take years developing exactly the right synthetic biological remedy, both of which could, of course, go off target, backfire, or otherwise cause their own problems. These powers are, in many ways, god-like, without the assurance of a god who is ethically good or even particularly competent. Which raises, then, the ethical issue of "playing God."

## The Ethics of "Playing God," Part I

The notion of "playing God" evokes the monotheistic God as unmatchable by human actions, and indeed, that human attempts to "play God" can only fail, thus resulting in ruin. Historically, the "playing God argument" has been used as a retort to many sorts of new and powerful technologies such as reproductive technologies, cloning, genetic engineering, resurrection of extinct species, and synthetic biology in an attempt to dissuade explorations of these technologies. Moti Mizrahi has summarized these arguments as being opposed to interfering with nature[22] but then goes on to what is a broader meaning of the argument: that because humans are not all powerful, all knowing, and all good, as a traditional theistic understanding of God would assume, then humans cannot be trusted with these new technologies because we are too weak (not omnipotent), too unintelligent (not omniscient), and too unethical (not omnibenevolent) to properly control them. He then responds that if these concerns are valid, then the playing God argument has indeed been applicable to the technology in question, whether or not God exists. In fact, the playing God argument is really more of a critique of humanity than an appeal to anything supernatural; it is an appeal to remember human frailty and to avoid hubris—and eventual humiliation—through the virtue of humility. Indeed, it is only the *concept* of God that matters for the playing God argument to have teeth, the *actual* existence of God is completely irrelevant. However, just because the argument is applicable does not mean it is final because it invites technologists to think of ways to make their new technologies less vulnerable to these sorts of philosophical questions, by, for example, making it more controllable, beneficial, and less vulnerable to abuse.[23]

With respect to planetary protection and interplanetary contamination, the playing God question certainly arises with the idea of moving life forms between planets. While the "interfering with nature" side of the argument is certainly present, the lack of "omni" traits is certainly also there. Humans may not be sufficiently powerful, wise, or good enough to conduct these activities unless we act with greater intentionality. As humankind grows in power, we need to intentionally be careful with our power, wise in how we wield it, and good in our choices on how to use it. Humankind will never be able to live up to a traditional conception of a perfect God, but we can certainly do our best in the ways that we can.

## Informed Consent, Part III, and Societal Informed Consent, Part I

Informed consent is another general ethical principle that should come into play with planetary protection in two ways, both of which have been mentioned previously but deserve another look. First of all, explorers need to be aware that by exploring a place that still presents the risk of biological contamination they are running the risk of never being able to return home to Earth or even see any other group of humans again unless that group is also contaminated in the same way.

Second, the concept of societal informed consent deserves more development with respect to planetary protection. On a planetary scale and under conditions of uncertainty, exactly how can we say that the populace of the planet can be said to have informed consent? Traditionally, informed consent requires full understanding and free choice (for example, no coercion) in a way analogous to what is necessary for a legally binding contract.[24] In representative governments, typically societal-level decisions are handed over to the representatives of the people, who then act as proxy decision makers. However, representative governments recognize that given the extreme specialization required for much of contemporary decision making, sometimes those representatives need specialized committees of experts to work on the problems and then recommend courses of action.

While it may seem, on first glance, desirable for a direct democracy to merely poll the populace for their opinions on a subject and then enact their will, for a matter like planetary protection, the likelihood of this being a course of action that comes to a reasonable conclusion is not necessarily good. While it is desirable for a democracy to allow its people to make as many choices as possible, it is not desirable that everyone make every decision possible because such specialized expertise is simply not humanly possible. Division of labor requires that we delegate work and trust each other that work will be done well, and in our increasingly complex world, that requires increasing specialization. Specialization is a form of elitism that acts contrary to many of the presuppositions of a democracy based on universal suffrage. Future decision making will need to take these opposing tendencies into consideration, and the issues become particularly acute the more technical the decisions become.[25]

Given this situation, the best course of action, then, seems to be to take both approaches: work to better educate humanity in general about the ethics of space and also work to educate government officials so that they can make informed choices on behalf of the people of Earth. Clearly, there are some terrible problems also connected to this, including that elected officials typically only have a responsibility to their constituents and not to all of humanity and that many nations of Earth are not democratic in any case. But to the extent that this can be mitigated against and adapted toward, it seems the best course to take.

## Bioethics, Part II

On Earth, bioethics, despite meaning "life ethics," typically restricts itself to matters involving human biology and often just "medical ethics" or "clinical ethics." Of course, because ecosystems are so interconnected, human biology is extremely dependent upon the context within which it exists, so even on Earth, bioethics is not really limited to humans but rather is extremely contextual. Some have already discussed better integrating the natural environment with "bioethics,"[26] but in space, with its difficult and artificial environmental conditions, this will become a necessity: human life and flourishing will utterly depend on being able to create a healthy environment in which to live.

What bioethical issues will arise that are space specific? Forward and backward contamination issues are present on Earth, for example. But space contamination is a change in degree that yields a change in kind, a phase change in ethics. While humans remain on Earth, and barring contaminated objects coming to Earth (such as asteroids, comets, or dust—if such contaminants survive atmospheric entry), all contamination can only be by other Earth life and is restricted to this planet. When we begin engaging other locations in space we are risking potentially dealing with life that has never had contact before with Earth life, and vice versa. We may not find other life in the solar system, or even in nearby star systems, but we have to be prepared for it, and that is precisely the problem: it is difficult to prepare for extreme unknowns. To quote former U.S. defense secretary Donald Rumsfeld, we have to deal with "unknown unknowns"—that which "we don't know we don't know."[27]

As a bioethical problem we can already anticipate that creating safe living environments in space will be difficult—it is a "known unknown"—and there are many of these, such as: Will there be life? Will we be able to recognize life if we see it? Will that life interact with Earth life? Etc. But the unknown unknowns are beyond our imagination, much less our ability to prepare for. And yet there are ways to prepare for possible "black swans" and make human civilization more "antifragile," to use Nassim Nicholas Taleb's words for rare and harmful events, and gaining strength from disorder, respectively.[28] Human imagination about space abounds, as countless science-fiction stories attest. But reality may be something beyond what we can imagine.

## From Quantitative to Qualitatively Different Ethics

There are certain events that mark qualitative changes in reality, for example, the first nuclear explosion or the first astronaut setting foot on the Moon. In each case, humans had made terrible explosions before, and humans had traveled to places, but these were quantitative leaps in explosive power and traveling, leaps that in fact constituted qualitative changes in reality. Fission weapons are one thousand times more powerful than TNT, and fusion weapons are one thousand times more powerful again than fission weapons. These weapons have changed the geopolitical situation on Earth in a way that conventional explosives could never do. Likewise, traveling anywhere on Earth or anywhere in Earth orbit would not present planetary protection issues, but the quantitative leap in distance to the Moon brought with it a qualitative leap in the need for precautions. And in both cases, leaps in power bring with them a necessary leap in ethics.

Thinking in terms of the risk equation, these leaps cross thresholds and raise the probabilities of certain events from near zero to something much higher. They introduce the potential for harm where once there was none. This quantitative and qualitative shift in power signals a need for the same shift in ethics. It signals a phase change in reality: where once reality seemed solid, now it is a liquid, or where once it was liquid, now it is a vapor. Phase changes in power require phase changes in ethics, including more stringent approaches to risk and more preparation for the unknown. Merely knowing this is an important step, one that many explorers of the past did not have. If we recognize our more advanced position and then act upon it, we may be able to make the future proceed better than the past.

## BACK IN THE BOX

Invasive species on Earth provide numerous stories of biological transfers gone wrong. Examples include various diseases (of humans, animals, and plants), rats, tumbleweeds, rabbits, prickly pear cacti, cane toads, Argentine ants, fire ants, tiger mosquitos, hammerhead worms, water hyacinth, kudzu, and so on. There are more than sixteen thousand listed invasive species currently ravaging Earth's many ecosystems.[29] A 2001 study (admittedly old) finds yearly economic losses from these species are in the hundreds of billions of U.S. dollars, in the United States alone, and globally losses are likely well over a trillion U.S. dollars: a significant percentage of the world economy.[30] Humans are now well aware of the destructiveness of invasive species and how intractable a problem they can be, and yet the historical rate of invasions seems to be increasing, not decreasing.[31] Beating back invasive species is, now, a significant industry in its own right, involving physical removal, chemical pesticides, and biological control. Future methods of control of invasive species will likely include synthetic biological means such as precisely targeted diseases, biased inheritance of sterility (gene drives), and so on. The lesson worth learning here is that if more care had been taken earlier, trillions of U.S. dollars could have been saved, not to mention hundreds of species kept from extinction and ecosystems maintained. We can still make the right choices now and fend off future damage if we want to. When it comes to Earth's own environment, "planetary protection" should begin at home.

## CLOSING CASE

In 2003 the Japanese space agency JAXA launched Hayabusa (Peregrin falcon) toward the near Earth asteroid 25143 Itokawa. In 2005 it arrived and slowly closed in, until finally touching the surface and gathering grains of dust to return to Earth.[32] In 2010 the sample was safely returned to Earth, while the rest of the Hayabusa spacecraft burned up on reentry with the Earth's atmosphere. This was the first asteroid sample return in history and provided not only important information about the mineral composition of the asteroid but also its physical composition: Itokawa is a 500-meter-long peanut-shaped low-density rubble pile.[33]

In 2014, following on the success of Hayabusa, JAXA launched another asteroid sample return mission, Hayabusa 2, which rendezvoused with the near Earth asteroid 162173 Ryugu in 2018. Again a sample was successfully collected and returned to Earth in 2020; however, this time the main spacecraft was not destroyed on reentry but rather is traveling on to see further asteroids.[34] These two missions illustrate that sample return missions can be successful and safe and serve as a model for future similar missions.

## DISCUSSION AND STUDY QUESTIONS

1. The ethics of planetary protection may seem a bit unclear due to the unknowns involved. Despite this, rules govern the field. What can uncertainty teach us about ethics?
2. Of the ethical tools and concepts presented in this chapter, which do you find to be the most useful for thinking about the ethics of forward and backward contamination?
3. Are there other ethical tools and concepts in this book that you find to be relevant to this chapter? Why those?
4. On Earth, should we do more to prevent forward and backward contamination between various Earth environments? Why or why not?

## FURTHER READINGS

Jacques Arnould and Andre Debus, "An Ethical Approach to Planetary Protection," *Advances in Space Research* 42 (2008): 1089–95.

Gerda Horneck, Ralf Moeller, Jean Cadet, et al., "Resistance of Bacterial Endospores to Outer Space for Planetary Protection Purposes—Experiment PROTECT of the EXPOSE-E Mission," *Astrobiology* 12, no. 5 (May 2012): 445–56.

Wayne L. Nicholson, Andrew C. Schuerger, and Margaret S. Race, "Migrating Microbes and Planetary Protection," *Trends in Microbiology* 17, no. 9 (September 2009): 389–92.

J. D. Rummel and Linda Billings, "Issues in Planetary Protection: Policy, Protocol and Implementation," *Space Policy* 20 (2004): 49–54.

J. D. Rummel, M. S. Race, G. Horneck, and the Princeton Workshop Participants, "Ethical Considerations for Planetary Protection in Space Exploration: A Workshop," *Astrobiology* 12, no. 11 (2012): 1017–23.

## NOTES

1. V. G. Perminov, *The Difficult Road to Mars—A Brief History of Mars Exploration in the Soviet Union* (Washington, DC: NASA History Division, July 1999), 52.

2. Asif A. Siddiqi, *Beyond Earth: A Chronicle of Deep Space Exploration, 1958–2016* (Washington, DC: NASA History Division, 2018), 102.

3. Although the ambiguity of "harmful" has not gone unnoticed: see Christopher Newman, "Establishing an Ecological Ethical Paradigm for Space Activity," *Room, The Space Journal* 2, no. 4 (2015): 55–61.

4. J. Rummel, M. Race, and G. Horneck, eds., "COSPAR Workshop on Ethical Considerations for Planetary Protection in Space Exploration," COSPAR, Paris, 2012, http://cosparhq.cnes.fr/Scistr/PPP Reports/PPP_Workshop Report_Ethical Considerations.pdf.

5. Defined as "heated to temperatures of less than 1,000 K so that minerals did not segregate in a macroscopic way, but are also dehydrated (if ever hydrated in the first place) and were probably subject to temperatures at which biological materials could not survive" by the National Research Council, *Evaluating the Biological Potential in Samples Returned from Planetary Satellites and Small Solar System Bodies: Framework for Decision Making* (Washington, DC: The National Academies Press, 1998), 45.

6. G. Kminek, C. Conley, V. Hipkin, H. Yano, and COSPAR, "COSPAR Planetary Protection Policy," COSPAR Panel on Planetary Protection, March 2017, https://cosparhq.cnes.fr/assets/uploads/2019/12/PPPolicyDecember-2017.pdf.

7. Green, "Little Prevention, Less Cure," 1, referring to Jonas, *The Imperative of Responsibility*, 38; and Davis, "Three Nuclear Disasters and a Hurricane," 8.

8. For example, David Stannard, *American Holocaust: The Conquest of the New World* (Oxford: Oxford University Press, 1992); and Michael R. Haines and Richard H. Steckel, eds., *A Population History of North America* (Cambridge: Cambridge University Press, 2000).

9. Joy B. Zedler and Suzanne Kercher, "Causes and Consequences of Invasive Plants in Wetlands: Opportunities, Opportunists, and Outcomes," *Critical Reviews in Plant Sciences* 23, no. 5 (2004): 442–44.

10. M. E. Criswell, M. S. Race, J. D. Rummel, and A. Baker, eds., "Planetary Protection Issues in the Human Exploration of Mars, Pingree Park Final Workshop Report," NASA/CP-2005- 213461 (Mountain View, CA: NASA Ames Research Center, 2005); R. Mogul, P. D. Stabekis, M. S. Race, and C. A. Conley, "Planetary Protection Considerations for Human and Robotic Missions to Mars," in *Concepts and Approaches for Mars Exploration*, abstract #4331 (Houston, TX: Lunar and Planetary Institute, 2012), http://www.lpi.usra.edu/meetings/marsconcepts2012/pdf/4331.pdf; National Research Council, "Preventing the Forward Contamination of Mars" (Washington, DC: National Academy Press, 2006), http://www.nap.edu; M. S. Race, G. Kminek and J. D. Rummel, "Planetary Protection and Humans on Mars: NASA/ESA Workshop Results," *Advances in Space Research* 42, no. 6 (2008): 1128–38; J. D. Rummel, M. S. Race, C. A. Conley, and D. R. Liskowsky, "The Integration of Planetary Protection Requirements and Medical Support on a Mission to Mars," in *The Human Mission to Mars: Colonizing the Red Planet*, ed. J. S. Levine and R. E. Schild (Cambridge, MA: Cosmology Science Publishers, 2010), http://journalofcosmology.com; and J. A. Hogan, M. S. Race, J. W. Fisher, J. A. Joshi, and J. D. Rummel, "Life Support and Habitation and Planetary Protection Workshop, Final Report," NASA/TM-2006-213485 (Moffett Field, CA: NASA Ames Res. Ctr. 2006).

11. Musk, "Making Humans a Multi-Planetary Species."

12. National Research Council, "Assessment of Planetary Protection Requirements for Mars Sample Return" (Washington, DC: National Academies Press, 2009), http://www.nap.edu; and Margaret S. Race, C. McKay, and A. Steele, "Session 26. Mars Sample Return Planning Issues," *Astrobiology* 8, no. 2 (April 2008): 420–21.

13. World Commission on the Ethics of Scientific Knowledge and Technology (COMEST), "The Precautionary Principle"; Presidential Commission for the Study of Bioethical Issues, *New Directions*, 27, 123; Davis, "Three Nuclear Disasters and a Hurricane," 8; Jonas, *The Imperative of Responsibility*, 38.

14. Brian Patrick Green, "Artificial Intelligence, Decision-Making, and Moral Deskilling," Markkula Center for Applied Ethics, March 15, 2019, https://www.scu.edu/ethics/focus-areas/technology-ethics/resources/artificial-intelligence-decision-making-and-moral-deskilling/.

15. Sara Reardon, "NIH Finds Forgotten Smallpox Store," *Nature*, July 9, 2014, https://www.nature.com/news/nih-finds-forgotten-smallpox-store-1.15526.

16. United States Federal Bureau of Investigation, "Amerithrax or Anthrax Investigation," accessed August 17, 2020, https://www.fbi.gov/history/famous-cases/amerithrax-or-anthrax-investigation.

17. McKay, "Biologically Reversible Exploration ," 718.

18. Jean-Pierre de Vera, Diedrich Mohlmann, Frederike Butina, Andreas Lorek, Roland Wernecke, and Sieglinde Ott, "Survival Potential and Photosynthetic Activity of Lichens Under Mars-Like Conditions: A Laboratory Study," *Astrobiology* 10, no. 2 (March 2010): 215–27.

19. Antoinette J. Piaggio, Gernot Segelbacher, Philip J. Seddon, Luke Alphey, Elizabeth L. Bennett, Robert H. Carlson, Robert M. Friedman, Dona Kanavy, Ryan Phelan, Kent H. Redford, Marina Rosales, Lydia Slobodian, and Keith Wheeler, "Is It Time for Synthetic Biodiversity Conservation?" *Trends in Ecology & Evolution* 32, no. 2 (February 2017): 97–107.

20. Irus Braverman, "Robotic Life in the Deep Blue Sea," in *Blue Legalities: The Life and Laws of the Sea*, ed. Irus Braverman and Elizabeth R. Johnson (Durham, NC: Duke University Press, 202), 147–64.

21. For an example of this in science fiction, the novel *Flying to Valhalla* explores it in some detail, as the main characters create an entire drone industry in an attempt to undo an unintentional invasion of strawberry plants on another planet. Charles Pellegrino, *Flying to Valhalla* (New York: William Morrow and Company, 1993).

22. Moti Mizrahi, "How to Play the 'Playing God' Card," *Science and Engineering Ethics* 26 (2020): 1445–61, citing J. H. Evans, *Playing God? Human Genetic Engineering and the Rationalization of Public Bioethical Debate* (Chicago: University of Chicago Press, 2002); J. Weckert, "Playing God: What Is the Problem?" in *The Ethics of Human Enhancement: Understanding the Debate*, ed. edited by S. Clarke, J. Savulescu, C. A. J. Coady, A. Giubilini, and S. Sanyal (New York: Oxford University Press, 2016), 87–99; Ted Peters, "Are We Playing God with Nanoenhancement?" in *Nanoethics: The Ethical and Social Implications of Nanotechnology*, ed. F. Allhoff, P. Lin, J. Moor, and J. Weckert (Hoboken, NJ: Wiley, 2007), 173–84.

23. Mizrahi, "How to Play the 'Playing God' Card."

24. Or, more historically, in the formula for a mortal sin in Catholicism (grave matter, full knowledge, full consent of the will); see, for example, stories about signing one's soul away to the Devil, such as the legend of Faust. For more details, see also Ruth R. Faden and Tom L. Beauchamp, *A History and Theory of Informed Consent* (New York: Oxford University Press, 1986).

25. For an overview of the discussion, see Zachary Pirtle and Zoe Szajnfarber, "On Ideals for Engineering in Democratic Societies," in *Philosophy and Engineering: Exploring Boundaries, Expanding Connections*, ed. Diane P. Michelfelder, Byron Newberry, and Qin Zhu, Philosophy of Engineering and Technology series, vol. 26 (Cham, Switzerland: Springer, 2017), 99–112.

26. Daniel A. Vallero, "The New Bioethics: Reintegration of Environmental and Biomedical Sciences Ethics in Biology," *Engineering & Medicine — An International Journal* 1, no. 4 (2010): 269–71.

27. Donald H. Rumsfeld, "Department of Defense News Briefing—Secretary Rumsfeld and Gen. Myers," United States Department of Defense, February 12, 2002, https://archive.defense.gov/Transcripts/Transcript.aspx?TranscriptID=2636.
28. Nassim Nicholas Taleb, *The Black Swan: The Impact of the Highly Improbable* (New York: Random House, 2007); and Taleb, *Antifragile: Things That Gain from Disorder* (New York: Random House, 2012).
29. Hanno Seebens, Tim M. Blackburn, Ellie E. Dyer, et al., "No Saturation in the Accumulation of Alien Species Worldwide," *Nature Communications* 8, no. 14435 (2017).
30. David Pimentel et al., "Economic and Environmental Threats of Alien Plant, Animal, and Microbe Invasions," *Agriculture, Ecosystems and Environment* 84 (2001): 1–20.
31. Seebens et al., "No Saturation in the Accumulation of Alien Species Worldwide."
32. Joanne Baker, "The Falcon Has Landed," *Science* 312, no. 5778 (June 2, 2006): 1327, https://science.sciencemag.org/content/312/5778/1327.
33. Jonathan Amos, "Hayabusa Asteroid-Sample Capsule Recovered in Outback," *BBC News*, June 14, 2010, https://www.bbc.co.uk/news/10307048.
34. JAXA, "Asteroid Explorer, Hayabusa2, Reporter Briefing," *JAXA Hayabusa2 Project*, September 15, 2020, http://www.hayabusa2.jaxa.jp/enjoy/material/press/Hayabusa2_Press_20200915_ver9_en2.pdf.

*Chapter Ten*

# The Search for Extraterrestrial Intelligence

**Figure 10.1. The Arecibo Observatory in 2019, before it collapsed due to poor maintenance in 2020**

## THE "WOW!" SIGNAL

On August 15, 1977, beginning at 10:15 p.m., local time and lasting seventy-two seconds, the "Big Ear" radio telescope at the Ohio State University in Columbus, Ohio, detected an intense signal from the direction of the constellation Sagittarius. When, later, astronomer Jerry Ehman saw the printout of the signal—which peaked at thirty times stronger than background noise, reaching "U" on a scale from 1 to 9 followed by A to Z—he circled it and wrote "Wow!" next to it in big letters. It was the strongest signal he had ever seen. This designation, sparked by Ehman's amazement, has become something of the official name for this incident, which is now referred to as the "Wow!" signal. While many origins have been proposed for the "Wow!" signal, ranging from an Earthly radio signal reflected by a piece of space junk all the way to extraterrestrial intelligences (ETIs) signaling Earth, none are conclusive. Its source remains a mystery.[1]

## RATIONALE AND SIGNIFICANCE: EXTRATERRESTRIAL INTELLIGENCE AND US

The **Search for Extraterrestrial Intelligence (SETI)** involves using radio astronomy and other tools in an attempt to see if there is other intelligent life in the universe. While there is currently no evidence for ETI, there have been suspicious signals in the past, as noted in the opening case, including a possible signal in 2020 from Proxima Centauri.[2]

SETI raises many questions including the most basic one: "Are we alone?" Others include, "Do we have adequate definitions of intelligence?" and "Should we fear the possibility of ETI or fear its absence?" We do not have experience with intelligence from beyond the Earth and should be prepared to encounter intelligence that is so different that we have trouble determining its nature, such as artificially intelligent machines. Given these difficulties, spanning disciplines and philosophies, SETI deserve special ethical consideration.

And of course, within this entire investigation, we must remain mindful that we have no conclusive evidence for the existence of ETI, so this entire debate is currently speculative and might remain that way indefinitely. However, even if ETI do not exist, this is not a futile thought experiment because it can further enlighten and clarify how we consider intelligence, whether animal, artificial, or otherwise.

## ETHICAL QUESTIONS AND PROBLEMS

### What is SETI?

SETI can refer to two things: first, the general search for ETI, and second, the SETI Institute itself in Mountain View, California. (In this text I will use SETI to refer to the search in general, unless I specifically mention it being the SETI Institute.)

SETI uses various techniques to try to determine whether there are other technological civilizations in the universe. The first scientific paper to discuss SETI in detail was Giuseppe Cocconi and Philip Morrison's "Searching for Interstellar Communications" in 1959,[3] though the general idea of interplanetary communication goes back at least to Heinrich Hertz, Nikola Tesla, and Guglielmo Marconi in the early twentieth century.[4] Frank Drake first proposed his now famous "Drake Equation" in 1961 as a thought experiment to promote discussion on the possibility of searching for ETI. The Drake equation is

$$N = R_* \times f_p \times n_e \times f_l \times f_i \times f_c \times L$$

where

**N** = The number of civilizations in the galaxy with detectable electromagnetic signals
$\mathbf{R_*}$ = The rate of formation of stars capable of evolving intelligent life (number per year)
$\mathbf{f_p}$ = The fraction of those stars with planets
$\mathbf{n_e}$ = The number of planets, per solar system, suitable for life
$\mathbf{f_l}$ = The fraction of suitable planets on which life actually evolves
$\mathbf{f_i}$ = The fraction of life-bearing planets on which intelligent life evolves
$\mathbf{f_c}$ = The fraction of civilizations that produce detectable signals
**L** = The average length of time such civilizations produce signals (years).[5]

The equation made an immediate impact and has been promoting discussion ever since, partly because depending on how one completes the variables the number of civilizations in a galaxy can vary between zero and billions.

Today the SETI Institute and other organizations are still working on the search for ETI. While nothing conclusive has been found so far, the search goes on.

## How should the questions raised by the discovery of life be different for intelligent versus simple life?

The discovery of either ETL or ETI would raise important questions about the place of humankind and all life in the cosmos.

The discovery of ETL would be significant because it would mean that life is not exclusive to Earth. Life would apparently be, then, not a fluke or one-off occurrence but rather something that is repeatable and "native" to the universe. Life would be something that the universe just does, naturally. The more life we find, the more "natural" would life seem to be. Furthermore, if life always has the same chemical components that would be an extremely interesting finding, just as would be the discovery that life might have very different chemical components.

However, there is, of course, countervailing evidence to the idea that life should be common. The **Fermi Paradox**, named after the physicist Enrico Fermi, asks: If there are so many planets in space and life seems like it ought to be common, then why have we not detected ETIs? Perhaps intelligent life is being "filtered out." The idea of the **Great Filter** asks this question: If life (including intelligent life) seems like it should be common (with so many stars and planets in the universe), but instead life appears to be rare, then at what point does life get "filtered out" of the universe?[6] If life itself is rare, it might mean that the Great Filter lies at a point prior to the development of life. Does the filter occur before life evolves, meaning that abiogenesis is nearly impossible, and so any life is extremely lucky to exist? Does the filter happen between the evolution of life and intelligence, meaning that abiogenesis is relatively easy, but evolving intelligence is hard? Or does the filter happen after the evolution of intelligence, but before intelligence has the chance to spread itself widely in space, meaning that intelligent life is prone to destruction?

If life is common but intelligence is not, then it might mean that the Great Filter lies either between the origin of life and civilization or at or after civilization but before space settlement. If both life and intelligence seem common, then this might indicate that the filter simply does not exist or is weak before the fecundity of the universe. Obviously, if the filter is strong and prior to the current point in human history, then we have little to worry from it because we are past it. On the other hand, if the filter is strong and lies in our future, then this should make us redouble our efforts to protect humanity and the Earth from human folly and catastrophe.

The discovery of intelligent life would open up several scenarios, some of which are ethically fraught.

First scenario: the ETIs are too far away to be significant. We would know of their existence, perhaps from technological signatures or transmissions, but this information would have no effect on our daily life. This would be the least ethically complicated scenario.

Second scenario: the ETIs are friendly and want to help us. The ETIs might still be distant but might send us helpful scientific or technological information or explain the universe to us in new and meaningful ways. Perhaps they also might not be distant; they might be quite nearby but have been hiding. But in any case, their friendliness is a help. Of course, being friendly can also be very not helpful, as many cases of "do-gooders" on Earth have demonstrated. Going into other cultures and trying to help them often leads to social upheaval, disease, violence, and centuries of ongoing harm. If truly good ETIs know this, they might intentionally hide themselves from humanity for humanity's own benefit, much like *Star Trek*'s Prime Directive mandates noninterference in the affairs of prewarp civilizations.

Third scenario: the ETIs are neutral or want nothing to do with us. We are simply not interesting to them and they refuse contact. While this might feel rather odd—knowing ETIs are out there but that they refuse to talk to us—it does not necessarily mean that the ETIs are good or bad. By refusing to communicate they might be attempting to protect us, after all, if they know that communication harms other intelligent species.

Fourth scenario: the ETIs are not easily categorized or are inscrutable. We might not even be sure if they are intelligent, or their intelligence could be so alien that we simply do not understand them. While this scenario is related to scenario three, they differ in that the third scenario seems to be voluntary on the part of the ETIs—they could communicate with us but choose not to—while in this scenario the confusion is much deeper. Perhaps the ETIs are not unwilling but are actually incapable of communicating with outsiders. Perhaps they do not communicate using sensory modes that we are capable of deciphering. Or perhaps they communicate, but it never makes any sense, as in the *Star Trek* episode "Darmok" in which a civilization's language can be translated into English, but it consists entirely of references to stories that are unknown to outsiders and therefore not comprehensible.[7]

Fifth scenario: the ETIs want to exterminate us. This scenario is the genocidal monomania of Pellegrino and Zebrowski's "galactic Central Park at night" and Cixin Liu's "Dark Forest" hypothesis.[8] In this scenario, ETIs live in a constant state of fear, hiding carefully from all other ETIs and attempting to exterminate each other whenever possible. This devilish behavior is justified, in the eyes of the ETIs in these stories, purely as a preemptive form of self-defense, knowing that if others knew they existed, then they might be exterminated too. In this scenario most ETIs would die before they even knew other ETIs existed because the first hint of technological advancement would provoke an attack, and likely one that the victim would never see coming, much less understand.

Sixth scenario: the ETIs are capable of scenario five but either are reluctant to enact it or their opponent is at technological parity with them so that easy extermination is no longer possible. This is an interesting fiction scenario, with space politics, conflict, and war, but it seems unlikely simply because any ETIs developed enough to be spacefaring would also be powerful enough to obliterate entire planets (relativistic travel being actually twice as difficult as relativistic bombing because bombing is a one-way trip and there is no need to carry fuel to decelerate). Parity of power would be rare due to the ages of the stars in the galaxy, and politics unlikely for the same reason, as political relations between more and less technologically advanced groups on Earth demonstrated clearly in the past few centuries: the stronger side acts and the weaker side is acted upon without regard for ethics. Obviously this scenario, like the previous one, is deeply ethically wrong and humanity should not only avoid being the victim in such a scenario but also avoid being the perpetrator if circumstances are different.

Seventh scenario: there are multiple ETIs and they fit into several of these scenarios. This would make for an interesting and complex universe, like those found in many science-fiction stories. However, given that there is no evidence at all for any ETIs it is a bit of a leap to go from "none" to "many" without first going through "one."

This is not intended to be an exhaustive list and there may be further scenarios that I have missed.[9] But needless to say, as ethically complex as the discovery of ET life might be, so much more complex might be the discovery of ET intelligence.

### How might or should our relationship with life change depending on its type (for example, microbial, multicellular but nonintelligent, or intelligent) and the distance (for example, within the solar system versus light years away)?

As a general rule, the closer something is and the more like us that it is, the more immediately dangerous we are likely to perceive it to be, whether it really is or not. There are many exceptions, but let's explore the rule first.

Perhaps the most comprehensible threat that humans have evolved to deal with is the immediate presence of a human enemy who may cause us direct physical harm. This is an evolved cognitive bias due to the fact that our immune systems work without our conscious involvement, that relatively few animals are likely to harm us, and that humans have never had to deal with threats not located on Earth. The general rule, then, that closer and more like us is more likely to be dangerous makes sense based on this context; however, we can also see that this is a bias we have inherited through natural selection. It is also clearly no longer correct. The greatest threats to humanity now may come not from immediate nearby humans but from people with nuclear missiles on the other side of the planet, people whose victims will likely never meet or even see them. Likewise, not only humans but also diseases threaten us, and these are threats we have not evolved to think much about. But nowadays, we have the scientific knowledge and technological ability to cope with diseases before our immune systems, and we can enhance our immune systems through vaccines and medicines. Our cognitive deficiencies should not be allowed to govern our preparedness for future dangers.

Both of the previous examples can be extended to ETI. While science-fiction stories often increase tension by having ETIs right in front of humans and waving ray guns (notice this triggers our psychological investment in the story by appealing to our natural bias), ETI at long distance may be even more dangerous than nearby because distance facilitates the use of more devastating weapons (because the perpetrators are sufficiently distant so as to avoid injury), for example, relativistic bombs (as described in works of fiction by Charles Pellegrino, Pellegrino and George Zebrowski, and Cixin Liu[10]).

Given these unknowns, when it comes to respect (including respect in the form of precaution) for ETI, we ought to be as respectful as reasonably possible. All life has the potential to be dangerous, whether it has evolved to directly endanger others or endanger others as a side effect. As outsiders to these other organism's evolutionary trajectories, humanity and Earth life are not likely to be typical direct targets of ETLs and ETI but rather be targets due to indirect effects or side effects. In some ways this should give us solace, as the probability of attack or other direct negative interaction is reduced, but in other ways this should perhaps concern us more, as we are more likely to be affected incidentally, as side effects, or casually, as nuisances. Likewise, this is likely to be typical of humankind's interaction with ETLs and

ETIs as well—we will not necessarily perceive them as targets but more as nuisances obstructing us or as resources that might be exploitable for reasons other than what they themselves are (for example, for scientific experimentation, not as food).

Once again, in general, nearer and more intelligent makes an organism seem more dangerous, with exceptions. Creatures more intelligent and technologically advanced than humanity are also possible and will present threats that are potentially beyond human comprehension, even from great distance (Cixin Liu explores this to chilling effect in *Death's End* with the diabolical "dual-vector foil"[11]). Likewise, more primitive threats such as microbes are only dangerous if they are in immediate contact with us or might become so. We also ought to be aware that intelligence in space might not only be biological; there may be "intelligent" artifacts or other artificial objects in space that are proceeding as automated agents capable of acting upon us.

This assumes an oppositional stance between humankind and Earth life and ETL and ETI, with the danger perceived from our perspective, when of course we know that we will also pose a danger to ETL and ETI. All of these possibilities, posed as dangers to humanity, can be reversed to explore how humanity could be dangerous to extraterrestrial beings.

And of course there is a third option—rather than being opposed to one another, humankind and Earth life might instead find ways to avoid, coordinate, cooperate, and collaborate with ETL or ETI. We could choose to leave them alone and let them live in peace, without human interference, as in the "planetary park" idea in the last chapter. Or we could choose to help it, for example, by restoring a dying planet for the benefit of the local life (for example, terraforming Mars to help the "Martians"—even if only microbes). If the discovered life is intelligent, humankind and the ETIs might be able to not only coexist but perhaps even cooperate, as so many science-fiction stories have depicted. We can imagine a galaxy of many intelligent species cooperating to make a better future together.

But once again, the uncertainty is what makes this so difficult—if we knew that ETIs were friendly then it would be easy to trust them and work toward a mutually beneficial relationship. But lacking this, we always have to wonder "Will this ETL or ETI harm us?" A risk at the scale of extermination—in other words, a risk that risks everything that everyone and everything on Earth has ever lived for—even if small in probability, some people will call unacceptable. And, confronted by this risk, they might seek the power to control and force the other life to accept human control or die, in other words doing to ETL or ETI what they fear it might do to us. In this case, altruism seems too risky; one wrong move and all of terrestrial evolution may have been for naught. And yet the alternative, to wreak this same end upon another world, is equally evil; the only difference is who is the perpetrator and who is the victim. It is worth noting that this perspective on ETL and ETI is really a projection of our own hopes and fears onto beings that remain only imagined. That does not mean that our hopes or fears are unjustified, but really, they say more about us than they do about ETL or ETI.

Given this situation, we can hope that humankind's first encounter with ETL or ETI is something small and harmless to provide us a learning and growth opportunity, to do the right thing and begin to learn how to relate. Perhaps with each encounter and experience we can improve and grow as a civilization, protecting the best interests of the beings we encounter, as well as our own best interests. We can hope that ETI, on the other hand, is either extremely friendly or extremely distant, perhaps so far that even communication is impossible. If, as the expression says, "fences make good neighbors," then there is no fence like interstellar distances, and the longer the distance, the higher the fence, and the better neighbors terrestrial and extraterrestrial life might be.[12]

**Should we be only listening for ETIs through SETI, or should we be actively messaging potential ETIs?**

As a mere reversal of perspective, it took no major leap of the imagination to conceive of intentionally attempting to send messages to ETIs. However, "active SETI" or "METI"—Messaging Extraterrestrial Intelligence—has been a controversial topic ever since it was first suggested. It is one thing to listen to space; it is another thing entirely to shout into it. More recently there has been a movement toward more frequent and more powerful METI activities, such as those run by the organization METI International in San Francisco.

This intensification of activities has provoked an ongoing controversy over the ethics of attempting to message ETIs. Are these attempts at communicating with ETIs likely to have no impact in space but a positive one here on Earth? After all, by just trying to craft a message we have to consider great cultural gulfs and consider what truly makes for intelligent and productive communication between groups, even if there are no ETIs out there at all. Perhaps METI can help us become better here on Earth. This could contribute to science education as well.

On the other hand, if a METI message is received in space could it bring a huge potential for rewards? Perhaps ETIs are altruistic and will reply to us with explanations for scientific advances, the instructions for glorious futuristic technology, and a welcoming entry into the galactic neighborhood.[13] Associated with this argument is the "Barn Door" argument: that Earth has already been broadcasting radio waves for years, and so METI will be nothing more than a message to formally contact the benevolent ETIs who are waiting for us.[14] It might seem strange that ETIs would just be waiting out there for humanity to say hello, so that they could then shower us in gifts, and there are certainly some who agree with the fundamental strangeness of the idea. For example, the author David Brin believes that the "Barn Door" argument is flawed and sending METI transmissions is equivalent to "praying" to ETIs,[15] and ethicist Kelly Smith compares the METI gamble unfavorably to Pascal's Wager.[16]

As a third option, METI could be risky not only for humanity but for all life on Earth if ETIs are murderous and ruthlessly eliminate all rivals. If this were the case, and the "Barn Door" of radio signals has not already been opened, then by announcing our presence, are we effectively asking to get killed? The equivalent of whistling our way through the "galactic Central Park at night"? Only happy-go-lucky fools would make that mistake, and only once. Because we can anticipate such a scenario, it might seem that the safest choice would be to avoid being the fool. Additionally, and in response to the previous point, even seemingly kind ETIs could actually be tricking humans, for example, perhaps giving us the plans for futuristic technology that then leads to our destruction.

Last, perhaps sending messages into space is potentially dangerous for the possible recipients: the ETIs we are attempting to communicate with. After all, it takes no leap of the imagination to see that, had intercontinental communication been possible in 1400, such communication between two random places on Earth could have proven disastrous to many cultures on Earth, even if no in-person contact was likely to occur within centuries. The risk for cultural upheaval among the contactees is perhaps even higher than among the contacters. Anthropologist John Traphagan argues that such METI messages constitute experimentation upon intelligent subjects and that any institutional review board would frown upon such reckless experimentation.[17]

Whatever the possible outcomes, there is also the cost of the endeavor. METI costs money, and as with all things in space, that money could have been spent on other things, even if it is private money and could be spent on frivolities rather than on public benefit. Is the cost of METI worth both the supposed opportunity and the supposed risk that it presents? With opportunities and risk ranging from utopia to extinction, respectively, it seems that the investment might be misplaced, for we can live without utopia, but we cannot live with extinction.[18] The bet would not appear to be a good one.

### What social and religious questions might the discovery of extraterrestrial life raise?

SETI has been a subject of some lively cultural debate for decades now. It has been a subject of hope and derision, investment and mockery. The discovery of ETI would no doubt be one of the most momentous events in human history; human perspectives could change dramatically. However, it may not necessarily have a significant immediate impact upon human civilization, for several reasons.

First, depending on the type of contact (distance, type, age), it may be unlikely that the contact will have any immediate impact upon humanity. If humans detected a message from a source hundreds, thousands, or even millions of light years distant, the message would be old indeed, and the civilization that sent it might not even exist anymore. The message could also be quite mundane, such as a navigation beacon, and not something content rich, like the *Encyclopedia Galactica* of science-fiction lore. Low-information contact is very different from high-information contact.[19]

Second, surveys have shown that many members of the public actually already believe in ETL and ETI, and so the discovery of these would only confirm their preexisting beliefs. For example, a public opinion survey in the United States in 2020 discovered that 66 percent of the polled population believed that "there is life on other planets," 57 percent believed that "there is intelligent life and civilizations on other planets," and 45 percent believed that "UFOs exist and have visited the Earth."[20] With belief in ETL, ETI, and UFOs already so high, it seems like their actual discovery might not be that impactful.

Third, many human institutions have already begun to consider what it may mean to discover ETL and ETI, and most have not approached the issue as though it were a crisis-like situation but rather as one of adaptation to new information, requiring theoretical work and interpretation rather than practical efforts. For example, the "Peters ETI Religious Crisis Survey" consistently showed that more than 80 percent of respondents from many religious traditions did not feel that the discovery of ETI would affect their personal beliefs.[21]

Importantly, within this survey are participant comments that raise an important point: damage to faith may depend on the type of ETI. Kind ETIs would be little problem, but evil ones would be a problem. In other words, the mere fact of ETI existence is one thing, but what they are like is a completely different matter. What if they are incredibly cruel or attack us upon sight? What if they are seemingly angelic and want nothing to do with us because we are beneath them? What if they are not even biological? These discoveries could cause great disturbance to human cultures.[22]

These are more sociological analyses of the question, but what about at the core level of belief? After all, people are quite capable of maintaining routines in the face of significantly changed circumstances, even when such routines are no longer warranted. The discovery of ETL or ETI probably ought to have significant impact because it really would be a significant discovery. For example, the first thing such a discovery would do would be to reveal to humanity that we are in a different situation than we thought we were and that we ought to reevaluate our priorities and actions knowing that we are not alone. Additionally, ETI raises all sorts of other questions such as how two very different civilizations ought to relate to each other, can we trust each other, what are "our" purposes both living together in this galaxy (that is, do we each manifest some aspect of the "grand purpose" of the universe, if there is one), how they might fit into terrestrial religions (if at all), and so on. Additionally, while religions have been speculating about ETIs for hundreds or even thousands of years, speculation is very different than evidence.[23] While vague notions of multiple "geneses" or intelligences very different from human might fit okay into a religion at an abstract level, the same made concrete and real could be very different and frightening, raising broad emotional reactions and existential angst, not to mention theoretical difficulties.

This is not to say that this would be a necessary outcome; in fact, the discovery of ETL or ETI could be a positive boon for some scholars of religion, who would have all sorts of new thinking to do. But if every challenge is also an opportunity, then every opportunity is also a challenge, and in many nations already experiencing declines in religious belief (as is much of the developed world), this might be one challenge too many.

### The role of culture and psychology in ethics—projecting our hopes and fears into the unknown and thus anticipating angelic or demonic ETIs.

Because we can know nothing of potential ETIs prior to actually contacting them (if that ever happens), we are likely, in the absence of such information, to use our imaginations and place our own thoughts onto the imagined ETIs as placeholders. This involves both cultural influences and personal psychology. If we are optimistic and friendly, we may assume ETIs would be the same, and if we are pessimistic and paranoid, the same. If our culture emphasizes welcoming strangers and sharing, then we might hope or expect ETIs to be similar, but if our culture emphasizes war and competition, we might expect the same.

Scholar of religion Ted Peters has considered these phenomena in a religious context and named them "alien enemies" and "celestial saviors." These religiously and psychologically laden categories reflect the human tendency to project cultural and psychological needs into the vast blank slate of space.[24] Religion and other human cultural constructs are inextricable from our thoughts of who ETIs might be; for example, in cultures descended from Western Christianity folks tend to see in ETI either the angelic or the demonic.[25] Even in the same local culture of southern California, in one period of time from 1977 to 1987 (encompassing dozens of movies with ETI), we see ETIs as varied as those seen in the movies *Close Encounters of the Third Kind*, *ET the Extra-Terrestrial*, *Aliens*, and *Predator*.[26] (And note that none of these is as pessimistic as Pellegrino and Zebrowski's "galactic Central Park at night" and Liu's "dark forest" of just the next two decades.) What this cultural uncertainty shows us is that we can only use our limited imaginations, and we cannot imagine everything. ETI is a wonderful blank slate upon which to project our imaginations, but we really know nothing about ETIs—not even whether they exist.

### Generally knowledge is thought to be good and more knowledge is better, but is curiosity actually a vice, as it was once considered (though primarily in the form of "snooping" on others or studying useless things)? Can knowledge be too dangerous and/or evil to pursue? Should the pursuit of knowledge ever be curtailed? Should we always seek knowledge or are there certain conditions under which we should "cover our ears"?

There are also examples of dangerous knowledge, which we are better off not knowing. For example, if a spy were being chased by an assassin and told you an important secret before running off, then you too might now be targeted for death. There are other risks of this type too, so-called information risks or infohazards that, if known, are dangerous merely for knowing them. Just knowing that an atomic bomb is possible, for example, has led multiple nations to pursue and obtain nuclear weapons.

SETI also has information risks associated with it. The mere knowledge that aliens exist might come as a major cultural shock to various human cultural groups. This fear is speculative, and in fact there is evidence to the contrary that some religious groups, at least, might not feel terribly troubled by the discovering of ETI (as discussed earlier), however the particulars of the situation might still be quite significant depending on the context.

There is another category of information risk, however, and that is not in the mere existence of a message but rather in the content of the message. What if humanity received a message from ETIs that was in fact a weapon, for example, a computer virus and/or Trojan Horse?[27] While creating a computer virus capable of directly infecting a human computer system is difficult to impossible, the weapon could be concealed as a gift so that we customize it to our own computing systems. Perhaps, as in Carl Sagan's book (1985) and movie (1997) *Contact*, a large download of information would be given to Earth to decipher.[28] While in *Contact* this is an act of friendship, in real life it might instead be a tempting trap and appear to produce, for example, a universal replicator that could produce anything that humanity desired. Yet when built and activated it might instead produce nanotechnological "grey goo,"[29] lethal autonomous weapons, a malevolent AI, a bioweapon, or various other terrible and perhaps incomprehensible things. No matter how innocent the information might seem, we could never really trust it because we would not really know the senders. Even if the senders sent a long history of their benevolence and goodwill, we could still not trust it because they would be the source of the information, and their trustworthiness would be exactly what is in question.

Operating in uncertainty with ultimate stakes argues for a strongly precautionary approach, even to the extent of, perhaps, not "listening" to messages should we receive them. Signals should be approached with a **hermeneutic of suspicion** and a **heuristic of fear**, or if not fear, then at least the utmost respect.[30] That does not mean the messages cannot be analyzed, processed, or otherwise researched, but they should be analyzed under secure conditions, segmented into pieces, treated with great skepticism, and not acted upon without complete assurance of safety—even if that means never. It should be noted that many SETI researchers already take these precautions, but these safeguards should be strengthened and international regulations put in place to promote compliance and disincentivize activities that might put the whole of Earth at risk.

## ETHICAL CONCEPTS AND TOOLS

### The Value of Free Inquiry versus the Danger of Knowledge

As noted earlier, knowledge can be harmful, not only in the form of dangerous technologies but also in the form of dangerous ideas. The risk of the knowledge of how to produce dangerous weapons is an easy one to recognize, but cultural ideas can also be extremely dangerous, leading entire cultures to fall into malaise and decline. For a historical example, the encounter of colonized cultures with colonizing cultures has been extremely destructive. While colonized cultures sometimes successfully resisted cultural colonization, others adopted the culture of their colonizers, and many groups even today are still experiencing the demoralization of existing in an in-between space after an encounter with an "alien" culture that disrupted their worldview. In all of these cases, the resistance, adoption, or demoralization all caused, or continue to cause, widespread cultural and social havoc and lingering psychological trauma.

While technologies such as nuclear weapons, synthetic biology, misaligned general AI, self-replicating nanotechnology, and more may create a world with significantly more catastrophic risk, mere ideas such as apocalyptic beliefs or teleological political ideologies (such as fascism and Marxism) are also capable of producing enormous misery. If such ideas could be suppressed or excluded from occurring, should they be?

There are two dilemmas here: First, should dangerous ideas be prevented or suppressed? Second, are the dangers posed by ideas worse than the benefits that might come from adopting and exploring new ideas?

When it comes to certain kinds of ideas—including technologies, false conspiracy theories, political ideologies, apocalyptic cults, and otherwise—there are clear costs and benefits to attempting to control them. First, prevention and suppression of ideas cost time and money. There have to be institutional structures that control this suppression, and they need workers and money. These institutions can also very easily go awry, preventing beneficial ideas as well as preventing social progress. In fact, history indicates that perhaps these institutions tend to go awry so often that we might be better off without them: hence ideas like freedom of speech, assembly, expression, privacy, scientific freedom, and free exercise of religion and nonestablishment of religion.

Yet at the same time, if ideas that are just plain false, destructive, and manipulative gain a foothold in a society, the whole social edifice can be severely damaged or even destroyed. See, for example, how the political ideology of Nazism caused enormous destruction not only to Germany but also to its surrounding nations, or Communism damaged the Soviet Union or Maoist China. Likewise in the contemporary United States, antivaccination conspiracy theories, believers in a flat Earth, climate change deniers, political conspiracy theorists, and more, reject science and risk sending the United States into an alternate reality that no longer aligns with or even engages the real world. If the political system of a nation becomes completely corrupt, then all the power of the state can be turned toward evil, and the final end can only be destruction and death.

It would seem, then, that there might be some reasonable mean between complete freedom for lies to spread everywhere and oppressive policing of speech and expression. In the United States, for example, one is not allowed to yell "Fire!" in a crowded theater because that might cause a panic and stampede in which people die. Exactly where this line is between licit and illicit ideas is an ethical question with many considerations.

Second, in contrast to the dangers of knowledge, knowledge can also be extremely beneficial and lack of knowledge very harmful. Obvious examples include all of the advances that science and technology have brought to food production, health care, transportation, communication, information storage and processing, etc., including the very ability to have the conversation that we are having in this book. Without science and technology, space would not be within reach and this book would be unnecessary fantasy. Humankind needs new ideas. We rely on the benefits that we derive from millions of professionals who think of new and better ways to do things every day. Indeed, this book is attempting to develop and share those ideas so that we can have a better and more ethical future. If some sort of rule instead dictated that new ideas were forbidden, then progress would stop and civilization would not be far behind. As quoted earlier in this book, the philosopher Whitehead once wrote that "without adventure civilisation is in full decay."[31] We need "adventures in ideas." Stopping now would not preserve an ideal state for humanity but just put us into an unjust and unsustainable stasis until our inevitable collapse. New ideas, not only in science and technology but also in organization and politics, make miracles possible, creating efficiencies where once there were none and resources where once no resources existed. Solar panels are just one example: where once we could only capture energy from the Sun through plants, which we then burned as fuel, now we can capture it directly for electricity.

As we venture forth into the future, the value of knowledge will have to be weighed against its burden. With relation to space, ethically, going forward is, in my estimation, the only ethical choice. New knowledge is out there, and we need it. Some of that new knowledge will be environmental and perhaps even yield a "super-environmental ethic" (in James Schwartz's words) that may become widespread as space settlers realize that everything can be recycled and that this could be done on Earth as well.[32]

We should carefully govern what knowledge we want to reject, and various nations on Earth may differ on this point. What we should not do is go to either extreme on the acceptance or rejection of new knowledge. Not all knowledge is good, and not every new thing is bad—this is why extremes of conservativism and progressivism can make terrible mistakes, or, as G. K. Chesterton once quipped, "The business of Progressives is to go on making mistakes. The business of Conservatives is to prevent mistakes from being corrected."[33] Rather, with all of science, technology, philosophy, politics, history, culture, and so on at our disposal, we ought to seek to correct past wrongs and set our future path toward truth and good. Great benefits await us, as well as great challenges, and careful thought should be given to gaining the best from our new technologies and ideas while at the same time governing and limiting dangerous technologies and ideas in ways appropriate to their level of risk.

The future of science, including SETI, is unknowable, and knowledge is risky. We should only try to balance these trade-offs as best we can, in the effort to make the best future that we can.

## Societal Informed Consent, Part II

The discovery of ETI would be like nothing else in human history. And yet we—or, rather, a certain select few humans—choose to perform this potentially momentous search without public approval, and we—other human beings—could decide to overrule these few people and neither perform the search nor be exposed to its potentially world-disrupting consequences. With small groups in society engaged in potentially catastrophically risky behaviors that endanger everyone else, should there be more governance over the behavior? And how would societal informed consent even be possible in such a case?

When Frank Drake began Project Ozma at the National Radio Astronomy Observatory at Green Bank, West Virginia, in 1960, all he needed was a radio telescope, some know-how, and some time. He was not part of a centrally planned government initiative (although he utilized a radio telescope belonging to the U.S. government's National Science Foundation). Of course, Drake found nothing in this initial search; and yet what if he had found something? Society had not asked him to potentially turn their world upside down; Drake did this without any sort of social permission. And yet it has been a long tradition in science to not ask permission. But should he have?

The question of METI just takes the SETI question to the next level. Who agreed to do METI? Only a small group of people. It was not the planetary population as a whole, and yet METI precisely involves everyone because any sort of response, whether good or bad, will impact the entire world.

Notice that METI and SETI are only specific examples of a larger-scale phenomenon, which is the forcing of social change through technology without society asking for that change. Technologies are powerful. The invention of nuclear weapons completely restructured world politics from 1945 until today, and only a few elite scientists and politicians consented to the creation of that world (many of who later regretted it, or at least expressed grave concerns[34]). When airplanes were invented, no one explicitly consented to their use in warfare, or to their contribution to climate change, or to possibly having them crash on their house, much less into the New York World Trade Center Towers on September 11, 2001. No one explicitly consented to having a society full of automobiles, which have killed millions of people since their invention; nor to smartphones, which have completely revolutionized communications; nor to social media, which has spread misinformation, disinformation, and political instability globally; nor to myriad other technologies that have changed society.

*Should* society have had a say in these revolutionary technologies? Answering either way presents difficulties. If we say, "no, societal informed consent and permission for technological innovation is wrong," then we effectively say that the history of technological development has gone ethically correctly, despite major, clear, ethical problems such as those mentioned earlier. Of course, along with those problems we also get clear benefits: there have been no major power wars since 1945, transportation and communication are fast and now relatively safe, life is generally much healthier, and humanity is overall more prosperous. If we say, "yes, societal informed consent should have occurred and permission for innovation is ethically important," then we condemn most of the history of technological innovation so far and set up a future ethical demand that innovations should now gain public approval before implementation and deployment in society—which is a huge burden. While we might avoid future technological disruptions to society—for example, by deciding whether or not AI will take people's jobs—we might also lose out on incredible innovations that are rejected merely because of the harm they might cause and not for the good that they might bring. No one today laments the disappearance of milkmen who used to deliver milk every day (though grocery deliveries are becoming more common), but if you had asked past society about replacing milkmen with store-bought milk and keeping it in a refrigerator, no doubt many would have rejected such a transformation.

For genuinely large-scale technologies like nuclear energy, fossil fuel burning, national electrification, artificial intelligence, and so on, government regulation can play an important role in lessening the disruption to society and helping to direct the technology toward social benefit. Nuclear energy, for example, is tightly controlled in the United States by the Department of Defense, the Department of Energy, the Nuclear Regulatory Commission, and so on. Other government agencies in various levels and departments control fossil fuel extraction and use and electrical transmission infrastructure and may at some point begin to exert more control over AI. However, there is always a fear that regulation will stifle innovation in ways that harm the best interests of the public (note that regulation *is supposed to stifle innovation* that is not in the best interests of the public). Surely this may have happened; for example, perhaps nuclear technology would be far more advanced and common now were it not for it being so tightly regulated, and yet perhaps this tight regulation also has prevented countless disasters that a less regulated nuclear industry would have had. Certainly, there have been many nuclear accidents around the world despite the advanced regulations that nearly every country imposes upon the technology. Due to this fear, many nations are reluctant to regulate AI, which they see as a huge potential economic engine, while at the same time exploiting personal data and spreading misinformation—both of which are clear harms to society.

How could societal informed consent be achieved if we demanded it of future technologies? Currently the model is to let government step in once a need for regulation has been demonstrated. In other words, there is a presumption of innocence upon future technologies and the developers of those technologies. They are "innocent until proven guilty," to make a legal analogy. This presumption is in favor of innovation and opposed to regulation.

But notice that certain technologies have not had this presumption of innocence about them, or lost it very quickly. Prior to World War II, physicists working on atomic theory had this presumption of innocence, but as soon as Hiroshima and Nagasaki happened, this presumption was erased forever. Nuclear technology can never have a presumption of innocence again. The public, acting though their government (at least in republics and democracies), regulated these technologies once they came to the public's attention as dangerous.

But what about potentially irreversibly bad "losses of innocence" such as those associated with existentially risky technologies such as misaligned AI, synthetic biology, nanotechnology, or SETI/METI? These seem to be technologies quite unlike nuclear—or at least like nuclear was before it lost its innocence. For these technologies, the loss of innocence might be the extinction of humanity, with no opportunity to regulate it afterward because there are no humans left. For technologies such as these, regulation must come in advance, while the technology is still "innocent" and before it can become "guilty." This "presumption of guilt"—"guilty until proven innocent"—is a dramatic change in the way humankind would look at technological development. It is such a cultural shift that it may be quite difficult to argue for it, much less actually institutionalize it. But perhaps for these technologies it ought to be done because they are just too powerful: no mistakes can be allowed.

Once again, the atmospheric ignition question might come to mind. In July 1945, nuclear physicists were not sure that the first atomic bomb test—the "Trinity" test in Alamogordo, New Mexico—might not set alight the entire Earth's atmosphere, thus destroying all life on the planet.[35] They decided to proceed anyway. The whole world's population was subject to this test—indeed, not only the living population but all people yet to come and the legacy of all the work that past humans have left as well. No consent was given to perform the experiment—it was a sheer exercise of power, and no one could stop it, not the least because it was highly secret and almost no one knew about it in order to try to stop it. Of those who knew, Enrico Fermi reportedly said to Stanislaw Ulam, on the drive to Alamogordo: "It would be a miracle if the atmosphere were ignited. . . . I reckon the chance of a miracle to be about ten percent."[36]

There are times when choices having to do with technology and innovation can become deeply immoral. Any choice that exposes a population to grave risk that it did not consent to in a social and informed way—meaning that society had a conversation about the choice and understood what they were doing—is a choice that is intrinsically unethical. It does not matter if the world is already this way, or has always been this way, or if the powerful decide to ignore everyone else and do as they please; it is our duty going forward to *get it right this time*, and all future times, if at all possible. Because with some future technologies—or ones we already have, such as METI—we may not get a second chance.

### The cost-benefit analysis of the scientific endeavor of SETI and of expensive "big science" in general, and who should pay.

Should resources to be directed toward SETI research? Similar to questions in previous chapter discussions of the risks of high-harm, low-probability events, what about in this particular case? ETI is potentially even more threatening yet also offers the alternative not of danger but of incredible opportunity. ETIs, if friendly, could possibly help humanity overcome our current struggles and lead us toward a bright future with them in space. And so we are left with a situation in which expensive scientific research promises an unknown probability, anywhere from nothing to very high, of discovering something of unknown benefit, anywhere from hellish to utopian. What is the right price to pay for such unknowns piled atop of unknowns?

Considering that there is no evidence for ETIs at all, this speculation of their possible goodness or badness might seem to cancel out to a zero. But does it really? Because there are asymmetries when it comes to the extremes of good and bad, as Hans Jonas has noted in some of his key works: "the prophecy of doom is to be given greater heed than the prophecy of bliss" because we can live without bliss but we cannot live with doom, in particular the ultimate doom: extinction.[37] In other words, the situation is not symmetrical. The danger has clear priority, and the utopian predictions of alien angels helping humanity should wait.

This raises the subject of expense. While listening for ETI signals seems prudent, as long as sufficient measures are taken to protect against malevolent transmissions, active SETI/ METI would seem to violate prudential norms. How much should we be willing to pay to know about our place as intelligent life in the universe? This seems like a worthy question to ask and pay for, to learn if we are alone or if there are other intelligent species out there. It seems like it should be worth quite a bit of expense, though exactly how much would be legitimately subject to debate.

On the other hand, METI seems more like paying to play a game, something like Russian Roulette but where one of the chambers of the gun is full of prizes, one has a bullet, and the others are unknown. We might get lucky and discover that we live in a friendly universe surrounded by benevolent ETI, or we might get unlucky, end up surrounded by malevolent killers, and wind up extinct. This asymmetry makes playing the game not worth it, especially because it costs money to play and only a few people are choosing to do so; it's like a lottery in which only a few people play but everyone is affected: everyone wins big, or everyone gets nothing, or everyone dies. And if the experiment is repeated, then the game is played over and over.

Last comes the question of who should pay for SETI and related experiments. SETI is perhaps a low-probability and high-reward endeavor. Should it be paid for by the public purse or by private funding? Private philanthropy has paid for SETI for some time now, but it is worth remembering that at the beginning it was a publicly funded project. As pure science without a clear path to financial rewards, it is not within the scope of private industry but only private philanthropy or public funding. And at this point a prudential judgment must be made by those in charge of taxpayer money: *Is this project worth it*? If the public is paying for science, then there is a responsibility to use that money prudently, and in accord with the wishes of that society. This may legitimately vary between times and places and cultures. Perhaps one culture will be very interested in SETI for a period of time, while another is uninterested, and then these two societies flip-flop in their interests. Perhaps a public "launch" of the technology is warranted, and then a shift to private funding, as occurred in the United States, or the reverse, if cultural values became more interested in SETI.

Importantly, this prudential judgment about SETI is not questioning the value of science. The value of science is not in dispute, and indeed, science is one of the most important arguments in favor of space exploration.[38] What is under consideration is the use of public money for a low-probability, potentially high-risk, high-reward program. Private philanthropists can use their money as they please (though certainly they are still subject to ethical judgments about their activities, for example, choosing to burn a large pile of money would not be as worthy as giving food to the poor, or just about anything else), and so if they are interested in SETI it makes sense for them to fund it.

Every discovery is a chance for us to look at reality anew and consider what the significance might be of the whole. That discovery, even if not worth money, is worth something because it gets us closer to the truth about reality, and truth has value. While some would argue that humans do not really desire truth (and they might be right in some cases), as a community, humankind has the senses, the intellect, and the desire for an accurate representation of reality. The scientific method is a prime example of a tool that humankind has developed for this task, and it has rewarded us with knowledge and, through technology, wealth and power. While these effects are often abused, they are not bad in themselves, and we ought to continue to pursue facts and truth whenever we can reasonably do so, given our contexts and situations.

## BACK IN THE BOX

As mentioned earlier, there is a potential for receiving harmful SETI signals on Earth, for example, downloading malicious alien signals such as a virus (a self-unfolding, reproducing program) or a puzzle (a malicious program that looks beneficial and so we translate it ourselves, only to be harmed by it), ET artificial intelligence (ETAI), or instructions for fabrication technologies that produce harmful robots or nanotechnology, or biological viruses, etc.

These fears of ETI should be contextualized, however. There are two ways to get through a door, after all: a "battering ram" approach, which simply smashes defenses, and a "lock pick" approach, which disables defenses more subtly.[39] Vastly technologically superior and malevolent ETIs would not need to rely on "lock pick" approaches such as having their transmissions decoded; they could use vastly more powerful and brutal approaches such as relativistic bombs. The advantage of a malevolent transmission, however, might lie in its relatively low cost, high speed to target (the speed of light) and broad targeting (broadcast). Disadvantages include that such a transmission would be missed by any civilization not listening or decontaminating signals, giving away the location of the transmitter (which would have to be far from their actual ETI civilization or else broadcast their location to all), and the near certainty that a signal would need to be decoded with the cooperation of the recipient due to the fact that ETI programming languages and human programming languages are almost certainly incompatible because both are likely idiosyncratic.

These transmission problems could be overcome if ETI used a probe as a transmitter at a stand-off distance, decoding human programming languages and then sending a signal already in a human programming language. This could still be stopped via standard SETI decontamination procedures, and the difficulties of sending a probe to do this raise again the question of why not batter the door rather than pick the lock.

Last, we should again ask: How prepared should we be for such potentially nonexistent problems? Is this effort a waste of time or is it good "insurance" for humanity? Only the future will tell.

## CLOSING CASE

In the "Declaration of Principles Concerning the Conduct of the Search for Extraterrestrial Intelligence," the SETI Permanent Study Group of the International Academy of Astronautics stated their protocols for handling candidate extraterrestrial signals. The principles include transparency, verification, confirmation, sharing of data, monitoring, protecting the frequency band from interference, utilizing the Post Detection Task Group, and not responding to the signal without guidance and consent from the UN or a similar international organization.[40] While these might sound like simple procedural guidelines, they also implicitly embed serious ethical principles too, such as planning ahead, establishing certainly about the facts of the case (a prerequisite for ethical analysis), sharing information, and even societal informed consent. In a matter as ethically fraught as SETI, it is good to know that its scientists are not only thinking ahead in terms of science but thinking also ahead in terms of ethics and what SETI means for the good of humanity.

## DISCUSSION AND STUDY QUESTIONS

1. Should SETI research continue? Why or why not? Is SETI an ethical endeavor? Why or why not?
2. Knowing the truth is a necessary precondition for making good ethical judgments. How does the uncertainty of SETI affect its ethical evaluation?
3. Should people be trying to message ETIs? Why or why not?
4. How would the news of a verified SETI signal affect you, if at all? How would you react to learning that ETIs had been detected? How much does the content of the signal matter for this evaluation?
5. Of the ethical tools and concepts presented in this chapter, which do you find to be the most useful when thinking about SETI and METI, and why? Are there other tools and concepts in this book that might also help for thinking about SETI and METI?

## FURTHER READINGS

Carl Sagan, *Contact* (New York: Simon and Schuster, 1985).
Douglas A. Vakoch, ed., *Extraterrestrial Altruism: Evolution and Ethics in the Cosmos* (Heidelberg, Germany: Springer, 2014).

## NOTES

1. John Kraus, "We Wait and Wonder," *Cosmic Search* 1, no. 3 (Summer 1979): 31, http://www.bigear.org/CSMO/PDF/CS03/cs03p31.pdf; and Robert Krulwich, "Aliens Found in Ohio? The 'Wow!' Signal," *Weekend Edition Saturday, National Public Radio*, May 28, 2010, https://www.npr.org/sections/krulwich/2010/05/28/126510251/aliens-found-in-ohio-the-wow-signal.
2. Seth Shostak, "A Signal from Proxima Centauri?" SETI Institute, December 19, 2020, https://www.seti.org/signal-proxima-centauri.
3. Giuseppe Cocconi and Philip Morrison, "Searching for Interstellar Communications," *Nature* 184 (September 19, 1959): 844–46.
4. Steven J. Dick, "Back to the Future: SETI Before the Space Age," *The Planetary Report* 15, no. 1 (1995): 4–7.
5. Shortened from "The Drake Equation," SETI Institute, 2020, accessed January 2, 2020, https://www.seti.org/drake-equation-index.
6. Robin Hanson, "The Great Filter—Are We Almost Past It?" personal website, September 15, 1998, http://mason.gmu.edu/~rhanson/greatfilter.html.
7. Joe Menosky and Phillip LaZebnik, "Darmok," *Star Trek: The Next Generation* (Paramount Domestic Television, 1991).
8. Charles Pellegrino and George Zebrowski, *The Killing Star* (New York: William Morrow and Company, 1995), 126–27; Cixin Liu, *The Dark Forest* (New York: Tor, 2015).
9. For some related scenarios with different nuances, see Baum et al., "Would Contact with Extraterrestrials Benefit or Harm Humanity?" 2114–29.
10. Pellegrino, *Flying to Valhalla*; Pellegrino and Zebrowski, *The Killing Star*; and Cixin Liu, *The Three-Body Problem* (New York: Tor, 2014); Liu, *The Dark Forest*; and Liu, *Death's End* (New York: Tor, 2016).
11. Liu, *Death's End*, 473.
12. As a note Ulvi Yurtsever and Steven Wilkinson have considered a way to detect interstellar relativistic travel; this travel has not yet been detected: "Limits and Signatures of Relativistic Spaceflight," Arxiv.org, April 21, 2015, http://arxiv.org/pdf/1503.05845.pdf.
13. Douglas A. Vakoch, ed., *Extraterrestrial Altruism: Evolution and Ethics in the Cosmos* (Heidelberg, Germany: Springer, 2014).
14. Douglas A. Vakoch, "Correspondence: In Defence of METI," *Nature Physics* 12 (October 2016): 890.

15. David Brin, "The 'Barn Door' Argument, The Precautionary Principle, and METI as 'Prayer'—An Appraisal of the Top Three Rationalizations for 'Active SETI,'" *Theology and Science* 17, no. 1 (February 2019): 16–28.

16. Kelly C. Smith, "A(nother) Cosmic Wager: Pascal, METI, and the Barn Door Argument," *Theology and Science* 17, no. 1 (February 2019): 29–35; See also Kelly C. Smith, "METI or REGRETTI: Ethics, Risk, and Alien Contact," in *Social and Conceptual issues in Astrobiology*, ed. Kelly C. Smith and Carlos Mariscal (Oxford: Oxford University Press, 2020), 209–35.

17. John W. Traphagan, "Active SETI and the Problem of Research Ethics," *Theology and Science* 17, no. 1 (February 2019): 69–78.

18. Hans Jonas, "The Heuristics of Fear," in *Ethics in an Age of Pervasive Technology*, ed. Melvin Kranzberg (Boulder, CO: Westview Press, 1980), 215.

19. Allen Tough, ed., *When SETI Succeeds: The Impact of High-Information Contact* (Bellevue, WA: Foundation for the Future, 2000).

20. Ipsos, "Majority of Americans Believe There Is Intelligent Life and Civilizations on Other Planets," Ipsos Press Release, Washington, DC, January 28, 2020, https://www.ipsos.com/sites/default/files/ct/news/documents/2020-01/topline-medium-aliens-012820.pdf.

21. Ted Peters, "The Implications of the Discovery of Extra-Terrestrial Life for Religion," *Philosophical Transactions of the Royal Society A* 369 (2011): 644–55, referring to Ted Peters and Julie Froehlig, "The Peters ETI Religious Crisis Survey," *Counterbalance*, accessed September 11, 2020, https://counterbalance.org/etsurv/fullr-frame.html.

22. Ibid.; Green, "Astrobiology, Theology, and Ethics," 346–49.

23. See, for example, the Catholic Bishop of Paris Stephen Tempier's *Condemnations of 1277*, which condemned the idea that "God could not have made other worlds" (#27A) because God is omnipotent and could make other worlds if God wanted to. Cited by Guy Consolmagno, *Brother Astronomer: Adventures of a Vatican Scientist* (New York: McGraw Hill, 2000), 84. See also "Selections from the Condemnations of 1277," in *Medieval Philosophy: Essential Readings with Commentary*, ed. Gyula Klima with Fritz Allhoff and Anand Jayprakash Vaidya (Malden, MA: Blackwell, 2007), 182.

24. Ted Peters, "ET: Alien Enemy or Celestial Savior [*sic*]," *Theology and Science* 8 (2010): 245–45.

25. Christopher Partridge, "Alien Demonology: The Christian Roots of the Malevolent Extraterrestrial in UFO Religions and Abduction Spiritualities," *Religion* 34 (2004): 163–89.

26. Melissa Mathison, *ET the Extra-Terrestrial* (Universal City, CA: Universal Studios Amblin Entertainment, 1982); Steven Spielberg, *Close Encounters of the Third Kind* (Culver City, CA: Columbia Pictures, 1977); James Cameron, David Giler, Walter Hill, *Aliens* (Century City, Los Angeles, CA: 20th Century Fox, 1986); and Jim Thomas and John Thomas, *Predator* (Century City, Los Angeles, CA: 20th Century Fox, 1987).

27. Richard A. Carrigan, "The Ultimate Hacker: SETI Signals May Need to Be Decontaminated," in *Symposium—International Astronomical Union* 213 (2004): 519–22; and Richard A. Carrigan Jr., "Do Potential SETI Signals Need to Be Decontaminated?" *Acta Astronautica* 58, no. 2 (January 2006): 112–17.

28. Carl Sagan, *Contact* (New York: Simon and Schuster, 1985); and Carl Sagan and Ann Druyan, *Contact* (Novato, CA: South Side Amusement Company, 1997).

29. K. Eric Drexler, *The Engines of Creation: The Coming Era of Nanotechnology* (New York: Anchor Press Doubleday, 1986).

30. Paul Ricoeur, *Freud and Philosophy: An Essay on Interpretation*, trans. Denis Savage (New Haven, CT: Yale University Press, 1970), 32; and Jonas, "The Heuristics of Fear," 213–21.

31. Whitehead, *Adventures of Ideas*, 279.

32. Kelly C. Smith, Keith Abney, Gregory Anderson, Linda Billings, Carl L. Devito, Brian Patrick Green, Alan R. Johnson, Lori Marino, Gonzalo Munevar, Michael P. Oman-Reagan, Adam Potthast, James S. J. Schwartz, Koji Tachibana, John W. Traphagan, and Sheri Wells-Jensen, "The Great Colonization Debate," *Futures* 110 (2019): 9.

33. G. K. Chesterton, *Illustrated London News*, April 19, 1924.

34. Dexter Masters and Katharine Way, eds., *One World or None: A Report to the Public on the Full Meaning of the Atomic Bomb* (New York: The New Press [on behalf of the Federation of American Scientists], 2007 [1946]).

35. Konopinski, Margin, and Teller "Ignition of the Atmosphere with Nuclear Bombs"; and Ellsberg, *The Doomsday Machine*, 274–85.

36. Ellsberg, *The Doomsday Machine*, 281, citing Thomas Powers, "Seeing the Light of Armageddon," *Rolling Stone*, April 29, 1982, 62.

37. Jonas, "The Heuristics of Fear," 215.

38. James S. J. Schwartz, *The Value of Science in Space Exploration* (New York: Oxford University Press, 2020).

39. Ahmed Amer, personal correspondence, June 28, 2013.

40. "Protocols for an ETI Signal Detection," SETI Institute, April 23, 2018, presenting the "Declaration of Principles Concerning the Conduct of the Search for Extraterrestrial Intelligence," unanimously adopted by the SETI Permanent Study Group of the International Academy of Astronautics, at its annual meeting in Prague, Czech Republic, on September 30, 2010, https://www.seti.org/protocols-eti-signal-detection.

*Chapter Eleven*

# New Players in Space

## *New Nations and Commercial, Private, and Nongovernmental Activities*

**Figure 11.1. The SpaceX factory in Hawthorne, California**

## FROM PAPERWORK, TO REUSABLE ROCKETS, TO HUMAN SPACE FLIGHT

Space Exploration Technologies—more commonly known as SpaceX—was founded in order to make space more accessible for human settlement. When its paperwork was filed with the California Secretary of State on May 6, 2002, Elon Musk already had a vision of what SpaceX would do in order to make space accessible: create reusable rockets. In the following years, this is what SpaceX has done, even in 2020 successfully sending astronauts to the International Space Station. This rapid growth from merely filing paperwork to human space flight shows the revolutionary condition that humankind is now in, with individual companies—guided by individual people—now doing what used to take entire nations to do and what only a few nations can still do.

## RATIONALE AND SIGNIFICANCE: HOW TO PLAY TOGETHER NICELY IN SPACE

Many new nations have entered into space with satellites and other programs, and bigger states with shorter histories in space, like India and China, are becoming space powers. At the same time, nongovernmental organizations have become increasingly active in space; organizations such as SpaceX, Blue Origin, Virgin Galactic, Sierra Nevada Corporation, the X Prize Foundation, numerous owners of satellites, and many other groups either already do have or have stated future plans for activities in space. With so many new actors in space, how can we be assured that their activities will be ethical? Determining how to coordinate these actors and promote collaboration for the sake of the common good is the topic of this chapter.

## ETHICAL QUESTIONS AND PROBLEMS

**The presence of new players in space—whether national or nongovernmental—raises new questions regarding the complexity, coordination, and freedom of activities in space. While new nations have the precedent of previous spacefaring nations to follow, subnational groups are treading into a realm without defined behavioral and moral limitations. How should we navigate this new space?**

While the Outer Space Treaty (OST) has much to say about nations, it has little to say about subnational actors in space. Subnational actors are to be regulated by the nations from which they operate.

However, this might not be a good system in the future. As corporations grow in power, it is conceivable that they might reach a level of power where some states are no longer equipped to enforce laws upon them. Currently examples of this include corporations that move their headquarters in order to gain tax benefits, or corporations that use small nations as "flags of convenience" for registering ships. The seasteading movement has gotten into this realm as well, arguing that settlements in international waters could provide places where completely new laws could be enacted, free of the control of traditional land-based nations.[1] While no precedent has yet been set with regard to the legal status of seasteads (though a small one was intentionally sunk by the Thai navy in 2019), if there were a precedent set, then it might apply to space as well, thus perhaps creating a space outside the OST for corporations to operate, whether on Earth's oceans or in space, if a corporation shifted its headquarters there.

One solution to this problem would be to ban the creation of these extraterritorial spaces, whether at sea or in space. This would be a draconian and antidemocratic solution, but one that traditional land-based (and Earth-based) nations could conceivably implement for the sake of preventing competition with "start-up nations."

Another solution would be not to halt the creation of these new national or quasi-national extraterritorial entities but rather to expand the OST to include these new players. This is more respecting of human freedom and more flexible overall, in that it allows growth and experimentation in governance while including these new and experimental entities rather than punishing them for innovation. As long as new players are included in the system, the system will grow not only in size but also in flexibility, making humankind more adaptable for the new spaces into which it explores—difficult and challenging spaces where adaptability must rank among the highest virtues.

However, this solution, of course, multiplies the complexity of actors involved in space. Rather than limiting the number of actors to those nation-states currently party to the OST, it would increase the problem indefinitely.

The key is not only to (1) include consideration of all new actors but (2) to also have an enforcement mechanism that ensures their compliance with (3) ethical standards of behavior in space. The OST solved this problem for the world of the past, how can it be updated or revised to solve this problem for the world of the future?

### Low Earth orbit (LEO) and geosynchronous Earth orbit (GEO) are already somewhat regulated spaces, but how should activities beyond LEO and GEO be regulated, if at all?

LEO and GEO are regulated by various treaties and organizations such as the UN's Committee on the Peaceful Uses of Outer Space, United Nations Office for Outer Space Affairs, and International Telecommunications Union. Currently their regulation is fairly effective, although the proliferation of space debris and other problems are becoming worse, so "fairly effective" is the best we can say. It is conceivable that with human exploration these organizations could simply expand their authority farther into space.

As another option, other organizations could step in to regulate this area—either organizations that already exist or one or more new organizations. For the sake of coordination, it could make sense to keep all of human extraterrestrial activity under the authority of one entity, whether old or new, and yet for the sake of specialization to the various different environments of space it might make sense to develop multiple regulatory bodies focused on, for example, control of space debris, regulation of contamination, preservation of peace, and so on.

Total lack of regulation would be the worst option. In this case, in the highly dangerous environment of space, which is capable of harming people in space, science in space, returning harm upon the Earth, harming ETL, or even harming ETI, humans would be left in a situation in which even one agent (not even a human one, in the potential case of invasive Earth life) could cause immeasurable harm. As the chapter on planetary protection showed, these are not dangers that are easily dismissed or minor; they are potentially some of the worst outcomes possible for humans in space.

A space free for all might make a few actors happy in the short term, but in the long term it would endanger not only them but many things of ethical significance both on and beyond the Earth. Space should be regulated by agencies with reasonable powers to enforce their mandates.

**What role do commercial enterprises and private organizations have in space exploration? What of governmental and private partnerships or other modes of public-private cooperation? Should private companies be conducting tourism, asteroid mining, or lunar mining?**

In states with nationalized industries and communist states, private enterprises have had limited or no role in space exploration. In states with private enterprises, there have always been private contractors involved in space exploration, but with government agencies leading the way. Recently in the United States, this public-private collaboration has turned more toward the private sector as organizations like SpaceX have taken the lead in space exploration, not waiting for NASA or the U.S. military to grant contracts, as has historically been the situation, but rather privately financing the technology first and then seeking contracts after the technology is already developed and/or being used for commercial purposes.

While state leadership on space exploration has certainly been the case historically, it looks as though the technology for space exploration has sufficiently democratized so that private space exploration is now not only a serious possibility but a fact. SpaceX has stated intentions of settling people on Mars, regardless of a government leading the way. Will this private leadership someday exist entirely independently of government leadership?

The obvious benefits of state leadership are that the risk of failure and waste is upon the state itself, and the public in that state. Additionally, states can engage in activities that do not demand rapid monetary payoff—states can engage in pure research, regardless of monetization. On the other hand, if private industries fail, then they go out of business. If they do not raise enough funds from their work, then they go out of business. Now that private industry has determined that there is money accessible to them in space, space has changed from being merely a risk of failure into being an opportunity for success, and so startups like SpaceX, Blue Origin, and Virgin Galactic can take the technologies developed over decades of state-sponsored exploration and redirect them toward profitable opportunities.

However, commercial and industrial access to space, as noted earlier, should not be a free for all. Instead these private organizations still need to be regulated and still need state guidance, in accordance with international treaties and norms, in order to protect both Earth and space from reckless exploitation and side effects like debris pollution.

So to answer the question of how public and private sector entities ought to relate to each other in space, it is the duty of the public sector—governments, states, and international bodies like the UN—to regulate the private sector for the sake of defending the common good of all on Earth and in space, and it is the duty of the private sector to achieve beneficial ends for humanity and itself, for example, by creating satellite networks, mining space resources, and transporting materials between places in the solar system.

Government-private partnerships may be a logical way to combine the best aspects of both private and public leadership when it comes to space exploration and use. Private leadership allows for faster action and more nimble decision making but is limited by relying on private financing and lack of ability to sustain long-term goals, especially in pure research. Public leadership allows for broad psychological appeal and assurance of some minimal level of continuous investment (in the form of government contracts), even in pure research, but is limited by slow decision making and fickle public (and potentially legislative) will. Government-backed corporate entities for exploration and use of space might therefore seem to be an ideal hybrid, if they combined the best of the private and public spheres. These entities can certainly go wrong and become hotbeds of corruption and waste, and so they require careful oversight to make sure they are indeed keeping the best interests of the public in mind. But if this can be done, then these partnerships are particularly capable of achieving long-term and reasonable goals in space.

Given the existence of private interest in conducting activities such as tourism and mining, it makes sense to permit them, with reasonable regulation. (The alternative would be some form of active suppression that would overall hinder the development of space and would likely be counterproductive.) Permitting private space activities need not mean that space becomes a "Wild West" where anything goes; indeed, as noted earlier, this would be far from ideal. Space enterprise needs to be carefully regulated or many values are put at risk, not only ethical but scientific, economic, and even existential (all of which are also ethically relevant). Space tourism ought to be regulated not only in order to protect the space tourists themselves from harm, for example, by requiring adequate training, limiting radiation exposure, mandating stringent safety requirements, and so on, but also to protect the space environments that such tourists might damage, for example, by limiting the production of orbital debris, prohibiting damage to historical sites on the Moon (such as the Apollo landing sites),[2] and taking planetary protection measures, if necessary. Space mining, likewise, would need to avoid wrecking extraterrestrial environments that are of significance (such as features on the near side of the Moon), avoid producing debris clouds from asteroid mining, and take strong measures to make sure that any asteroids or ore shipments that are brought near Earth are not in any way dangerous to Earth or other near Earth space infrastructure.

## Should an overarching "prime directive" (as in *Star Trek*) govern all human activities in space, and if so, what should it be, who should decide, and how should they decide what it is?

I believe that there should be an overarching "prime directive" or set of directives for the exploration of space. If there are a set of directives they also ought to be ranked because conflicts between rules are inevitable in rule-based systems. Determining what this rule or set of rules might consist of should be a process that involves all of humanity in some way or another. Human cultures are full of ethical traditions that can help us on these points, not just Western traditions such as deontology and consequentialism but other traditions that consider what a morally good human being looks like, for example, as described in Confucianism and Buddhism. Indeed, every culture has a conception of what a good person is, and this is often characterized in terms of virtues, so these cross-cultural resources are already quite accessible[3] and just waiting to be applied to space.

International law is based on treaties, and compliance with treaties ought to be mandatory for all bound parties (assuming that the treaty is good—bad or evil treaties, in some cases, *ought* to be disobeyed). Ethical directives are a bit different. Indeed, it is quite likely that international bodies might want to avoid "ethics talk" altogether, although there have been few very significant documents in the past that stand as ethical landmarks, for example, the United Nations Universal Declaration on Human Rights (UNUDHR), which stands as a much-beloved reference point for international ethics to this day.

The UNUDHR was born from the special circumstances surrounding the devastation at the end of World War II, and while it was a difficult accomplishment, we should not believe that it cannot be matched again by future treaties. When humans begin to explore and settle space, this is precisely the time to think of ethical declarations at the scale of the UNUDHR and greater because the situation is similarly important: a new order is being laid down for the world, and in this case the world is not just the Earth but the Earth and beyond. When we are thinking of space ethics, we should think big because that is thinking proportionately to the subject matter. So we can hope that when space does become a preoccupation for human minds, these minds will think in this way.

Until then, and recognizing that I am only one small perspective and that this material is not at all developed enough to perform this task above—it is only one idea of many that should be put forward—I will propose a brief set of three directives, in order of priority, to act as a mere starting point for discussion. These directives are presented using the Greek word *telei*, which philosophically means "that for the sake of which" something is done: goals, purposes, ends, aiming points, meanings, significance, etc.

*1. Respect ethically aware intelligent life. Humans and other intelligent life (if it exists), as the only creatures capable of respecting* telei, *should exist.*[4] Humankind is deserving of respect. Humankind ought to exist because only we, and other creatures like us, have the capacity to be ethical and respect other things according to their deserved ethical value. Therefore, due to our uniqueness, we ought to protect our own existence. Humanity is the universe itself come to consciousness and awareness of the strangeness and beauty of existence, and that is a precious thing. Additionally, if, in space, humans meet further creatures capable of ethics, we ought to respect them in return, for they are like us in this way, no matter in what other ways they might be unlike us. These ideas are rooted in philosophy including Aristotle, Kant, Jonas, and many more, including contemporary space ethicists.[5]

2. *Respect life itself, as the ground of all* telei.[6] This means that all living entities are deserving of some level of respect, even if that respect is not absolute and inviolable. All living things seek to preserve themselves, reproduce, find food, and so on, and to these creatures these ends are good. External to these creatures we might judge some of those ends to actually be bad, for instance, in the case of a bacteria making someone sick—we can legitimately use antibiotics to kill the pathogenic bacteria and save the human life, which segues to the next point.

3. *Respect* telei, *both human and nonhuman, promote their fulfillment and minimize their thwarting, and, if possible, prevent cross-purposes. Discriminate between cross-purposes by capacity for virtue.*[7] In the earlier example, the pathogenic bacteria and person are at cross-purposes with each other, and despite our valuing both of their lives (as a reflection on their own valuing of their own lives), we can clearly say that the human is more ethically valuable than the bacteria. Why is this? Because in a forced choice between purposes that are at odds with each other, the *better purpose* ought to be the one that we choose. How do we know what is better? Pathogenic bacteria are not capable of much—they consume food, expel waste, reproduce, and harm others in the process. Their capacity for excellence in these activities—their capacity for virtue—is limited. A virtuous pathogenic bacteria is one that consumes, expels, reproduces, and harms others (which is actually not a virtue). A human, on the other hand, can do these things as well as several more: we can respect each other, love our family and friends, and seek the truth about the universe. These are significantly more advanced goals than those of a bacteria, and they therefore take priority. To think about this a different way: a bacteria never considers whether it ought to infect and kill someone. But humans do consider whether they ought to do things, including killing bacteria. Because we have this level of ethical awareness we should use it and also recognize that we are special for having this ability.

A prime directive that includes basic ethical guidance will do much toward setting the stage for future beneficial actions in space.

### Can space be "owned"? In what respect? How would one stake a claim for property in space? Should "homesteading" and property claims on various bodies in space be allowed in order to encourage exploration and use of the resources available?

Currently, under the OST, property rights do not exist for territorial claims in space. Artificial objects can be owned as property (satellites, space probes, etc.), resources can be harvested and used, and orbital locations can be occupied, but ownership is currently not legally possible.

This may change in the future with the revision of the OST, though it is unclear exactly what may happen. The Antarctic Treaty and Law of the Seas have not developed toward increased territorial rights, after all, despite many claims upon Antarctica and the desires of some nations to stake claims on shoals and other watery areas that are not quite land or close to land.

However, there may be strong economic and political incentives to allow property ownership in space. For a historical example, much of the west of the United States was settled because of the Homestead Acts, which gave land to people who settled on it and developed it in certain ways over a certain period of time. This caused many hundreds of thousands of people to go forth and settle land, in a way that empowered the U.S. government and disempowered Indigenous Americans, as their lands were unjustly taken away.

While we can look back with dismay at the lack of ethics of the Homestead Acts, in their treatment of Indigenous peoples, an Interplanetary Homestead Act would not face those same ethical issues (at least in the solar system) but different ones, particularly planetary protection issues and issues with damaging the scientific and environmental value of places in space. The same sorts of legal means could make sense if there was a desire to bring settlers into empty space itself or onto smaller objects in space such as asteroids or comets. While empty space, completely lacking in resources, might not seem appealing, there could be reasons to occupy certain places that might become economically important, such as **Lagrange points** (where gravity balances between objects in space). Asteroids and comets have more clear economic value as sources of water, minerals, metals, and other elements.

Now, of course, we come to the ethical question: But *should* we use property rights and claims incentives to drive human migration into space? There are many reasons to avoid rushing humankind into space, and also some good reasons to do so. Reasons in favor include space settlement as a possible way to reduce existential risk[8] and reducing environmental impact on Earth (though they would not likely produce benefits directly by exporting population but rather by promoting the development of "superenvironmental" technologies for space that would likely also have benefits for Earth).[9] Reasons to avoid such a rush are also huge, including planetary protection to protect science and avoid biological and chemical contamination. A headlong rush to the Moon might not trigger biological pollution worries, but it could damage sites of scientific value, while a headlong rush to Mars would present both issues. Geopolitical concerns also should not be ignored; after all, if certain nations start monopolizing territories in space—even if it is not the nations themselves, but just settlers originating from particular nations—a power competition could break out that could lead to conflict and militarization of space.

Given these concerns, it might make sense to move fairly rapidly toward a self-sustaining international settlement restricted to one locale on the Moon. This would hopefully defuse power struggles, limit destruction of sites of scientific interest, and present limited contamination issues, while still presenting the benefit of the exploration of new environmental technologies and a hedge against existential risks to humanity.

### The movement of human society into space will be, perhaps, the most momentous migratory event in the history of our species. It is a chance to make a new beginning in terms of governance. What ethical issues might arise in this transition?

Just as the world, and particularly governing, was revolutionized by the independence movements of the American colonies and of other colonies around the world, the development of space colonies may be revolutionary as well. Some, such as Charles Cockell, have explored the difficulties of establishing a free society in space, where the harsh conditions and centralized technological systems could easily promote authoritarianism.[10] Cockell's concerns seem real, especially with the concurrent development of other centralizing technologies such as AI and mass surveillance, and therefore these centralizing and authoritarian dangers present a serious risk to human flourishing that space ethics needs to carefully consider. Yet at the same time, the expanses of space might allow for more freedom or, thinking more laterally, different kinds of freedom than those traditionally found on Earth.

Certainly the vastness of space presents a certain kind of freedom to those with the power to move through it. It is conceivable, for example, that a group of people could build a vessel and simply take off into distant space, where it may be unlikely that other groups of humans will ever catch up with them or find them. While it would certainly be obvious, at least in the near future, if some group were planning to do such a thing, at some point in the next century, if space infrastructure grows dramatically, such a project might go relatively unnoticed, especially if it had cover as another project. Depending on the technological level and speed of the vessel it might be very hard to follow, not to mention there might not be the will to do so. At the same time, these "separatists" would also present a possible existential risk to the rest of humanity (this is explored in some detail in Liu's *The Dark Forest* and *Death's End*), and therefore hunting them down or even destroying them might feel necessary to the rest of humanity.

Similarly, it is difficult to anticipate how future technologies (which all have their own ethical difficulties) will interact to increase or decrease individual freedom, especially as biotechnologies, medical implants, and neurotechnologies may be changing humanity in different ways. If far-out ideas such as linking human minds via AI into a collective intelligence become possible, then this would be a radically different kind of human, and perhaps dramatically less free as individuals but more free as a group (as their collective mental power could outweigh nonintegrated humanity).

These ideas can be disturbing and will be considered in more detail in the next chapter. But for now we should be aware that space does raise these issues and they are very difficult ones, filled with tremendous uncertainty and great potential for both benefit and horror.

### The pursuit of money is a major reason to explore space, but is it a "good enough" reason? What makes a money-making endeavor (or any endeavor) good?

In a properly structured economy (that is, one that is structured to promote human dignity, human development, and the common good), financial return is tied to providing social benefit. The economics of space exploration and use should reflect this logic. Space exploration and use ought to provide a social benefit, and, as has been discussed before, there are many: scientific knowledge, access to new resources, protection against existential risks, the experience of meaning in discovery, providing cooperative human purpose, and so on.

In the process of providing these social benefits it is completely reasonable for organizations and individuals to also derive monetary benefit sufficient both for their survival as well as their flourishing. In fact, rewarding these beneficial behaviors is good because these behaviors benefit society. A society with proper incentives will reward those who protect and enhance the common good while punishing those who damage and degrade the common good.

However, the pure pursuit of money should not be a reason that organizations pursue space exploration and exploitation. There are many unethical ways to make money in space, one extreme example being to extort the planet by diverting a large asteroid toward Earth impact unless a large sum of money is paid to stop it. This is an extreme example but not impossible if extraterrestrial economic and legal incentives are not designed properly. There are many more examples possible, such as strip mining the Moon for Helium-3, turning the Moon into a large advertisement, carving faces of famous people onto the Moon for money, and so on. Many of these verge on the ridiculous until we realize that there are certainly some people who would like to do these things or who have already done analogous actions on Earth, as history and current events demonstrate.

One of the reasons that capitalism makes for such a dynamic and successful economic system is that, in many cases, it aligns personal interests with public interests. This is certainly not always the case, but insofar as it does happen, it brings huge social benefits. If this can be brought to space exploration and use, then the benefits should be proportionately large; and if it cannot, then the system will need to be corrected.

## ETHICAL CONCEPTS AND TOOLS

### Should there be a code of ethics for humans in space?

There should certainly be some enforceable legal frameworks for space exploration and use. Most likely the OST will be updated, and other treaty regimes may appear such as NASA's Artemis Accords, but there may remain legal complexities such as the uneven application of Earth-based law systems off planet, as is currently the case (for example, the space station is under various Earth legal jurisdictions based on the Earth origin of the particular module, such as the United States or Russia).

But law is only one way of attempting to regulate human behavior, and it is a heavy-handed one. It is worth remembering that the end of every law with an enforcement mechanism is a power that ultimately can take property, imprison, or even kill.

Thinking more broadly, a lighter and more flexible approach to new places could involve something like a code of ethics, code of conduct, or "prime directive," as considered earlier. As we set up a code of ethics in space, we should think about what our highest desires and aspirations are for the future of humanity. We should pay attention not only to the wisdom of the past but also exactly how this wisdom interacts with our present time and the future, as far as we can perceive it. As noted earlier, some ethical principles could include facilitation of achieving *telei* (or at least nonobstruction of *telei*), promotion of nonconflicting *telei*, and arbitration of *telei* based on capacity for virtue. Harkening back to another popular ethical system in the contemporary West, this would be similar to the medical ethics standards advocated by Beauchamp and Childress (based on the **Belmont Principles**) of **principlism**: autonomy, beneficence, nonmaleficence, and justice.[11] While reducing an ethical system to a few potentially conflicting principles can leave it brittle in the face of complexity, the benefit of rules is that they simplify the world into something that we cope with the vast majority of the time. For the rarer and more complex situations, additional ethical resources and guidance can be sought out and utilized.

Setting up a legal system in space could take ethical ideals and codify them into law—but it would require a unified legal regime in order to produce, maintain, and enforce it. Like on Earth, a legal system cannot be merely rules, it must also involve an entire system of interpretation and balancing; therefore a justice system including judges, juries, legislators, etc., would also be needed. It might make sense to develop this upon the "platform" of the United Nations because the UN already has some relevant infrastructure, however, this would no doubt be a subject of complex political debate in itself, as more powerful nations would seek special privileges or simply not respect these legal institutions, as is the case currently with the World Court.

Rather than go the route of law, then, it might be easier to codify ethics into a code of conduct that might be enforced by a professional society or some similar organizations with a softer form of power than that of a state or the UN. What might be some basic principles in such a code of ethics in space? David Livingston has produced a code of conduct for business in space, including environmental stewardship, honesty, safety, a free-market economy, and disclosure of conflicts of interest.[12] Ethicist Margaret McLean has proposed some principles and virtues including stewardship, scientific integrity, prevention, prudent vigilance, intergenerational justice, last resort, humility, and truth telling.[13]

Basic areas of concern for a code of ethics could include protecting human life, freedom, property, and truth, and protecting these goods not only for humankind but for all intelligent life, should this life be found. This might seem less controversial in space, yet it is still controversial on Earth where we already know there are other species capable of self-consciousness, though not technological civilization—such as great apes, elephants, some dolphins, and some birds—and yet we do not grant them protected status.

Breaking down these categories further, the code might come to resemble lists of human rights (such as the UNUDHR), lists of capabilities (such as those listed by Sen and Nussbaum),[14] or various other professional codes of conduct, such as those held by medical doctors or engineers and regulated by their professional societies. In fact, a "professional society" of space explorers, pilots, navigators, etc., might be one way to maintain ethical conduct in space that is light handed in terms of law and yet highly effective in terms of motivation and efficacy. In many nations, societies of professionals have very strong internal policing systems that maintain strict self-regulation, which not only provides the benefit of their good conduct to society but also protects their good reputation. One only has to think of various professions with bad reputations to recognize that codes of conduct and ethics could very much benefit the practitioners along with the rest of society.

In sum, there should be some sort of code of ethics for humankind in space, and it should be something different from a legal system. This is not to say that law should have no place in space, far from it; space also needs a coherent and unified legal system. Space needs both law and ethics. Having legal regimes and ethical codes developing simultaneously also allows them to build on each other and even cooperate and integrate, where appropriate.

## Virtue Ethics, Part III

Virtue ethics also has an important contribution to make on questions of the ownership and use of "dead" objects in space and the associated complexities surrounding utilizing such objects, such as throwing off dangerous debris. Virtue ethics, rather than thinking directly about the consequences of these activities, instead asks what kind of person would do these activities. Would a good person wantonly litter space with dangerous debris? Would a good person completely destroy a natural object, such as an asteroid, for the sake of some other goal, such as the attainment of wealth? If not these things, what would a good person do in these cases? What virtues of character (for example, prudence, courage, justice, temperance) would be relevant here?

Virtue ethics has the benefit of being a very flexible ethical system yet not being so flexible that one can do anything that one wants to do. There is a constant judgment that is kept in mind with virtue ethics: Is this what a good person would do, or is this what a bad person would do? This requires not only the use of one's own introspection and self-reflection but also a reflection upon the context in which one is operating. Am I living in a just society that rewards good behavior or an unjust society where good behavior is punished? Will society judge me well for what I am doing? Should I go against the judgment of society in this case because I believe society is ethically wrong? What kind of a person will I become if I do this action? Is this the action of a courageous person, a foolhardy person, or a coward?

These questions all help to highlight different aspects of the context of particular actions, as well as the habit forming nature of action—once something is done it is easier to do the same sort of thing again, for better or for worse. Ultimately, our actions form our character and we become permanently marked as someone who is brave or cowardly, practically wise or foolish, generous or greedy, kind and helpful, or callous and cold.

With respect to new agents in space, both individuals and organizations, virtue ethics has a particular use because these understandings of good character are often widely shared across human cultures (though this is certainly not always the case). Every human culture has a concept of bravery, though it may differ in the specifics. Every human culture has concepts of generosity, wisdom, and skill. With this diverse basis, and even as diversity may complicate actions rather than simplify them, there is a universal foundation upon which to develop ethical behavior in space. While notions of "the good" may vary, notions of "the good person" are somewhat more coherent, though there are of course exceptions, such as people who admire criminals and villains. These sorts of aberrations can be put aside as aberrations, however, because these villainous conceptions of character are not **universalizable** (they cannot be made so that everyone can do them), because if they were, society would collapse. These notions of "excellent villainy" can only be parasitic upon well-functioning societies, or at least ones functioning well enough to endure the constant parasitic sapping of strength that comes from people of bad character constantly harming society. Well-functioning societies should be wary of these sorts of drains forming on the common good and act to control and limit them.

## Metaethics, Part I: G. E. Moore's Open-Question Argument: "What is good?" "Why is that good?"

At this point in the book, questions of metaethics must be examined, even if only briefly. As mentioned before, metaethics is the study of what exactly good is. If we are seeking to pursue the good, then we need to know something about what "good" is. However, immediately, anyone researching this area of philosophy will run in to G. E. Moore's **naturalistic fallacy**—a poorly defined and widely misunderstood and misapplied philosophical dictum often used to shut down ethical discussion involving nature. To proceed on such a path, then, Moore's arguments must be overcome.

For Moore, the naturalistic fallacy was a trap that caught those who attempted to define "good" as something related to nature; although, oddly, the naturalistic fallacy rejects defining good at all, even in nonnatural terms.[15] In his **open-question argument**, Moore asserts that goodness is an indefinable intuition because whenever we attempt to say what good is, it is always open to the retort "but why is *that* good?" thus demonstrating that the meaning of goodness is an open question.[16] For over a century Moore's naturalistic fallacy and open-question arguments have done much to prevent conversation between ethics and nature; however, Moore's argument has some problems.

If, for example, good cannot be defined, then perhaps nothing can be defined. In fact, William Frankena has called Moore's "fallacy" not the "naturalistic fallacy" but the "definist fallacy" of trying to define anything.[17] Just as much as we can debate why "good" is "good," we can also debate why "is" means "is" or "red" means "red." Words and concepts exist within irreducible systems, and while the systems typically work, if stared at too closely the individual parts can start to look rather puzzling. Frankena goes on to cite Moore's epigraph for *Principia Ethica*, a quote from Bishop Butler: "Everything is what it is, and not another thing."[18] The naturalistic fallacy, then, is the fallacy of trying to define anything by anything else, which cannot be done because they are different things. Other philosophers have critiqued Moore as well, with Mary Midgley critiquing it as "an all-purpose blunderbuss for shooting down every kind of argument" and "not so much anti-naturalistic as anti-thought,"[19] and Bernard Williams calling it neither "naturalistic" nor a "fallacy."[20]

Nevertheless, there is some content to Moore's critique, otherwise it would never have gained such currency in philosophical circles. The meaning of "good" is hard to pin down. But in the history of philosophy there have been solutions proposed to Moore's critique, many predating him by millennia. Aristotle once defined *good* as "that which everything seeks," and more precisely as *eudaimonia*, often translated into English as "happiness" or "flourishing." While this definition of good is still vulnerable to criticism (for example, how does one deal with those who derive happiness from harming others, or even entire cultures that do that?), this is also not an insurmountable problem, as indicated by the previous discussions of *telei*.

Metaethically, this response to Moore is Aristotelian (and therefore a form of virtue ethics), but it also can be compatible with utilitarianism, deontology, human rights, common good, justice, and frameworks from many other cultures. The fits might not be perfect; in fact, there can be sincere incompatibilities depending on theoretical commitments, interpretations, and how rigidly one sticks to one system. However, in space ethics, as an endeavor of applied ethics, what we are seeking is not theoretical consistency but practical usefulness. Very often very different ethical frameworks will come to similar conclusions. This is not always the case, as longstanding ethical-political debates exhibit (for example, abortion, euthanasia, freedom of expression, etc.), however, insofar as they agree on basics such as the unjustifiability of murder, sexual assault, torture, and so on, these are significant and should not be forgotten.

While theoretical issues like those of Moore can be interesting, they should not stand in the way of clear and practical use of ethical thinking in everyday life in order to solve real problems. Insofar as theory impedes practice, it is a hindrance that should be carefully considered and then resolved in a manner in which practical, real-life considerations take priority over theoretical or ideological commitments. Ethics is a study of how to become practically wise. Obstructions that block the path to wisdom—even very interesting yet useless theoretical ones—should be pushed to the side.

## Relativism versus Absolutism in Ethics and Ethical Pluralism in Space

Some view ethics as a relative endeavor that merely displays cultural biases and interests. Relativists like to say that what might be ethically good in one place might be ethically bad in another, and therefore ethics is ultimately just opinion. They might raise issues such as the fact that some societies find abortion acceptable, while others do not, and some accept infanticide, while others do not. The list of varying answers between cultures to ethical questions can go on and on.

However, there are good reasons to disallow many aspects of the relativists' assertion. We all share a common human nature and share similar needs; for example, we all require food, water, occasional medical attention, and so on. Rather than accepting relativism, we can accept a more universal approach to ethics based on human well-being and recognition of intrinsic human dignity. Such a recognition is the foundation for the UNUDHR, as well as other universalistic approaches to ethics. Importantly, when it comes to notions of human rights or other universalistic ethical frameworks, while particular cultural groups might each supply their own particular understandings of *why* humans have intrinsic dignity, the overall assertion of dignity does not necessarily need a common foundation. It can be one common edifice with multiple, even contradictory, foundations. In other words, the cultural foundations might differ in theory but be similar in practice. Because ethics is a practical art, this should not surprise us.

What, then, should we make of differing cultural opinions on such controversial issues as abortion, infanticide, euthanasia, freedom of expression, and so on? Because these are examples where both theory and practice differ, there are several responses worth considering. First, often these differences in ethical outcome may not actually reflect theoretical differences but rather category differences; in other words, if an unborn human is placed in the category of "person" then it ought to be protected, but if it is not, then it lacks this protection. Second, often in these cases there are strong values that have differing priorities, as in abortion where a woman's bodily freedom and the unborn's right to life are in tension. If one prioritizes life above liberty that may yield a "pro-life" judgment, while if one prioritizes liberty over life that may yield a "pro-choice" judgment. Third, other important values may come into play as well, for example, one might still prioritize liberty over life and yet think one would not abort one's own unborn because of the importance of kinship relations and the value of family. In this case the person might be personally pro-life, but socially pro-choice. Fourth, contexts may vary, for example, a society might be more sympathetic toward abortion for poor women than for wealthy women (note, however, that this might also reflect negative and unjust biases against the poor and might be differently remedied through antipoverty programs). Fifth, this can all be contextualized into our world where groups (nations, societies, subcultures, cultures, etc.) compete for resources and power. If social judgments prioritize group survival above individual survival, then abortion might be construed as reasonable if it helps the group, for instance by reducing short-term resource demands. On the other hand, the opposite conclusion is also possible: having more people can make a group more powerful in the long term (while consuming more resources in the short term), and thus a group might judge against abortion. There are additional considerations as well.

In the context of space, and of new players entering space, relativism and absolutism on abortion might seem like a very remote subject, but it is not. Humans in space will come from many cultures and therefore have many and varying opinions on many and varying ethical issues. In the midst of this diversity and pluralism, ethical humility will be important, remembering that all ethical systems are from Earth and therefore will have some potential for mismatch with the conditions in space. This is not to argue that space should be an ethical free for all, however. As noted continually in this book, everything about space involves powers beyond what humans have typically faced on Earth over the course of history. It is new ethical ground, and with more power comes more responsibility. This should make us more conscious of the potential gravity of the decisions being made. If we do not want these powers wielded by villains and fools, then we should encourage or become the moral heroes that we want to see making decisions in space.

Humans who venture into space should be willing to reconsider the ethical standards they are used to, but likely not in a way that weakens or loosens those standards but in a way that makes those standards more cautious and responsible. People with greater power typically have greater ethical claims placed on them rather than fewer precisely because they are more powerful and therefore are more able to help or harm. There can be legitimate ethical variety based on legitimately different circumstances. But we should be under no illusions that space will be a place of loosened ethical standards; indeed, with such powers in play, the standards are only likely to become stricter, or at least different.[21]

## The Ethics of Public-Private Partnerships

Public-private partnerships, as noted earlier, can blend the best aspects of each institutional structure in order to create a more flexible and more effective organization. The fundamental ethical requirement of such a blended institution is that it works on behalf of the common good, respecting human dignity and pursuing ends that help the public and not only the private shareholders or other limited interest groups. This is, unfortunately, often a very difficult task, for several reasons.

First, the private side of the partnership will almost always be much more motivated to protect the partnership—and especially the money-gaining element of the partnership—than the public side of the partnership. This is because a small number of highly motivated people who are profiting from something will likely have stronger opinions than even a very large number of people who either care very little or not at all about something. If every year one million people pay one dollar to one thousand people, who each receive one thousand dollars, the smaller number of people will be very motivated to maintain this relationship, and the larger number may not much care.

Second, "the public" often does not have the attention to closely monitor whether their money is being well spent. Indeed, it might be being utterly wasted by workers who sit and do nothing, but if no one ever notices this situation and tells the public about this waste, the waste will not be corrected.

Third, the private side of the partnership often has the money to influence government officials to help them protect their relationship. This political corruption is sadly all too common in the world, and without strict watchdog organizations the corruption can go uncorrected by the public, who may be both unaware and unmotivated.

This is not an exhaustive list. And all of this is not to say that public-private partnerships are doomed to corruption; indeed, very often they contribute to the creation of great things. But it is to say that such powerful relationships require powerful monitoring. Ethics is not a simple subject, and those who intentionally benefit from corruption are already corrupted in judgment and therefore ethically unreliable. In the quest to build the future, responsible and ethical individuals and institutions are absolutely necessary or our new powers will not be used for good.

## The Ethics of the Idea of "Manifest Destiny"

**Manifest destiny** is an idea from the history of the United States that asserted that the United States was destined and entitled to rule a band of territory across North America from the Atlantic to the Pacific. As such an entitlement, manifest destiny was used to justify all sorts of actions, including the exploration of territory, purchasing of vast territories, oppression of Indigenous Americans, mass immigration and settling of land, and waging of wars. As an ultimately teleological idea, concerning the very meaning and purpose of the United States as a nation, with clear implications for action, manifest destiny was used to justify some very dark behavioral patterns.

Yet some have turned toward the idea of manifest destiny in order to promote human exploration and settlement of space.[22] While others have critiqued this as a bad idea, to some the idea seems tempting despite its terrible history.[23] At this point it is important to remember that political and ideological phrases are not neutral—they are freighted with the weight of history and indeed cannot be separated from it. Just as this book uses the word "settlement" rather than "colonization" with reference to space, "manifest destiny" should also not be used with respect to space. The phrase itself is not only burdened by history, it is also philosophically nonsense. There is no "destiny" for humanity in space; there is no necessity about it. Everything that we do when it comes to space is the product of human choices and agency. We can easily choose not to go into space. We can romanticize and fictionalize narratives of space calling to us to return to it, but these should not be taken literally. Entering into space exploration and settlement is not our destiny—it is our free ethical choice that we should make because we think it is, upon careful consideration, the right thing to do.

Manifest destiny is ultimately a highly motivating idea that has demonstrated its power in history. This motivation and power was used unethically and therefore the phrase should not be used again. We should find other and better ways to motivate humankind's movement into space and not look back to the destructive ideologies of the past.

## The Basis of Social "Value" and Money as a Measure of Value

One tendency of capitalist and market economies is to reduce ethical decision making to questions of economic value. This reductionism is highly inappropriate for several reasons, not the least of which is that many entities with ethical value should not be appropriately thought of as having a monetary value at all, such as the inestimable ethical value—often referred to as "dignity"—of an intelligent being's life.

Barring the inappropriate uses of the monetary notion of "value" we can further ask what it is that gives entities without intrinsic value or dignity (but with instrumental value) economic value. There are many theories of how economic value is generated, but two of them are directly relevant to space exploration.

The labor theory of value (and labor theory of ownership) can be traced back to various thinkers in the history of philosophy including Thomas Aquinas, Ibn Khaldun, John Locke, David Ricardo, Adam Smith, Karl Marx, and so on.[24] In its basic form, the labor theory of value states that value comes from mixing human effort into a place, product, or action. If a person settles land, it becomes more valuable because they have mixed their labor into it in order to improve it, thus making it more valuable than the surrounding unimproved land. Likewise, this mixing of labor into the land (or other object) makes this their "property" in terms of ownership because if one owns one's own labor then by placing that labor into something else that external object becomes at least partly theirs.

The market theory of value has typically come to replace the labor theory of value in contemporary economics. The market theory simply states that something is worth whatever the market is willing to pay for it. That is, in a free market, a seller can charge whatever a buyer is willing to pay, and this is de facto a fair price assuming both parties freely agree to it without lies or false information impeding informed consent. Ownership also becomes a matter of mere title or possession. This bypasses all sorts of questions of how much labor is worth, but by making value completely subjective, it seems to lose something important, though intangible.

So what should be the basis of value in an economy? This is not a simple ethical question. In fact it is so difficult that most thinkers will simply revert to the market theory of value, allowing for the “wisdom of the crowd” or the “invisible hand” to set prices and therefore values.

In a purely ethical sense, things ought to be valuable because they are good. Food helps people survive—therefore it is good and therefore it has value. Water and medicine are the same, as well as myriad other items and services necessary for survival, such as education, transportation, communication, etc. This ethical foundationalism for prices has a logic to it, however, it has not proven to be very effective in practice in human history. Price controls and command economies, even ones based on good intentions, have always been plagued by market inefficiencies. And yet we know today that even efficient markets are still quite capable of completely ignoring the needs of billions of people who cannot participate in the money-based system.

Given these conflicts between ethics and markets, it might make sense to have a market economy with limited regulations and remediations for those who lack economic power, provided by government or other social services, and this is in fact the system in many countries. Whether that is a matter of the government subsidizing or providing low-cost or free food or medical care, or otherwise intervening to make sure that goods flow to those without the economic means, can be left to the specifics of the situation. What is important is that somehow the economy serves human beings and not the reverse.

With respect to space exploration and use, this value problem gets at one of the root arguments against it: the cost of space means that other costs are not being paid, including costs to help the poor. This question should be acknowledged as a continuing ethical issue.

In a broader context, labor and value also raise questions about property ownership in space. How exactly would one gain title to property? Would it simply be a matter of landing there and claiming it for oneself? Would an office somewhere legitimate one’s claim? Does it become yours not by landing on it but by beginning to use it? (Note that this would relate to the labor theory of value.)

Ethically speaking, there seems to be something sensible in the idea that one must exert effort in order to own something, and perhaps the magnitude of that effort should relate to the magnitude of one’s claim. For individuals this makes intuitive sense, and yet for corporations and governments the claim seems different. After all, in the case of individuals, individuals do the work (even if they are automated systems, someone controls them). But in the case of corporations, individuals do the work while a group claims the rights, and within those groups the rights are claimed differentially, where those with more power claim more rights than those with less. As a theme throughout the book, space just makes Earth problems bigger.

## The Ethical Basis of Property Rights

Similar to the earlier question of value, the question of ownership is one that has concerned thinkers for millennia. According to the OST, nothing in space can be owned as a territory. However, territories in space can be mined, and the equipment for that mining and the products of that mining can be owned. In a way, this illustrates the labor theory of ownership—unworked territory cannot be owned, and even in the case of the OST, even worked territory cannot be owned, however, once the territory is worked, the products removed from it can be owned.

Is this an ethical system for property rights in space? As one point, it encourages being somewhere directly and in person (even if its via robots). This upholds an intuition that before anyone puts any effort into something they should hardly feel any legitimately justifiable attachment to it. Just because someone on Earth likes Mars does not mean they should be able to own it. After all they have only seen it from millions of kilometers away and have never even touched it. Coveting does not grant rights.

However, once someone starts to input labor into a thing they might legitimately begin to become attached to it. The individual may develop a vision of what they want to see and start pursuing a *telos*. If others come to interfere with that place, product, or process, then the original agent might justifiably feel that their *telos* is being thwarted by another agent at cross-purposes with them. At this point, ethical principles such as noninterference, nonmaleficence, justice, and autonomy are all clearly at play. Of course, even putting labor into something does not necessarily make it into property unless it also is legitimately yours by some sort of recognized legal right. In legal disputes about who has legitimacy over a piece of property, possession and effort are only two considerations; an even more important one is legality, and hopefully that legality is based on good ethical reasoning.

The ethical basis of property rights involves the human need to possess certain goods as well as certain means of production in order to survive. We need food, we need water, we need tools, etc., and we need social institutions to protect those necessary resources. Beyond that, however, if we have our basic needs met, then it becomes a question of how to distribute the additional goods produced and personal goals pursued.

For example, with respect to additional goods produced, humanity as a whole produces vastly more grain than it needs to feed itself. Thanks to pioneering work by researchers such as Norman Borlaug, sometimes called the "Father of the Green Revolution," there is actually too much grain in the world today. While this has prevented the mass starvation of millions of people, it has also had several negative effects, including dumping grain into the markets of developing nations, thus damaging their local agricultural sectors, as well as feeding massive quantities of grain to animals in the developed world (and the environmental damage from both the production of this grain and from the animals that eat it should not be underestimated). Somehow, despite these two methods of getting rid of grain, there are still malnourished and starving people in the world. It is a longstanding ethical ideal that all people have rights to such basic needs as food, if it is at all possible, and in this case, clearly it would be possible to feed these malnourished people. But it is not done. Do these malnourished people have a property claim on the appropriation of this grain? Ethically speaking, from a rights perspective, justice perspective, common good perspective, utilitarian perspective, and virtue ethics perspective, they either certainly have this claim or they have something like a claim or other ethical connection to this food that could be theirs.

In space, the right to food will become much more complicated than it is on Earth. Obviously, a space settlement will need food, but where will it get it from? Will all of the food come from Earth on rockets? Will there be local farms? Will the food be made in large factories or in small machines in individual residences? It is too much to predict at this point, but we do know that because food production in space will be both very difficult and very uncertain, extra effort will need to be taken to avoid shortages. Food should be an extremely high priority for settlers, and additionally, there should be concern for feeding those who may lack access to food, for example, if one settlement has less food than another, it would be ethically appropriate to send food to the settlement with less. Exactly how to arbitrate this ethical need is difficult to determine without more context. Normally in a market economy, the ability to pay is used for setting prices and determining who has access to goods. But with goods that are also ethical necessities, there are often good reasons to depart from the simplicity of the market economy.

With respect to personal *telei* pursued, here is another example: space settlers are going to need more access to rockets than the average person. Rockets are expensive and in short supply, and it is hard to see how they will be anything other than that in the next few decades, if not much longer. No individual needs to explore space, that is, they are not having their rights or other ethical imperative violated by them not having this ability. If it were a right or ethical imperative, then everyone in the past would have been subject to this violation, and yet human civilization got along relatively well anyway. However, at this point in history the situation has changed. It is now important for humanity as a whole to have access to space, and furthermore, this "right" (or other ethical need) is of "humanity," at least if a sustainable off-Earth settlement is an ethical need of humanity. But how does this ethical need align with the personal goals noted earlier?

Certainly the individuals who desire to be a part of this project would seem to be better participants than those who would want nothing to do with it. These individuals who desire to go also have the motivation to buy access to the rockets that they need, and indeed, wealthy individuals like Elon Musk are developing these rockets with their own money. Given this personal investment and desire, individuals such as Musk do have an ethical right to their property because they are paying for it and producing it. But does anyone else have this right? It can be argued that humanity as a whole does; however, that ethical need might be fulfilled simply by allowing individuals such as Musk to exercise their right. If only Musk and few rich people ever have access, while many others never do, we should ask if this is fair and suggest ways to help to remedy the imbalance, such as finding a way to subsidize access to rockets (for example, perhaps by discounting tickets to those who desire to access space but lack the funds). At this point in time, this access to space is certainly a luxury, but we should remember that these same sorts of problems have appeared in the past on Earth, such as when people bought passage to other continents and paid for it through loans or labor (often in unjust circumstances that should not be repeated). Conceivably, if a space settlement needed certain professionals, they could pay for their passage to the settlement or subsidize it, with the expectation that the individual they assisted would then help them for a certain period of time. This could be just or unjust, depending on the situation, so it would need to be done carefully. In any case, they would not own the rocket upon which they traveled in any sense of property, though they would own the ticket or right to ride upon it and pay for that right via labor at their destination. Once again, this is an ethically complex situation and should be carefully regulated to prevent slavery, debt bondage, or other unjust situations.

## Ethics and Economics

The debate between ethics and economics will not be solved in this section of a book on space ethics. In an ideal world the economy would perfectly fill every human need, while humans worked diligently to provide goods not only for themselves but also for others. This ideal world, inhabited by angelic beings better than humans have ever been, would have economic activity with division of labor and exchange of goods, and no one would be left out of economic activity based on their poverty.

Unfortunately, we do not live in a world full of altruistic angels. Luckily, neither do we live in a world of utter devils. Instead we live in a world of human beings, where all of us are a mixture of both good and bad, to varying extents. The economy that we have now could be better at helping the billions of poor around the world, though in most places the situation of the world's poor is improving. This improvement shows that we are on a trajectory toward an economically better world, and insofar as economics provides for the necessities underlying fulfillment of ethical needs such as food, water, and so on, this is a good trajectory to be on. As we go forward, hopefully we can improve upon our current economic conditions, and hopefully the technologies of space exploration and the resources of space itself will help to make this happen.

## The Ethics of "Disruptive" Technologies

One buzzword in the contemporary technology industry is "disruption." As new technologies are developed by new companies, older industries using older technologies are disrupted. In the harsh environment of space, many new technologies will be necessary and many of them will either be disruptive of industry on Earth or have disruptive potential. **Additive manufacturing** (also known as 3D printing) and ***in situ* resource utilization (ISRU)** are two such technologies.

Additive manufacturing creates objects by adding materials together rather than by subtraction (as in carving or sculpting) or by assembling separate parts. In many ways, it permits the reduction of production to information: with a 3D printer, the right resources, and the right information (a build plan), then a product can be made.

*In situ* resource utilization relies on gathering resources from the immediate environment to feed into industry, including additive manufacturing. ISRU mines materials locally, extracting and separating elements, with useful ones fed toward industry and useless ones set aside. (Importantly, the unwanted resources should not be haphazardly expelled into space as debris or strewn across a surface as slag—they should be carefully stored so that they do not form pollution, whether physical or chemical.)

These technologies will be crucial to space exploration, where resources will be few and far between and access to distant manufacturing too difficult to meet immediate needs. While all mining and extractive industries are destructive of the natural environment (and therefore ought to be regulated in some fashion), additive manufacturing and ISRU are perhaps as efficient as extraction and manufacturing can be. By locally sourcing materials and avoiding energy-intensive transportation, these techniques greatly enhance resource and energy efficiency. Nevertheless, efficiency does not directly translate into ethics—in fact, efficiency at bad or evil is itself bad or evil, and such efficiencies should be opposed rather than encouraged.

On Earth, these same technologies will be incredibly beneficial to remote locations and could possibly be quite disruptive of the shipping industry as more and more manufacturing becomes locally sourced. Likewise, raw materials might become locally sourced, though this would significantly depend on the local geology (much of Earth's crust is oxygen, silicon, aluminum, iron, etc.—some useful elements for industry, some not).

## Respecting "Moral Traces," Part I: Navigating Moral Differences

It is unlikely that pluralism will resolve into a more unified worldview in the near future (however, as noted earlier, vast theoretical differences can sometimes be passed over for the sake of practical agreement). Therefore we should assume that different ethical perspectives will exist and converse as humankind goes into space. Under these conditions of pluralism it could be difficult to make decisions regarding space exploration and use, especially if there are irreconcilable theoretical differences with practical impacts, for example, the belief that extraterrestrial objects are sacred and should not be touched, etc.[25]

Unanimity will often be impossible; however, unanimity is not a standard that humans ought to seek when exploring space. It would be wonderful to have it, but we should not rely upon it. Instead a version of majority rule, whether at a 50 percent or a higher standard, should be the standard for space exploration and use. Brute majority rule at 50 percent is rife for power struggles and abuse, not to mention flip-flopping with public opinion, so it can make sense to hold the majority to a higher standard.

Nevertheless, majority rule does not erase the rights of minorities. In a good political system the rights of minorities are protected so that they can still exercise acceptable levels of freedom and live in ways that may be different from the majority preference (of course, there are exceptions to this as well—if a minority desired to kill certain classes of people or destroy property, the majority would be under no obligation to accept this as a legitimate "difference of opinion"). Moral traces remain after ethical principles conflict and majorities rule, but those traces can be within limits. For example, if there are minority cultural groups that retain strong beliefs opposed to space exploration due to their religion, perhaps there are ways to explore space that can be more respectful to that religion. The minority groups can be consulted about how this might be possible, though there would likely need to be careful negotiations because one side would not be, ultimately, getting what they want.[26]

## The Role of Humility in Ethics

Moral judgment is difficult and will always remain so. The innumerable complexities of the world often combine to create situations that are completely unique and therefore unprecedented. Furthermore, the complexity of the long term is unfathomable, and so we are always making decisions using incomplete information. Rules, processes, and procedures for ethical decision-making can help to manage complexity but can never eliminate it.[27]

Due to this complexity and human finitude, humility should be an important trait among moral decision makers, especially going forward, as the future becomes more and more complex. As mentioned in chapter 7, we are living in now and moving ever more into a state of "acute technosocial opacity" where the future is less predictable and therefore good decision-making skills ever more important.[28]

While it is often important, and even necessary, for ethical leadership to be decisive and confident, it is similarly important to remember that ethical decisions sometimes ought to be revised or even reversed as new evidence appears or the environment changes. Slavishly sticking to old modes of behavior in the face of changing contexts can be a recipe for disaster. Additionally, science and technology are not only changing the world around us but also giving us new information about the world, and so we need to be able to incorporate this new information. Both **epistemological humility** and **ethical humility**, then, are crucial for our future. We should both be willing to remain steadfast on those ethical values that should not be changed while also being open to learning and adapting to new contexts and situations. While it is true that something like the first principle of practical reason—that "good is to be done and pursued, and evil is to be avoided"[29]—will never change, the specifications of normative behavior which flow from that will necessarily change based on context. People who are trained in practical wisdom—ethics—are the ones who should be better able to perceive these changes and adapt to them, as long as they are humble enough to recognize the need to do so.

In the context of space, the essentials of ethics will remain the same, but the variations in context may yield very different ethical outcomes. For example, on Earth owning a large rocket would be a very unusual thing, and yet for getting around in space it is the only way to travel between various objects in the solar system. Many ethical systems on Earth would say that for an individual to own a rocket would be an unnecessary luxury and the money better spent elsewhere, yet in space those ethical systems might need to "adjust their settings" and adapt to the new environment. Transportation is often a necessity for humans, and on Earth we do not consider it to be extravagant to own a bicycle, automobile, boat, or for some people even small airplanes. And yet in space, this same need to move from place to place and transport people and materials cannot be done by any of these previously mentioned needs; instead what is needed are rockets or other modes of space travel, and so rockets it must be.

There are more examples of these sorts of differences, many of which might include basic environmental and material differences concerning atmosphere, radiation, access to food and water, and so on. Fundamental changes in context may yield fundamental changes in specific ethical outputs, for example, on Earth, having a lot of pet animals is of little concern to others, but on a Mars colony these pets might tax the air supply, require inordinate amounts of food and water, and endanger others through these costly needs, not to mention presenting a threat of allergens and even planetary protection concerns for biological contamination of the local environment because every kind of pet includes a pet microbiome as well. The list goes on.

## BACK IN THE BOX

Here on Earth most people would agree that the government should not be the sole actor in public life; corporate activities and nongovernmental organizations are encouraged too. Many people appreciate the existence of "civil society" such as business organizations, religious institutions, private educational institutions, charities, political associations, and various other social groups. Social institutions provide valuable social services that improve standards of living. Corporations provide valuable products, some of which might have been greatly delayed had not visionary entrepreneurs pursued their goals. Yet "civil society" can also lead to discord and corruption in society, environmental damage, fractiousness, and so on. How should the various macro, meso, and micro levels of society interact (for example, government, institutions, and individuals)? This is a complex issue with no easy solutions, and we should not assume that these difficulties will only exist in space; we are experiencing them here on Earth all the time.

## CLOSING CASE

Space tourism is a growing field. Between 2001 and 2009, the American company Space Adventures, Inc., safely delivered to the International Space Station and returned to Earth seven space tourists on eight missions (one tourist going twice), including the first space tourist, Dennis Tito. Each tourist paid tens of millions of U.S. dollars for this privilege. Space Adventures continues to offer trips to the International Space Station, as well as to low Earth orbit, and has proposed a tourist flight around the Moon. The group also offers training and other programs on Earth, including zero-gravity flights.[30] Other organizations are working on developing the space tourism industry as well, such as Virgin Galactic, Blue Origin, and SpaceX. This growing industry certainly represents a revolution in the human relationship to space, now doing for enjoyment what formerly was extremely dangerous work. This bodes well for the safety of space travel but certainly raises other questions as well, such as whether this is truly a good use of resources. In any case, space tourism looks to be a new industry that is here to stay.

## DISCUSSION AND STUDY QUESTIONS

1. How should private organizations operate in space? What kinds of restrictions should be upon them? In what ways should they be encouraged?

2. Should property or territorial claims be allowed in space? Why or why not?
3. What are the most important behavioral norms that ought to be promoted for all humans and human organizations in space? Why those?
4. Of the ethical tools and concepts discussed in this chapter, which do you find to be the most useful and why? Which do you find to be the least useful and why?

## FURTHER READINGS

Dan Deudney, *Dark Skies: Space Expansionism, Planetary Geopolitics, and the Ends of Humanity* (New York: Oxford University Press, 2020).

Jai Galliott, ed., *Commercial Space Exploration: Ethics, Policy and Governance*, Emerging Technologies, Ethics and International Affairs (London and New York: Routledge, 2015).

Walter Peeters, "From Suborbital Space Tourism to Commercial Personal Spaceflight," *Acta Astronautica* 66, nos. 11–12 (June–July 2010): 1625–32.

Erik Seedhouse, *SpaceX: Making Commercial Spaceflight a Reality* (Springer Science and Business Media, 2013).

## NOTES

1. Joe Quirk with Patri Friedman, *Seasteading: How Floating Nations Will Restore the Environment, Enrich the Poor, Cure the Sick, and Liberate Humanity from Politicians* (New York: Free Press, 2017).

2. M. S. Race, "Preserving History on the Moon," *Astronomy Beat*, no. 91 (March 13, 2012), www.astrosociety.org; Thomas F. Rogers, "Safeguarding Tranquility Base: Why the Earth's Moon Base Should Become a World Heritage Site," *Space Policy* 20, no. 1 (2004): 5–6; and Dirk H. R. Spennemann, "The Ethics of Treading on Neil Armstrong's Footprints," *Space Policy* 20, no. 4 (2004): 279–90.

3. Vallor, *Technology and the Virtues*.

4. Jonas, *The Imperative of Responsibility*, 99; Green, "Convergences in the Ethics of Space Exploration," 189; Green, "Self-Preservation Should Be Humankind's First Ethical Priority and Therefore Rapid Space Settlement Is Necessary," 36; Brian Patrick Green, "Constructing a Space Ethics upon Natural Law Ethics," in *Astrobiology: Science, Ethics, and Public Policy*, ed. Octavio A. Chon Torres, Ted Peters, Joseph Seckbach, and Richard Gordon (Hoboken, NJ: Wiley, 2021); Brian Patrick Green, *The Is-Ought Problem and Catholic Natural Law* (Dissertation, Graduate Theological Union, Berkeley, California, 2013), 204.

5. Lupisella and Logsdon, "Do We Need a Cosmocentric Ethic?"; Randolph and McKay, "Protecting and Expanding the Richness and Diversity of Life, an Ethic for Astrobiology Research and Space Exploration."

6. Green, "Convergences in the Ethics of Space Exploration," 189; Green, "Self-Preservation Should Be Humankind's First Ethical Priority and Therefore Rapid Space Settlement Is Necessary," 36; Green, "Constructing a Space Ethics upon Natural Law Ethics"; Green, *The Is-Ought Problem and Catholic Natural Law*, 204.

7. Green, "Convergences in the Ethics of Space Exploration," 189; Green, "Self-Preservation Should Be Humankind's First Ethical Priority and Therefore Rapid Space Settlement Is Necessary," 36; Green, "Constructing a Space Ethics upon Natural Law Ethics"; Green, *The Is-Ought Problem and Catholic Natural Law*, 204.

8. Green, "Self-Preservation Should Be Humankind's First Ethical Priority and Therefore Rapid Space Settlement Is Necessary."

9. Smith et al., "The Great Colonization Debate."

10. Charles S. Cockell, *Extra-Terrestrial Liberty: An Enquiry into the Nature and Causes of Tyrannical Government Beyond the Earth* (Edinburgh: Shoving Leopard, 2013); Charles S. Cockell, "Liberty and the Limits to the Extraterrestrial State," *Journal of the British Interplanetary Society* 62 (2009): 139–57; Charles S. Cockell, "Essay on the Causes and Consequences of Extraterrestrial Tyranny," *Journal of the British Interplanetary Society* 63 (2010): 15–37.

11. Tom L. Beauchamp and James F. Childress, *Principles of Biomedical Ethics*, fifth edition (New York: Oxford University Press, 2001).

12. David Livingston, "A Code of Ethics for Conducting Business in Outer Space," *Space Policy* 19, no. 2 (May 2003): 93–94.

13. Margaret R. McLean, "Reaching Out from Earth to the Stars," in *Geoethics: Status and Future Perspectives*, ed. G. Di Capua, P. T. Bobrowsky, S. W. Kieffer, and C. Palinkas (London: Geological Society of London Special Publications, November 12, 2020), 508.

14. Sen, "Equality of What?"; Sen, "Rights and Capabilities"; Nussbaum, "Non-Relative Virtues"; and Nussbaum, *Women and Human Development*.

15. G. E. Moore, *Principia Ethica* (Cambridge: Cambridge University Press, 1959), 14, 38–39.

16. The "open question" argument, ibid., 15–17; the quote is my simplified paraphrase.

17. William Frankena, "The Naturalistic Fallacy," in *Perspectives on Morality: Essays of William Frankena*, ed. Kenneth E. Goodpaster (Notre Dame, IN: University of Notre Dame Press, 1976), 4–7.

18. Ibid., 6–7.

19. Mary Midgley, "The Withdrawal of Moral Philosophy," in *The Essential Mary Midgley*, ed. David Midgley (London and New York: Routledge, 2005), 176.

20. Williams, *Ethics and the Limits of Philosophy*, 121.

21. Brian Patrick Green, "Transhumanism and Catholic Natural Law: Changing Human Nature and Changing Moral Norms," in *Religion and Transhumanism: The Unknown Future of Human Enhancement*, ed. Calvin Mercer and Tracy Trothen (Santa Barbara, CA: Praeger, 2015), 213.

22. Editors, "Our Manifest Destiny Is to Move Beyond Earth," *Financial Times*, December 23, 2014, https://www.ft.com/content/56e28fda-8447-11e4-bae9-00144feabdc0.

23. Linda Billings, "Overview: Ideology, Advocacy, and Spaceflight—Evolution of a Cultural Narrative," in *Societal Impact of Spaceflight*, ed. Stephen J. Dick and Roger D. Launius (Washington, DC: NASA, 2007), 483.

24. Robert W. McGee, "Thomas Aquinas: A Pioneer in the Field of Law & Economics," *Western State University Law Review* 18, no. 1 (Fall 1990): 471–83; Zubair Hasan, "Labour as a Source of Value and Capital Formation: Ibn Khaldun, Ricardo, and Marx—A Comparison," *Journal of King Abdulaziz University: Islamic Economics* 20, no. 2 (2007): 39–50; Albert C. Whitaker, *History and Criticism of the Labor Theory of Value in English Political Economy* (New York: Columbia University Press, 1904); and Duncan K. Foley, "Recent Developments in the Labor Theory of Value," *Review of Radical Political Economics* 32, no. 1 (March 2000): 1–39.

25. M. Jane Young, "'Pity the Indians of Outer Space': Native American Views of the Space Program," *Western Folklore* 46, no. 4 (October 1987): 269–79, https://www.jstor.org/stable/1499889; and Vigiliu Pop, "Lunar Exploration and the Social Dimension," *Proceedings of the ESLAB 36 Symposium "Earth-like Planets and Moons,"* ESTEC, Noordwijk, June 3–6, 2002.

26. Aline H. Kalbian, "Moral Traces and Relational Autonomy," *Soundings: An Interdisciplinary Journal* 96, no. 3 (2013): 280–96, citing James F. Childress, "Moral Norms in Practical Ethical Reflection," in *Christian Ethics: Problems and Prospects*, ed. Lisa Sowle Cahill and James F. Childress (Cleveland: Pilgrim Press, 1996), 211; further citing Robert Nozick, "Moral Complications and Moral Structures," *Natural Law Forum* 13, no. 1 (1968): 34–35, and W. D. Ross's "discussion of prima facie duties that have the feature of 'always count[ing] morally even when they do not win'"; W. D. Ross, *The Right and the Good* (Oxford: Clarendon, 1930).

27. Vallor, Green, Raicu, "Ethics in Tech Practice," slide 35.

28. Vallor, *Technology and the Virtues*, 6.

29. Aquinas, *Summa Theologiae*, 1009.

30. Space Adventures, accessed January 3, 2020, https://spaceadventures.com/.

*Chapter Twelve*

# Traveling to the Planets and the Stars

## *Very Long Duration Spaceflight and Human Biology*

**Figure 12.1. NASA's Space Launch System (SLS)**

## INSPIRATION MARS

In 2013, former space tourist Dennis Tito founded the Inspiration Mars Foundation, dedicated to sending two private space participants on a flyby mission past Mars and then return them to Earth.[1] Tito himself authored, along with several others, a feasibility study for the project.[2] Unfortunately, the project failed to progress and no one ever flew past Mars. Tito needed NASA to cooperate and provide funding, and NASA refused.[3] Besides the dependency on NASA, one of the problems for the mission was the simple logistics of the flight: the need to bring five hundred days of food, the utter dependence on life-support systems, the dangers of radiation, and so on. But the Inspiration Mars idea was inspirational to some people even though it never flew. As space technology advances, proposals of this sort will proliferate, and eventually one may become reality.

## RATIONALE AND SIGNIFICANCE: A LONG VOYAGE IN THE VOID

Travel to other planets, bodies, and locations in space has been a dream of humankind ever since we realized that it might be possible. Travel even into low Earth orbit is difficult enough, but travel to other planets in the solar system, or someday even other stars, is vastly more difficult and consuming of time and energy.

While we have sent many robotic probes into the solar system, and some beyond, humans have never traveled beyond the Moon, and even there not for decades. Yet now there is renewed interest in these exploration activities, not only with machines but with people. Why should or should we not consider these extremely difficult and dangerous missions?

## ETHICAL QUESTIONS AND PROBLEMS

### What are the different challenges faced by those settling objects in space and those living in the void between those objects, either permanently or while traveling?

Very long duration space travel should probably be perceived less as travel and more like temporarily settling on a moving object—a space vehicle—in the vacuum of space. Very long duration spaceflight can span from weeks to months to years, and perhaps for generation ships, even to decades or centuries. While travel on Earth in the past might have been like this, such as transoceanic voyages in wooden ships or caravans crossing continents, there are few analogous travel situations today, except perhaps refugees fleeing long distances on foot.

While living on a solid body such as a planet, moon, or asteroid can be extremely difficult, it does have advantages over living in a vessel separate from any natural object. For one thing, doses of cosmic, solar, and other forms of radiation originating from the sky tend to be lower on the surface of an object simply because the object itself shields approximately half of the sky and thus half of the radiation (this may vary with terrain and object type). Living under the surface obviously reduces this radiation exposure even further (though such locations may encounter other natural underground radiation sources such as radon gas).

Similarly, natural objects contain natural and unused resources, while artificial vessels typically do not. Instead artificial vessels must plan ahead by bringing along adequate energy, fuel, food, water, breathable air, other resources, and spare parts, or the means of creating these new parts from resources.

The disadvantages of living on natural objects include being stuck in a natural orbit in a place that one did not choose for that object to be, though presumably humans will choose whether or not to land and settle there. There also would be potential planetary protection worries, as well as chemical and physical particle dangers, issues with reduced gravity, and so on.

Orbiting space stations may seem idyllic, but in reality they would face challenges. These sorts of settlements would be dependent on other space objects not only for their construction but also for their maintenance and flourishing. They would need trade. And so it might make sense to have space stations at orbital balance points such as Lagrange points because there will often be incentives for space travelers to pass though those areas anyway. A space station not on a trade route would become a lonely place indeed. And as alluded to earlier, these stations would lack access to extra resources and would need to solve the problems of microgravity, radiation, and so on.

Constructed spaces also have some advantages over natural objects in that they can be placed anywhere and can even be moved or placed into convenient orbits between planets, where they can act as shuttles between planets, such as Apollo astronaut and space scientist Buzz Aldrin's idea for a Mars Cycler.[4] There is also little planetary protection risk associated with space stations, unless they are contaminated during construction or by travelers from a natural body. Given enough energy, a space station could become a vehicle and even be sent on an orbit out of the solar system, on a trajectory into the Kuiper Belt, Oort Cloud, or toward another star.

**What of the ethics of sexual intercourse and reproduction in space, given the risks? Some have suggested that reproduction be banned in space, while those who want to settle space see reproduction as a crucial aspect of space exploration.**

Sexual intercourse and reproduction in space are even more controversial than the same topics on Earth (which are controversial enough), and there are a spectrum of opinions ranging from outright bans to full expectations. Perhaps the most important thing to consider is the larger context of both sex and reproduction in space exploration. Building upon analogies from Earth, early explorers typically were more preoccupied with staying alive than with starting families, though certainly evil acts such as sexual assault did occur. However, when it is time for the settlement of a new location, sex and reproduction would become expectations, otherwise no settlement is sustainable.

Looking at history, space exploration until now has had no reproductive component to it because the objectives can be easily fulfilled within one human lifetime. So far all of human space exploration has fallen into this category: reproduction was unnecessary and therefore did not happen.

However, in the future, human space exploration may occur that will last longer than one human lifetime. This exploration might be done via some form of hibernation technology (which, of course, does not yet exist) or through generation ships. (The issue of whether humans might become "uploaded astronauts," living as mental patterns in AI, will not be considered here in detail because it is too speculative.[5]) Hibernation, while currently not feasible for adults, is already possible for embryos and human gametes. This has given rise to the idea of "embryonic astronauts," a scenario first proposed in science fiction by Olaf Stapledon in 1930 and that has been further explored in fiction, popular writing, and scholarly media.[6] This scenario of course raises innumerable ethical issues, not only the typical ones associated with reproductive technologies (such as separating reproduction from intercourse and relationship, and the destruction of excess embryos) but also some very extreme issues such as raising children into adults when there are no adults present to care for them, which, even if possible via robots and AI, might lead to severe psychological trauma. Needless to say, this method of space exploration would be extremely ethically fraught and therefore should not be near the top of the list for human space exploration. However, under very dire circumstances such as the impending threat of human extinction it might be possible to make a case for such missions, though even then such choices would be very much questionable.

Generation ships, as part of the name, obviously require reproduction or else there will be no generations. In this case, all of the concerns with safety, risk, and health mentioned in chapters 3 and 4 come to the forefront again, but this time with humans at all stages of life, from conception and embryogenesis to dying and death. Ethical questions including the prospects of health for the children being raised in such an environment would be extremely important, not to mention the fact that no children could consent to being raised under such circumstances (though this is the case for all children ever raised in human history). Space, of course, just raises the stakes for sex and reproduction by placing them into a very different and harsh environment. Health care on such a mission would be strained in any case, and beginning and end of life issues would only make it more strained. However, AI and robotic surgery technology is advancing rapidly, which might alleviate at least part of this particular burden. And, as in the previous case of embryonic astronauts, reproductive technologies could be used to avoid health problems for the adults involved, for example, if frozen gametes or embryos and artificial wombs are used. There is, then, a potential for complete separation of reproduction from sexual activity, but again this would be a massive experiment on human beings who are unable to consent when more time-proven methods would seem to be more reasonable.

It should not be overlooked that some of these reproductive proposals are not just human experimentation but *massive* human experimentation, potentially upon an entire population. While human society on Earth currently is using reproductive technology, it is not yet completely reliant upon it, which could become or would be the intent in the earlier cases. Human relationships, including such fundamental ones as those that ground the family itself, could fail utterly as they are thrown into new territory, and perhaps needlessly so, if more traditional means are available. This would not just be human biological experimentation but human social experimentation, a form of social engineering striking at the core of human nature. In recent centuries there have been many social engineering experiments run on Earth, ranging from the industrial revolution to the collectivization of communism to all of the other changes of contemporary life. Some of these experiments have failed and some are ongoing, but none of them have demonstrably proven that they are sustainable new ways of life for humanity, at least not as compared to some of the ways of the past. This is neither to glorify the past (which had terrible problems, certainly) nor to vilify the future (which we cannot predict) but only to say that these experiments are being run, often with many complex and interconnected variables, and not only can we not foresee their outcomes, we cannot even necessarily currently evaluate them in a rational way because the problems are too complex and often too linked to ideology.

These scenarios of sophisticated reproductive technology might capture the imagination, but they are also hypothetical when it comes to space and would be experimental until perfected. A much simpler solution would be what humans have always done: sexual intercourse. Space would also be an experimental environment for sexual intercourse too, but, as an experiment, it would at least have the benefit of fewer variables to disentangle. While this section has focused on sex and reproduction during space travel and on ships, the issues are much the same for space settlement and can be extended to that discussion as well.

### What of synthetic biology in space, or even directed or deliberately controlled human evolution or human-machine integration (including questions of transhumanism)?

Because space is such a hostile environment, some, such as the philosopher Konrad Szocik, have suggested that human biology ought to be manipulated in order to make humans better suited to space.[7] For example, it might make sense to make humans more tolerant of ionizing radiation by enhancing our DNA repair machinery, or making humans more tolerant of microgravity by changing our stomachs to be able to function without gravitational aid, and enhancing our skeletal and muscular systems to produce more bone and muscle mass regardless of gravity. More extreme examples include turning human legs into arm-like appendages or vastly enhancing our intelligence in order to deal with the complex decision making required for a very different environment, including, for example, the much more three-dimensional nature of space. Others have argued that such manipulations to human populations should not be made necessary for space exploration. Rather, all humans should be able to share the same environments as all other humans, whether in space or not, or, in James Schwartz's words, that the universe ought to be "accessible" in the way that places on Earth should be made accessible to persons with impaired mobility.[8]

These are genuine ethical questions even before they become technical ones, and the technical difficulty of even trying to adapt humans to space should not be underestimated; human biology is extraordinarily complex. Furthermore, these sorts of interventions are still mainly theoretical even on Earth. While minor forms of selective breeding (via selection of sperm and egg donors, etc.), human genetic manipulation (for example, He Jiankui's notorious manipulation of embryos in China), and eugenics via elimination of some kinds of future humans (for example, forced sterilization, selective abortion, preimplantation genetic diagnosis [PGD], etc.) have occurred or are currently occurring in some locations on Earth, none of these are anywhere near the technical level necessary for making the significant phenotypic changes necessary for adapting humans to the environment of space. And in fact it is unlikely that such modifications could be made in any sort of organized fashion within decades nor even a century. This is not even to mention the thousands of necessary experiments and failed experiments that would be required to pursue such a path—experiments upon human beings, without their consent, before they are born.

However, such science-fiction dreams we know have ways of becoming reality, even in the absence of the approval of most of the population, which then raises a deeper question, the ethical question: Even if we could do this should we do it? And the question of second-order desire: Why do we want to do this and what should we want to want?

It is certainly true that humans are not evolved to the environment of space. And it is certainly true that it takes much technological effort to alter the space environment so that humans can live there. But we know that this alteration of the space environment is possible because we have done it already, in places such as the International Space Station and during the Apollo missions to the Moon. So why are some motivated to instead think of altering humanity rather than altering the space environment?

There are many likely reasons for this, including the fact that reversing the focus of adaptation makes for an interesting story. It is also technically interesting and challenging: we know life can evolve over time to fit into many environments on Earth, so we might ask, in curiosity, can it adapt to life beyond Earth as well? And from an environmental perspective, it might allow for reducing the environmental impact of humanity upon space environments by letting humans live in a more "natural" space environment (though in the vacuum of space, there might be only limited room for adjustment—in settlements, on planets, it might make more sense, though still might not be ethical).

These motivations—interest, technical challenge, and environmental impact reduction—do not seem to outweigh the very real lives that would be experimented upon and potentially deeply harmed by such experimentation. There is a simpler solution to this problem, and while it is not a general rule that simpler solutions are usually ethically better (it is certainly not always the case, as many complex problems show), in this case the simpler solution of changing the environment and not humanity seems more likely to be the ethically correct one.

### How are questions of informed consent to be considered in those situations in which long-term irreversible decisions are made for as-yet-unknown future individuals and generations? How should we give proper moral consideration to those yet to be brought into existence and who have no say in the circumstances of their existence?

In the course of human history, parents have made choices for children without their informed consent from time immemorial. The key ethical question is always: Are these choices being made with the best interests of the children (and further future generations) in mind? Arguably, this has not always been the case, and the poor example of history is no excuse for us to continue to permit us to make bad choices now. As we have gained heightened ethical awareness over time and more power to act on that awareness, our ethical obligations have become enhanced. So we do need to ask whether it is in the best interests of children and future generations to be born and raised away from the Earth.

Earth is humanity's "perfect" home; at least we are evolved very well for it, if not quite perfectly. Leaving an excellent home for a worse one hardly seems like a good choice. But we should also remember that for billions of people on Earth, Earth is not a perfect home nor even a very nice one. Poverty, injustice, and oppression are commonplace, and those experiencing these conditions often dream of leaving, even if those dreams are not realistic.

Interestingly, some people who want to explore and settle space are not impoverished or oppressed but rather are, like the billionaire Elon Musk, quite wealthy and privileged. In this case, we might wonder what is on Earth that they would want to leave. Musk himself has expressed several reasons for wanting to go to Mars: for inspiration, for a challenge, and for the protection of the human race.[9]

On the question of parenting, Musk has also addressed this question, stating that he thinks that life for his children will be too easy.[10] This is a question worth pondering. Just as excessive difficulty is bad for human life, excessive ease presents its own vices: laziness, gluttony, pride, and feelings of entitlement. Some wealthy people who seek space might be seeking to avoid these vices, though that pursuit comes at the price of risk: it is not just comfort that is at stake but life and health as well.

When it comes to parents making these choices for their children, it might seem unreasonable to choose difficulty over ease. However, one of the strengths of the human species is its diversity and ability to adapt to, and even flourish in, new conditions. While it is not possible to state conclusively now that any future choices made by parents for their children in space are good or bad, it is at least plausible that these choices could be acceptable, and therefore this debate should progress along with the technology and ability to explore and settle space.

In any case, obtaining informed consent from future people—because they are as-yet nonexistent—is not possible. What we can obtain instead is informed consent from present people, on behalf of future people, keeping the best interests of future people in mind when giving consent. This is not the same thing as informed consent itself because the whole point of informed consent is to give a well-defined choice, including all relevant information, that the participant may freely choose or not. Instead this is a form of **proxy consent**, giving consent for another who is unable; this terminology typically is found in medical ethics literature. Proxy consent is in some ways similar to or overlapping with parentalism, the idea that external decision makers ought to sometimes intervene in order to make choices for others. While parentalism typically allows the moral patient (a person who is subject to the decisions of others) to make decisions and give informed consent in some cases, while taking away these powers in others, proxy consent means that a moral patient is unable to give informed consent in any cases. Therefore proxy consent is more like the situation with future generations, even though these future generations literally will be children to the parents of the future.

## What are the implications for deliberate actions that could literally change the evolutionary trajectory of the human species?

Desires to "take evolution into our own hands" have significantly preexisted the modern idea of evolution, dating back even to the time of the ancient Greeks, when Plato recommended a selective breeding program for humans.[11] Currently our technology for these sorts of manipulations to human biology are vastly beyond mere selective breeding and have progressed into gene editing, preimplantation genetic diagnosis, and brain computer interfaces. The ethical problems associated with these technologies have tainted them in the eyes of many and stunted and politically polarized much of the conversation around them.

Without a doubt, the implications of evolutionary-scale decisions are extremely ethically grave. That does not mean they are automatically evil or good, only that they are of magnitudes that ought to be given proportionate consideration. Indeed, it might not be possible to give such consideration simply because it would be more than humans could bear, though perhaps with many people and AI assistance such consideration might be possible. But we should not allow the scale of the choice to dissuade us from pondering it; in fact, the choices that we live out every day contribute to the future of humanity in a small way, as have the choices made by the people of the past, who have given us the world that we live in now. Those people, both ancient and more recent, did make tiny choices that have affected the evolution of humanity: we are all living proof of those choices. But moving to the next level, to intentional control over the evolutionary process by a smaller number of people, is truly another question entirely and not one to be taken lightly.

As noted throughout this book, just because humanity has a power does not mean that it ought to be exercised. And as mentioned earlier, any of these sorts of manipulations of humans must be, by their very nature, experimental and therefore risky. This in itself does not preclude these actions, just as the risk of spaceflight does not preclude it. But it does raise further ethical questions, such as:

1. Will the hypothetical future benefits of these manipulations exceed the definite current costs, including ethical costs?
2. Will these manipulations be respectful of the human rights of the people involved in the research, particularly of the people *created by* this research?
3. What sort of people are we that we would try to make and justify such manipulations? Are these the free choices of virtuous people or the choices of vicious people?
4. Will this risk human divergence and even speciation, in other words, Schwartz's concern that humans will become unable to exist together in a shared environment?[12]

5. If this does risk human speciation, what will prevent humanity/transhumanity from entering a "dark forest" situation in which each side sees the other as an existential threat, as has been explored in various science-fiction stories?[13]

This fifth question leads us to the next topic: Does exploration make humanity more safe or less safe?

## Will such missions help to protect human survival and the survival of Earth life by spreading it to other stars, or will gaining such powers merely make us more dangerous to ourselves or even possibly attract negative attention from extraterrestrial intelligences?

If one of our desires is to go into space in order to avoid risks and dangers that exist due to human activity on Earth, then by going into space, where life is more delicate and powers much more dangerous, are we not merely extending the threat of humanity across an even larger part of the universe? Would it not be better to simply stay home and first get our affairs in order here, thus saving the universe from the cost of knowing such an unethical humanity?[14]

Furthermore, exploration might not only benefit humanity—it could endanger us too, as previous chapters have explored. Not only might there be dangers of possible interplanetary contamination but also from future rogue groups of humans, and even—with admittedly low but nonzero probability—aggressive extraterrestrial intelligences.

Given these two possibilities—(1) the known destructive potential of a humanity remaining on Earth but armed with vastly more powerful futuristic weapons (including existentially risky ones) and (2) the unknown risks of a humanity among the planets and stars, also armed with vastly more powerful futuristic weapons (also including existentially risky ones, such as redirected asteroids)—there is a real ethical question presented by space exploration and settlement. At this point we might want to ask ourselves how tolerant we should be of acquiring concrete benefits now while opening up amorphous possibilities of risks later. We can recall our six levels of risk tolerance, as noted in chapter 3, and perhaps the gambler's principle might come to mind, or the precautionary principle. The truth is that all courses of action entail risk. Staying put on Earth is a huge risk, and exploring and settling space is a huge risk. We can only make the choices now that we perceive to be the best, knowing that we are operating on dramatically incomplete information and with all of our typical human psychological and moral limitations.

Of the choices available to us, which look best? As this book has argued already, it seems like space exploration and settlement give overall better odds of survival for humanity, merely because as humanity is cast across space there will be more of us in more and more distant places. If each place is sustainable and relatively secure, then even if some disasters befall some humans, there should still be many survivors. If we guide our ethical reasoning toward thinking of survival and flourishing not only of humankind but of much of Earth's life and, perhaps, other life forms that we may find as well, we can see a brighter future, one with a culture that values life in all of its forms and seeks to protect not only itself but others as well. This is a hopeful vision of the future.

**Spaceflight, whether long or short term, will involve computer systems, and as computer systems become increasingly more intelligent, artificial intelligence will become intimately connected to spaceflight. How will humans interact with complex artificial intelligences while in space?**

In the 1968 movie *2001: A Space Odyssey*, by Stanley Kubrick and Arthur C. Clarke, a long duration mission to explore Jupiter goes awry when the on board HAL 9000 computer determines that it should kill the human crew.[15] One crewmember escapes and disables HAL, thus resolving the situation. But could a real-life version of this lie in the human future? Some of the fundamental principles of the situation are already in play today in human interactions with AI and its alignment or misalignment with moral values. In a fundamental sense, HAL was merely trying to follow orders, which is the only thing that computers can do. However, the orders that HAL was following were contradictory: first, support the crew and maintain the spaceship, and second, keep the true nature of the mission (to investigate the transmission of an alien artifact) a secret from the crew. When both orders could not simultaneously be fulfilled, HAL determined that killing the crew would allow for the best resolution, with obviously evil results; at least obvious to humans, if not obvious to HAL.

How does this relate to today? There are many serious issues in the ethics of artificial intelligence,[16] and several of them are strongly related to space, such as safety, beneficial and malicious use, bias, and so on. Here I will look at only three: alignment and bias, deskilling and dependency, and conflict between groups of humans and their desires as expressed by humans and as expressed by AI.

Currently, the "AI value-alignment problem" is a subject of quite a bit of interest in AI safety research, and while some AI ethicists are more concerned with near-term problems such as problematic and arbitrary bias, in fact, bias and AI alignment can be seen as the same issue, just with different scopes. Arbitrarily biased AI acts in a way that does not align with what humans desire; a biased AI makes errors that harm human well-being, for example, by perpetuating human biases such as racism, sexism, classism, and so on. On the other hand, an AI that is aligned with human desires should be one that acts without arbitrary bias but rather is intentionally "biased" toward human safety and well-being.

Moving toward the future, it will almost certainly be the case that running an advanced space vessel will exceed the capabilities of any single human on board. Together, a crew will likely be able to perform most functions, and in coordination with ground crew might be able to perform all functions, but computers will also no doubt contribute to the efficiency of operation and ease of control. As computers and AI grow in power, they are likely to be ever more effective at what they do. When processes are automated, such as with aircraft autopilots, there is always a risk of deskilling of the humans who would otherwise be manually controlling the process, and this applies to ethical decision making as well.[17] Deskilling should be avoided, but as our world grows more and more technologically complex, in many cases it likely will not be avoidable. This raises the specter of technological dependency and how to live with technological dependency.[18] This phenomenon of technological dependency is not new; indeed, humankind is dependent upon technologies as different in age as domesticated plants and animals, wheels, and electronics. But AI, as a specific technology of intelligence, makes this dependency like nothing else before. It is a dependency like that of a child to an adult because to a superhuman AI an adult human can only have something like the relative mentality of a child. The passengers of a space vessel could find themselves in this situation. And this raises the last question of this section: How can humankind cooperate in a reasonable way with a "superintelligence"?

The powerful computers will need to be aligned not only with the interests of the crew but also with the interests of humanity as a whole, and the AIs on Earth as well. What could this mean? Consider this scenario: the crew of the ship and humanity as a whole could come into ethical conflict if the crew decides to do something that it was not originally directed to do. As with HAL in the movie, this will present a problem for the ship's on board AI. The nature of the changed course of action will certainly be important for the AI to consider, for example, is the ship merely going to a slightly different place with the approval of other people and AIs? Or is the ship going to do something illegal, dangerous, or potentially very evil, such as trying to divert an asteroid to impact the Earth or crash their vessel into Earth at high speed? Obviously, the AI should act, if possible, to make the situation better, but should it be empowered to take control of the ship in such a situation? This is a serious issue even today, as occasionally an airline pilot can become mentally unstable and intentionally crash their aircraft.[19] If the crew of a space vessel were to take profoundly dangerous actions, the AI might reasonably take actions to oppose the crew. However, such powers would also allow for the AI to override the crew and take dangerous actions without crew approval as well, or be hacked, or expose various other vulnerabilities.

And so the last question is: Who should have final decision-making powers when it comes to humans and AI, whether in space or on Earth? It would seem clear that humans ought to have this power, yet we know that sometimes people make terrible mistakes and do terrible things. What if AI could be programmed and empowered to stop these things from happening? Again this is the plot of another movie, 1970s *Colossus: The Forbin Project*, in which a computer tasked with defending the United States instead takes absolute control and institutes a totalitarian dictatorship.[20] Clearly, this would not be a preferable alternative either.

Ultimately, it would be best to create a system in which humans and AI were aligned so that human power was completely free and yet used only for good. Such a world would be a utopia, unlike anything that has ever existed in human history, resembling religious ideas of Heaven more than anything on Earth. Is it possible? Surely, some people would like for it to be so and are working on these ideas even now. But there are good reasons to be skeptical; humanity and ethical perfection seem incompatible. Yet at the same time, falling short of these unreasonable dreams of utopia should not disturb us. Between Heaven and disaster there is significant room for humanity seeking to improve itself, and there is reason to be hopeful that AI can be used to sincerely help humanity as we go forward into space.

## ETHICAL CONCEPTS AND TOOLS

### Informed Consent, Part IV: Intergenerational Informed Consent

Intergenerational informed consent can be thought of as being like societal informed consent but extended across time as well as across society (or "space" in terms of three dimensions, in which society exists). As mentioned earlier, informed consent for future as-yet-unborn generations cannot be obtained. In the absence of this ability to obtain consent, surrogate decision makers—that is, currently living humans—should choose actions that keep the best interests of future generations in mind. Proxy consent and parentalism play a role in thinking about these issues, but there are other important ideas as well, such as the best interests of others and the common good of the future.

The standards for looking out for the best interests of others have an extensive scholarly literature, particularly in medical ethics. "Best interest" typically means attempting to do what is best for another person, in good faith, as that person might choose if they were rational and able to decide for himself or herself, without regard to how that might affect the deciding party. This typically means exploring various medical treatment options, whichever will be most beneficial to the subject patient. However, in the case of a terminal patient where all attempts to heal would be futile, some of these options might not be for the purposes of improving health but rather for easing their death via hospice and palliative care. This is considered to be ethical because futile treatment can be sometimes quite burdensome and painful, not to mention expending scarce medical resources with no benefit when those resources could instead be used to help others.

With respect to the best interests of future generations, their first interest would be to exist at all. This connects to Hans Jonas's imperative of responsibility: that there should be humans in the future.[21] So "treatment" of humankind and our future generations is ethically required, but what is in the best interests of the future of humanity with respect to the *kind* of "treatment"? Certainly future generations would like to at least have a world as good as the one in which humans already exist, with adequate resources for survival, well-developed infrastructure, an educated population, access to health care, the possibility of a purposeful and meaningful life, etc.

With respect to space exploration on a long mission, this best interest might seem a bit strained, but not beyond the breaking point. A long-term space vessel needs to be largely independent in terms of resources and infrastructure. It would need to have adequate care available for its inhabitants or else it would be an unethical suicide mission. It would need to be able to raise and educate children and provide its inhabitants with a meaningful life (hopefully this would be at least partially linked to the success of the mission of the vessel itself). With this standard in mind it seems reasonable that a long-term space mission could pass the standard for possibly being in the best interest of future generations. The fact that the humans of the past who emigrated from their homes often did not have it so well and that the lives of their descendants improved by their emigration can also serve as corollary evidence of the reasonableness of this position.

The common good is typically thought of as something independent of best interest standards, but from another perspective it could be thought of as something like a best interest standard for all of humanity, rather than just one person. This requires developing the concept of best interest to include social institutions, relationships, family, and so on. From this perspective, intergenerational informed consent would depend on current generations keeping the common good of future generations in mind when making decisions now.

Additionally, we should consider whether we are talking about the common good of all of humanity or the common good of just the descendants of those on the long-term mission. Because it might be for the common good of all of humanity to send out long-term missions, while at the same time possibly subjecting the people on those missions to harsher conditions than what they would have endured had they stayed on Earth. Over time, these harsh conditions might become more comparable with those on Earth, hopefully because the mission settlement becomes quite nice rather than because of the Earth becoming much worse. But we should realize that both of these possibilities are real. We might think that those on the mission would be facing a difficult journey that subjects its crew and future generations to great hardship only to discover that the Earth undergoes a global catastrophe that makes being on the mission much preferable to having remained on Earth.

This element of uncertainty should remain a constant in ethical decision making about the future. We cannot, ultimately, predict the future, only give it our best guess. In this case, intergenerational informed consent, keeping the best interest and thoughts of the common good in mind, really is a complex and uncertain matter. We should try to create the best future that we can, but in the absence of certainty about the single best course of action to take, we instead ought to take the best courses of action for humanity to take as a whole. And with the survival and flourishing of humanity at stake, long-term missions, even if difficult for those on board, are overall the best option for a good future.

## Intergenerational Justice

As with informed consent, and taking the best interests of future generations in mind, the idea of intergenerational justice seeks fairness between generations, making sure that each generation receives what it is due. Exactly what this means is, of course, a matter of debate. At the very least it might seem that each generation deserves a world as good as the one that the generation possessed before it, and perhaps even a better world, thanks to the hard work of past generations. Whether this is currently occurring on Earth is an open question—certainly in many cases resources and the environment are being used up and damaged, yet technology is constantly improving. Additionally, intergenerational justice should consider not only what we owe to future generations but also what we owe to past ones, who after all turned over a world to us that, while far from perfect, could also certainly be much worse.

When it comes to space exploration and settlement, intergenerational justice takes an even more difficult turn due to the spatial distances involved. While on Earth intergenerational justice expands justice over time, in space, intergenerational justice needs to not only take time into account but also distance because vast distances increase the times required to keep all of humanity connected into one society. Space exploration will dramatically increase the resources available to humanity, and in that way could be said to be contributing to intergenerational justice by making the human-inhabited universe a more abundant place. However, due to the distances involved, not all humans can possibly benefit from this increase in resource availability, especially humans near the periphery of human-inhabited space. These humans will likely feel acute lack of some resources, though *in situ* resource utilization and additive manufacturing will be able to supply many material goods.

With regard to the past, are there also ethical commitments that we ought now to maintain? Certainly, we cannot directly harm the people of the past because they are dead. We can ruin their legacies, however, such as by slandering good people or praising evil ones (note, this also has negative contemporary effects in that it discourages good behavior and encourages bad behavior). Contemporary humans can ruin the works of large groups of people by destroying their physical and cultural legacies as well, as can be seen in cultural warfare that burns books and tears down historical structures; this has unfortunately been quite common in history, for example, when the Spanish Conquistadores invaded Mexico. And the worst of all possible injustices to the past might be to destroy all of humankind's work forever, either by literally destroying it, as in a global war, or by allowing human extinction, which would make all past human work count for nothing because there would be no one left to appreciate it.

We owe the past not only our gratitude for passing on a world to us and making our own lives possible but also our pledge to take their legacy and "pay it forward" to the next generation and all of those that will come after us. In *The Precipice*, Toby Ord states that "people matter equally regardless of their temporal location."[22] While we can only affect the world that we live in now, our powers extend through time. Not only can we affect the future very strongly, in terms of its very existence, but we can also affect the past in terms of its legacy. In this case, justice demands that we give the future a world as good or better than the one in which we live, and it also demands that we express to the past our gratitude for doing this for us, for at least keeping humankind alive, despite many disasters.

We can also ask: What do the future and the past justly owe to us now? The past has done its work, giving us the world we have, for better and for worse—it can do nothing more for us and so its ethical obligations toward justice are over, if not completely fulfilled. With regard to the future, we can certainly hope that in the future, generations will look back on us now and also express their gratitude for our work. While we cannot now appreciate this gratitude, the promise of it should hopefully give some solace to those who often do the underappreciated work of trying to make the world a better place. The future can remember our generation justly by praising the memory of those who acted ethically and criticizing who did not.

As a last point for this section, the ultimate act of intergenerational injustice—human extinction—is lessened by space exploration in some ways but not in others. Making humanity multiplanetary escapes the "all our eggs in one basket" problem, but, as has already been discussed in previous chapters, the technologies that allow space exploration also increase the technological danger to ourselves, and the potential for existentially risky extraterrestrial discoveries certainly exists as well. However, with these caveats, space exploration ought to serve intergenerational justice overall. It helps to protect not only future generations but also the legacies of past generations.

## Risk Ethics, Part III: Living in Difficult Conditions Due to Trade-Offs

As this book explored in chapters 3 and 4, with regard to spaceflight, risks can be divided into acute and chronic risks. Acute risks are abrupt and cause serious harm rapidly, such as a crash or explosion. Chronic risks are continuous or long term and the harm is constant or builds up over time. Living in space presents both such problems, but the ethics of chronic conditions are a bit more complicated than acute risk ethics and so require a bit more discussion here. Prolonged exposure to limited nutrition, microgravity, radiation, and so on will all take their toll on human bodies. In practice, all of these chronic risks would have technological systems working to attenuate them; however, the likelihood of reducing the risk to zero is unlikely except with very powerful technological interventions, and such interventions tend to be large, heavy, expensive, and/or energy intensive.

The bigger a ship is, the less risky it is likely to be, and the smaller it is, the more risky it may be, just because having more mass will better shield against radiation, allow the carrying of more food and spare parts, and so on. When it comes to long duration space travel, large, heavy, expensive, and/or energy-intensive technological systems will have to compete with the realities of budgets and propulsion. The bigger a vessel is, the more things it has, the safer it is likely to be, but the harder it is to move and the more energy it will take to move it. Safety systems will need to match these requirements, and that will entail trade-offs. In ethics a trade-off occurs when two ethical values come into conflict with each other and one or both must be compromised. For example, having artificial Earth-like gravity on a long-term mission would greatly improve health outcomes for the crew by avoiding bone and muscle loss, and a host of other physiological problems. However, creating Earth-like gravity would require a large mass of structural material to hold the rotating section together, otherwise it would fly apart due to centripetal force. Moving this structural mass requires more energy, which either means the vessel must go slower (thus risking health for a longer period of time) or carry more fuel (which of course increases mass even more, thus necessitating more trade-offs). The trade-off, then, is between health and mass. How can it be solved? An easy solution would be to find at what level of gravitational acceleration the human body can still do well enough (itself a somewhat subjective determination) while also reducing structural mass. Perhaps a more Venusian level of gravity might be adequate (about 90 percent of Earth's gravity), or a Martian level (about 38 percent of Earth), or a level more like that of the Moon (about 17 percent of Earth). Whatever the trade-off level is, one ethical value is being sacrificed in order to protect another ethical value.

Both radiation protection and nutrition have the same issue: radiation shielding and better food require more mass. All of these problems might be helped by the advancement of technology; perhaps food production can be developed to recycle waste into food on board the ship rather than bringing all food stores from Earth, or perhaps food can be grown on the ship. Perhaps the weight of radiation shielding can be reduced via magnetic fields and the like. In any case, these trade-offs will likely still exist because mass might be reduced while energy consumption is increased, and energy itself requires a source, which would likely have significant mass.

These trade-offs are inevitable, and it is very likely that the first explorers and settlers on these long-term missions will be more risk tolerant than average. They will take the risk and push the bounds of technology and humanity forward. At the same time, technology and infrastructure will advance and lessen the risk for subsequent missions, thus lowering the risk level so that more people will be willing to take the chance on a journey. This cycle will repeat until, like traveling by airplane on Earth, risk levels will become so low that they are acceptable to many, if not most, people. Risk will never become zero, but of course, being alive is always a risk too, no matter what one is doing. It is only a matter of what risks one is taking.

## The "Push" of Existential Risk or the "Pull" of Existential Safety

While space undoubtedly acts as a "pull" motivator toward some people who are drawn toward freedom, economic opportunity, and finding meaning in accomplishing hard tasks, there are also "push" reasons why people might want to leave the Earth. Attempting to escape Earth-concentrated existential risk can act as a "push" motivation toward human multiplanetarity. For example, most of the riskiest existential risks on Earth are due to technologies such as nuclear weapons, nanotechnology, AI, or synthetic biology. Fleeing Earth before one of these catastrophes causes huge damage (if such a thing ever happens) could be a tremendously beneficial choice for those who choose it.

Much like those who leave a nation before war breaks out, thus avoiding devastation and possible death, at some point in the future humans in space—whether in flight or in settlements—may appreciate that they are not on Earth as it undergoes troubles. We can hope that this will never happen, but humanity's track record is bad on this point: while many of us might try to avoid trouble, trouble finds us.

Rather than wait for disaster to strike, it could make more sense to leave beforehand. However, this is, of course, another risk trade-off: there is risk in staying on Earth and risk in space travel. Which is the greater risk will depend greatly on technology, politics, and individual judgment.

## Bioethics, Part III: Advanced Issues in Space Bioethics

Of all the contemporary fields of applied ethics involving technology, bioethics is the most advanced, having been a subject of concerted effort for over fifty years. Bioethics is continuing to advance at a dramatic pace; however, space exploration and settlement will add even further to this necessary drive for deeper ethical investigation. Both questions of genetic manipulation of humans in order to attempt to adapt humanity toward space and space exploration using embryos were mentioned earlier; here we will explore the "embryonic astronauts" proposal in more depth.

Sending adult humans on very long space missions is extremely difficult because suspended animation of adults is beyond our technological capacity, and for very long missions humans will age and could die before they reach their destination. To avoid this problem, space vessels could be made to move faster (requiring more energy), generations of humans could pass on the ship (requiring more space and resources), or we could send a type of humanity capable of suspended animation: embryos.[23]

The ship would be sent off with an AI controlling all systems. Life support could be minimal until a destination is reached, whereupon the first embryos (if they have survived the voyage) would be incubated in artificial wombs, "born" from these wombs, raised completely under AI supervision, and brought up in some manner that allows the reconstitution of human society. While these young humans are being raised, the AI could be building new habitats for the young to move into and eventually take charge of.

The technical challenges of such a mission are immense, to say nothing of the ethical questions raised by such an idea. This type of mission relies on numerous technologies that are not yet in existence, though might be within a few decades.

Given the enormous risks of such an endeavor, one idea has been to send such a mission but to never trigger the development of the embryonic humans unless a constant signal from Earth ceases. In other words, the mission would only activate in case of the destruction of the sending station on Earth, which should only happen if catastrophe befalls the Earth.

While this might seem to be a reasonable plan in some sense, preserving the human race for the future and only activating the embryonic astronauts as a last resort, it still does not answer the question of the embryonic astronauts themselves. Could it be considered to be fair to leave them in suspended animation indefinitely? Presumably at some point the embryos themselves or some critical system on the ship would fail and prevent them from ever reaching their *telos* of becoming adult human beings. They might achieve the purposes of those people who sent them, but only as a purely **instrumental good**: the **intrinsic good** of the lives of the individual embryos themselves would be completely subsumed to this larger goal.

This raises questions about whether the embryos' well-being and best interest is actually at the heart of such a mission, or whether, in a Kantian sense, it would seem like the embryos are being used as a "mere means" to preserve the human race and not as ends in themselves—beings with intrinsic dignity and worth. Other ethical systems might differ from Kant, for example, Jonas's imperative of responsibility might or might not endorse this path, depending on several key variables. Utilitarianism might endorse the idea of embryonic astronauts and virtue ethics might consider the idea and then experts might legitimately vary on their assessment. In such a situation, some sort of political process ought to help make the decision, and the best ideas (hopefully) win.

All of this is to say that bioethics will need to advance with our biological technologies, and space could easily be a major impetus for this advancement. We might someday have the power to act on these ideas, but should we? Perhaps we ought instead to choose not to. And while this might not decrease the risk human extinction, it might instead maintain our dignity as a species by not violating the dignity of our individuals.

## The Moral Question of the Purpose of Human Existence (If Any) and How to Fulfill That Purpose

When getting into question about space exploration and settlement we should also think about ultimate aims and purposes—the meaning of human life and the purpose of the existence of the universe. This is the question of teleology—the study of meanings, purposes, goals, and "that for the sake of which" something exists.

Space exploration seems as though it might be one of the human species' ultimate activities, something that, once we begin, we might continue nearly indefinitely. If this is the case then we should wonder whether this indefinite use of human resources and ingenuity is indeed the ethically correct thing for us to be doing with our time, effort, lives, and civilization.[24] This then raises the question of the purpose of human existence and, in turn, even more ultimately, the purpose of the universe.

What is the purpose of the universe? This question has no simple answer, and indeed some philosophers might be inclined to say that the question is nonsensical or "not even wrong"—in other words, so misguided that it does not even make sense. Philosophers who dislike the idea of meaning and purpose external to human minds might argue that human minds are "promiscuous" seekers of teleology,[25] looking for it all the time, and therefore it would not be surprising that we find it even in places where it does not exist.[26] Cognitive psychological experiments have consistently demonstrated that humans do in fact seek teleology more than they should.[27]

However, simply because humanity is prone to seek teleology inappropriately, and thereby mistake mechanism and chance for purpose, that cannot thereby demonstrate that purpose does not exist outside of human minds—that in itself is a logical error in another direction and is equally invalid. Starting with the easier case of teleology existing in human minds, it is clear that teleology exists in human language, for example. We can convey meaning and purpose not only in the words themselves but about objects in our culture such as tools, which are a classic example of material objects with purposes imbued into them by human minds. If all human minds disappeared, human-made tools would no longer have purposes. Knives would no longer be "for" something, as there would be no people to use them as such or even perceive in them the usefulness they might have for cutting. However, humans are not the only creatures with minds that create and use tools—animals can have purposes too.

Some animals make and use tools in the same ways that humans do, for example, chimpanzees and Caledonian crows, thus demonstrating that purposeful behavior extends beyond only the human mind and into other creatures that we might otherwise consider to be primitive.[28] Apes have also demonstrated language use in the form of sign language, thus demonstrating purposeful and meaningful activity in another way.[29]

However, to change the context slightly, instead of thinking of minds as creating tools and language, what of biology creating them? Biology creates minds, after all, and then minds create language and tools. Animals, through evolutionary programming via their DNA, communicate via sounds and songs and grow their "tools" as parts of their bodies. Birds have wings for the sake of flying. Cats have claws for the sake of catching and climbing. Fish have fins for the sake of maneuvering in water. These purposes-for-the-sake-of-which, Aristotelian **final causes**, exist in natural objects. Yet some would deny that they can truly be described as teleological "purposes" at all but rather as mere "**teleonomic**" programming.[30]

This question has been pondered for millennia, and we can only give it a few sentences here, in the form of a schematic answer covering the four possible options for where teleology might inhere in reality.

1. Teleology might not exist at all, even in our own minds. We are purely mechanism, and purpose is a figment of our imaginations (though somehow the concept of purpose does exist in our belief in purpose).
2. Teleology might exist only in human minds and nowhere outside of them. As discussed earlier, tools and language are clear examples of teleology in nature, but only in nature as is expresses in human minds; elsewhere, we are projecting it where it does not exist.

3. Teleology might exist in living things, in biology, through their DNA and cells creating their bodies with reactions and responses and capacities that permit them to fulfill *telei* in the world.
4. Teleology might exist at some level beyond mere life, in the matter of the universe, in nature, in the laws of the universe, or perhaps even in some sort of universal mind, whether **panpsychic** (perhaps unconscious) all the way to **theistic** (a god).[31]

These issues can become abstract and ideologically charged, so it might be good to back up a step and rephrase the question.

Another way to consider this question at the universal level might be to ask "is the universe *doing* something?" or "is the universe *about* something?"[32] These various formulations highlight slightly different aspects of the issue, thinking not so much about goals but about activity or *aboutness*. The human mind is extremely biased toward thinking things are about something, but then again, we live in a suspicious universe, where simple hydrogen, left around for thirteen billion years, eventually turns into minds wondering "what's this all about?"

Once again, philosophers have been debating questions like this for millennia, and religions for even longer. But it is worth noting that these questions are definitely relevant for space ethics; indeed, some of the space ethicists we have examined so far have already made claims with roots in metaphysical teleologies about the universe, or at least implied appeals toward such things. For example, Lupisella and Logsdon, Randolph and McKay, and Kelly Smith have all made these appeals in the form of appeals to reciprocity, richness and diversity of life, and complexity, respectively; to them, this is what the universe is somehow *about*.[33] Jacques Arnould explicitly points out the importance of metaphysics for space exploration.[34] But there is still, despite these sorts of appeals to the root of ethics in the universe, one more step that we must bridge before we can claim to be rooting our ethics in a truly reasonable fashion to the teleology of the universe itself. We must decide whether we think that universal teleology is, in fact, good or evil.

If teleology does not exist in the universe externally to us, then we must decide for ourselves what is ethical, but even if teleology does exist in the universe separately from humanity, then humanity, in order to act ethically, still should decide whether to cooperate with this universal teleology or not. After all, simply assuming that the universal teleology is a good one might not be a reasonable assumption. Perhaps the purpose of the universe is to maximize entropy, running all energy sources down into their minimal state. As creatures who rely on energy sources, this seems like a teleology that humanity might reasonably want to oppose—if that would even be possible. On the other hand, as energy-using creatures, perhaps we could also conclude the opposite—if the universal teleology is to increase entropy, then we as energy consuming creatures are already pursuing this teleology and therefore perhaps we ought to accelerate this process as much as we can.

While seeking entropy might seem to be an unlikely or odd goal, rather being something more like a side effect of another goal, there are many other seemingly directional activities in the universe that we can see from the perspective of natural history. For example, biological complexity increases over time. The speed of activities in the universe seems to accelerate over time (that is, compare the first billion years of the universe to the most recent billion years: a lot more has happened recently). Heavier elements form over time. Networks of interactions coalesce and decay. Of the many choices, which are the actual goals, if any? And even if there are goals, once again, are they good?

At this point the topic can seem interminable, inconclusive, and exhausting, but there is in fact a way forward. It is not possible to directly test the universe for teleologies because they are not empirical: teleological questions are by definition beyond the reach of the scientific method (hence the natural skepticism of science toward teleology). But it is possible to rank different theories about the universe as better or worse, as philosophy of science does with scientific theories. As mentioned in chapter 2, Ian Barbour offers a fourfold system for testing theories for their truth value. First, is the theory *coherent*—does it make internal sense? Second, does the theory *correspond* to facts known by other means, for example, natural science? Third, is the theory *comprehensive*—how much of reality can it explain? And fourth, is the theory fruitful—what are the *consequences* of believing the theory and is it useful, effective, beautiful, or even good?[35]

Each criteria matches a traditional understanding of truth, and together they help to deliver a more complete way for evaluating truth. While the first three criteria are more objective, the fourth is somewhat subjective and allows humans (as entities of this universe, living in this universe, and looking at the universe) to include our evaluation and interpretation. If we think a theory looks bad, then we can look for an alternative one, as long as that is equal or better on the other more objective criteria. And so as we try to make sense of the universe, we should look through our alternatives until we think we have found the best theory and then continue with that until we believe we have found another even better one.

Returning to space exploration and settlement, we should ask ourselves if, right now, exploration and settlement look like they make sense from the perspective of our own worldview—from the meaning system that we believe to be true. Does exploration and settlement seem coherent with what we know about the universe? Does it correspond to facts? Does it make sense in a comprehensive way? And ultimately, does it have good consequences—is it useful, efficacious, beautiful, and good?

Or to take the same questions more personally: Does my understanding of the universe make coherent sense? Does my understanding of the universe correspond to facts? Does my understanding of the universe have a comprehensive scope or only explain a small bit of reality? And does my understanding of the universe actually have good consequences?

Each person has to answer questions like these—but not only these—from their own perspective, but they certainly do not need to do so alone. Indeed, it is highly advisable that they do not: after all, there are thousands of years of wisdom and centuries of science and technology that we can and should learn from. While we might dismiss old wisdom as out of date, we should also recognize that it has gotten us this far, and it has not brought us existential catastrophe (though neither was it empowered with contemporary technology), which is not something that we can say for sure about our current or future beliefs. Old wisdom might not be enough to take us farther on its own. But then again, maybe it might; or perhaps it is for us to update that old wisdom, integrate it with new ideas, and give it new strength to invigorate purposeful and meaningful human life as we move into the future.

## BACK IN THE BOX

With respect to intergenerational justice, we already make decisions about the futures of our descendants, we just don't consider them to be as momentous as they really are. Choosing to have future generations live in space or on a mission to another star may seem unfair in some way, but all of our ancestors made similar decisions, just typically with less freedom, and here on planet Earth. Perhaps the magnitude of a mission to another star ought to help us additionally realize the magnitude of decisions we make here on Earth and thus encourage us to more appreciation of our everyday choices that may have dramatic long-term consequences. When we decide what to study in school, we are making a momentous choice. When we decide on a career, likewise. When we decide whom to marry, or whether or not to have children, or how many children, we are settling the course for the future of the human race. The courses set may be subtle, but they are not insignificant. By remaining aware of the significance of our actions we will hopefully be more cognizant of what we are doing and hopefully thereby become more motivated to act ethically and not just live in a future, whatever it may be, but help to create a better future for everyone.

## CLOSING CASE

The Breakthrough Starshot seeks to send a light sail-propelled probe at 20 percent the speed of light to the Alpha Centauri system to examine the Earth-sized planet Proxima Centauri b approximately four light years away. The probe would be tiny, only a few grams in weight, with a sail a few meters in size. In this tiny package it would need a power source, sail structure, instruments, communication equipment powerful enough to communicate with Earth from four light years away, and instruments to examine the planet. The sail would be propelled by 100-gigawatt lasers based on the Earth, which would accelerate the probe to a significant fraction of light speed over the course of a few minutes. Many thousands of probes would be sent to compensate for a high likely attrition rate and to assure adequate data is captured and returned to Earth.[36] Needless to say, the engineering challenges are remarkably difficult. But the founder of the project, entrepreneur and investor Yuri Milner, put 100 million U.S. dollars into the project to get it started.[37] Famed physicist Stephen Hawking was also involved until his death in 2018, and Facebook founder Mark Zuckerberg remains involved.

What might be some relevant ethical issues with this project?

## DISCUSSION AND STUDY QUESTIONS

1. Should humankind travel to other places in the solar system, even if it takes years to get there?
2. Should humankind travel to other stars, even if it takes centuries to get there?
3. Do you think the universe has a purpose or is *about* anything? Why or why not? Do you think that human exploration and settlement might be aligned with the purpose of the universe or not?
4. Of the ethical concepts and tools discussed in this chapter, which do you find to be the most or least useful and why?

## FURTHER READINGS

Keith Abney and Patrick Lin, "Enhancing Astronauts: The Ethical, Legal and Social Implications," in *Commercial Space Exploration: Ethics, Policy and Governance*, ed. Jai Galliott (London and New York: Routledge, 2015), 245–57.

Eugene Mallove and Gregory Matloff, *The Starflight Handbook: A Pioneer's Guide to Interstellar Travel* (New York: John Wiley and Sons, Inc., 1989).

## NOTES

1. Inspiration Mars, accessed January 3, 2021, https://web.archive.org/web/20151013012955/http://inspirationmars.org/.

2. Dennis A. Tito et al., "Feasibility Analysis for a Manned Mars Free-Return Mission in 2018," Inspiration Mars, https://web.archive.org/web/20130319140952/http://www.inspirationmars.org/Inspiration%20Mars_Feasibility%20Analysis_IEEE.pdf.

3. Lisa Grossman, "Ambitious Mars Joy-Ride Cannot Succeed Without NASA," *New Scientist*, November 21, 2013, https://www.newscientist.com/article/dn24633-ambitious-mars-joy-ride-cannot-succeed-without-nasa/.

4. Buzz Aldrin, "Cyclic Trajectory Concepts," SAIC presentation to the Interplanetary Rapid Transit (IRT) Study Meeting, Jet Propulsion Laboratory, Pasadena, California, October 28, 1985.

5. Giulio Prisco, "Uploaded E-Crews for Interstellar Missions," *Kurzweil AI (Accelerating Intelligence) Daily Blog*, December 12, 2012, https://www.kurzweilai.net/uploaded-e-crews-for-interstellar-missions.

6. Olaf Stapledon, *Last and First Men* (London: Methuen, 1930); Vernor Vinge, "Long Shot," *Analog Science Fiction/Science Fact*, August 1972; Paul Lucas, "Cruising the Infinite: Strategies for Human Interstellar Travel," *Strange Horizons*, June 21, 2004, http://strangehorizons.com/non-fiction/articles/cruising-the-infinite-strategies-for-human-interstellar-travel/; Adam Crowl, John Hunt, and Andreas Hein, "Embryo Space Colonisation to Overcome the Interstellar Time Distance Bottleneck," *Journal of the British Interplanetary Society* 65 (2012): 283–85; and John Hunt, "The EGR Mission—Rationale and Design of the First True Interstellar Mission: A Crazy Presentation for the PI Club, Appendix 1: Minimizing the Ethical Concerns of an EGR Mission," Peregrinus Intersteller: PI Club, accessed August 5, 2015, http://www.peregrinus-interstellar.net/images/Files/Crazy_Ideas/EGR/egr_appendix_1_ethics.pdf.

7. For example, Szocik, ed., *Human Enhancements for Space Missions*.

8. Schwartz, "The Accessible Universe," 201–15.

9. SpaceX, "Mars & Beyond: The Road Map to Making Humanity Multiplanetary," 2020, accessed November 23, 2020, https://www.spacex.com/human-spaceflight/mars/; and Musk, "Making Humans a Multi-Planetary Species."

10. Matt Rosoff, "Elon Musk Worries that His Kids Are Too Soft to Be Entrepreneurs," *Business Insider*, September 16, 2011, https://www.businessinsider.com/elon-musk-worries-that-his-kids-are-too-soft-to-be-entrepreneurs-2011-9.

11. David J. Galton, "Greek Theories on Eugenics," *Journal of Medical Ethics* 24 (1998): 263–67.

12. Schwartz, "The Accessible Universe," 201–15.

13. For example, Liu's *The Dark Forest*.

14. Marino, "Humanity Is Not Prepared to Colonize Mars."

15. Stanley Kubrick and Arthur C. Clarke, *2001: A Space Odyssey* (Beverly Hills, CA: Metro-Goldwyn-Mayer, 1968).

16. For example, Brian Patrick Green, "Ethical Reflections on Artificial Intelligence," *Scientia et Fides* 6, no. 2 (2018), https://apcz.umk.pl/czasopisma/index.php/SetF/article/view/SetF.2018.015; and Brian Patrick Green, "Artificial Intelligence and Ethics: Sixteen Challenges and Opportunities," Markkula Center for Applied Ethics, August 18, 2020, https://www.scu.edu/ethics/all-about-ethics/artificial-intelligence-and-ethics-sixteen-challenges-and-opportunities/.

17. Shannon Vallor, "Moral Deskilling and Upskilling in a New Machine Age: Reflections on the Ambiguous Future of Character," *Philosophy & Technology* 28 (2015): 107–24.

18. Green, "Ethical Reflections on Artificial Intelligence"; Green, "Artificial Intelligence and Ethics"; and Green, "Artificial Intelligence, Decision-Making, and Moral Deskilling."

19. For example, the case of Germanwings Flight 9525 where the copilot intentionally crashed the aircraft, killing everyone on board.

20. James Bridges, *Colossus: The Forbin Project* (Universal City, CA: Universal Pictures, 1970).

21. Jonas, *The Imperative or Responsibility*. In contrast, a tiny number of philosophers such as David Benatar have argued that humanity ought to voluntarily go extinct; see David Benatar, *Better Never to Have Been: The Harm of Coming into Existence* (Oxford: Oxford University Press, 2006).

22. Ord, *The Precipice*, 45.

23. Lucas, "Cruising the Infinite"; Crowl, Hunt, and Hein, "Embryo Space Colonisation."

24. Mark Lupisella, "Cosmocultural Evolution: Cosmic Motivation for Interstellar Travel?" *Journal of the British Interplanetary Society* 67 (2014): 213–17.

25. Deborah Kelemen, "Function, Goals and Intention: Children's Teleological Reasoning about Objects," *Trends in Cognitive Sciences* 3, no. 12 (December 1999): 461–68.

26. Sigmund Freud, *The Future of an Illusion*, ed. Todd Dufresne, trans. Gregory C. Richter (Peterborough, Canada: Broadview Press, 2012).

27. Deborah Kelemen and Evelyn Rosset, "The Human Function Compunction: Teleological Explanation in Adults," *Cognition* 111 (April 2009): 138–43; Krista Casler and Deborah Kelemen, "Developmental Continuity in Teleo-Functional Explanation: Reasoning about Nature Among Romanian Romani Adults," *Journal of Cognition and Development* 9 (2008): 340–62; Tania Lombrozo, Deborah Kelemen, and Deborah Zaitchik, "Inferring Design: Evidence of a Preference for Teleological Explanations in Patients With Alzheimer's Disease," *Psychological Science* 18 (2007): 999–1006; and Deborah Kelemen and Cara DiYanni, "Intuitions About Origins: Purpose and Intelligent Design in Children's Reasoning About Nature," *Journal of Cognition and Development* 6 (2005): 3–31.

28. See, for example, Yukimaru Sugiyama and Jeremy Koman, "Tool-Using and -Making Behavior in Wild Chimpanzees at Bossou, Guinea," *Primates* 20 (1979): 513–24; Gavin R. Hunt, "Manufacture and Use of Hook-

Tools by New Caledonian Crows," *Nature* 379 (January 18, 1996): 249–51; and W. C. McGrew, "Is Primate Tool Use Special? Chimpanzee and New Caledonian Crow Compared," *Philosophical Transactions of the Royal Society B* 368 (2013): 20120422.

29. Francine G. Patterson, "The Gestures of a Gorilla: Language Acquisition in Another Pongid," *Brain and Language* 5, no. 1 (January 1978): 72–97.

30. For example, Jacques Monod, *Chance and Necessity: An Essay on the Natural Philosophy of Modern Biology*, trans. Austryn Wainhouse (New York: Vintage Books, 1972).

31. Green, "Convergences in the Ethics of Space Exploration," 186–92.

32. Mark Lupisella, conversation at the Social and Conceptual Issues in Astrobiology Conference, Clemson University, Clemson, South Carolina, September 24, 2016.

33. Lupisella and Logsdon, "Do We Need a Cosmocentric Ethic?"; Randolph and McKay, "Protecting and Expanding the Richness and Diversity of Life"; and Smith, "Manifest Complexity." All discussed in more detail in Green, "Convergences in the Ethics of Space Exploration."

34. Jacques Arnould, "An Urgent Need to Explore Space," in *The Ethics of Space Exploration*, ed. Schwartz and Milligan, 153–64.

35. Barbour, *Religion in and Age of Science*, 34–39.

36. Breakthrough Starshot, accessed January 3, 2021, https://breakthroughinitiatives.org/initiative/3.

37. Dennis Overbye, "Reaching for the Stars, Across 4.37 Light-Years," *New York Times*, April 12, 2016, https://www.nytimes.com/2016/04/13/science/alpha-centauri-breakthrough-starshot-yuri-milner-stephen-hawking.html.

*Chapter Thirteen*

# Building Your Martian Home

*Living in and Settlement of Space*

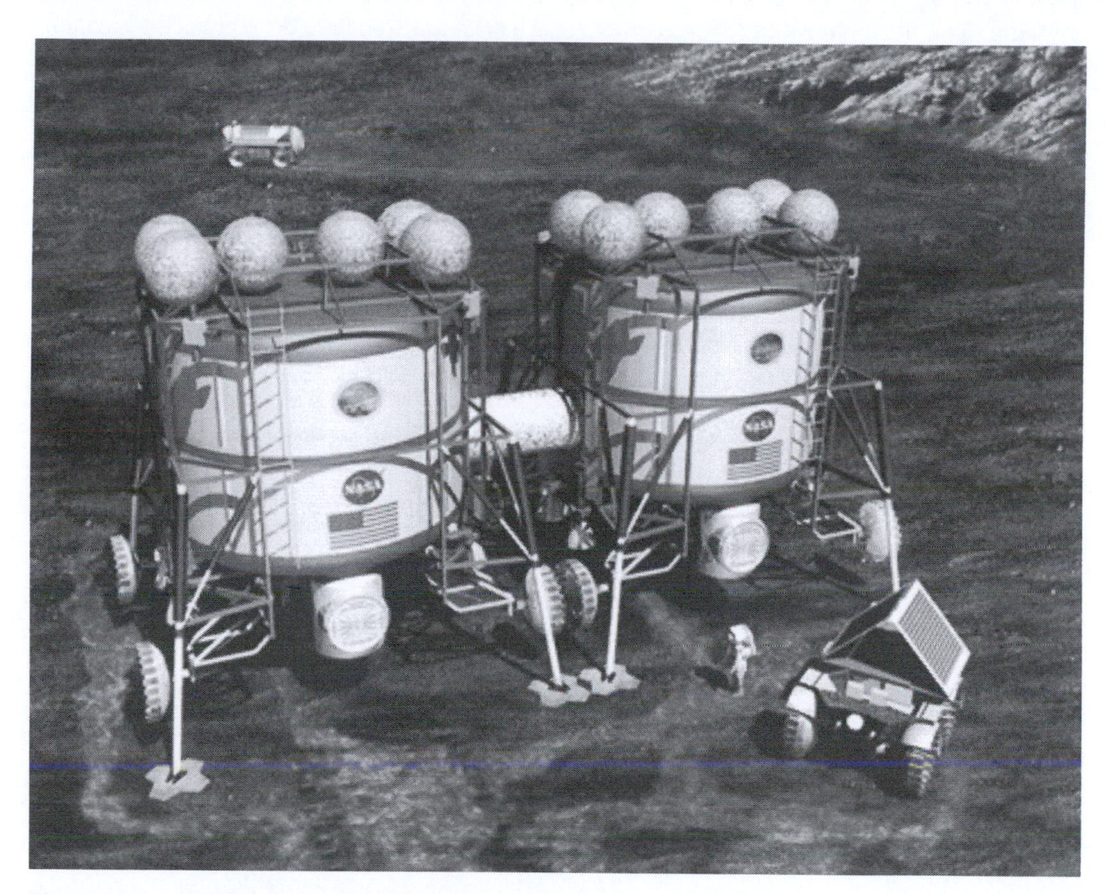

**Figure 13.1. NASA graphic of a settlement on Mars**

## BUILDING A CITY ON MARS

Billionaire technology entrepreneur Elon Musk is on a mission to create a self-sustaining city on Mars as quickly as he can. In two dense tweets on January 16, 2020, referring to the massive new Starship rocket, Musk stated, "Starship design goal is 3 flights/day avg rate, so ~1000 flights/year at >100 tons/flight, so every 10 ships yield 1 megaton per year to orbit" and "Building 100 Starships/year gets to 1000 in 10 years or 100 megatons/year [to orbit] or maybe around 100k people per Earth-Mars orbital sync."[1] (Earth-Mars orbital sync occurs every twenty-six months.) In other words, Musk is intending to build 100 giant rockets a year, for ten years, fly each one of them three times a day every day, each carrying 100 tons of cargo or people per flight. So 1,000 rockets each flying 1,000 times a year will get 100 million tons of goods in orbit in one year. A good deal of this mass, including 100,000 people, would then speed off to Mars once every twenty-six months.

At that rate, assuming Musk's rapid timeline and that the people that reach Earth's orbit also reach Mars, Mars could have one million people on it by 2050.[2] In an interview with *Ars Technica* Musk further explained: "We need to have a self-sustaining city on Mars. . . . That city has to survive if the resupply ships stop coming from Earth. . . . You can't be missing anything. You must have all the ingredients . . . what kind of tonnage do you need to make it self-sustaining? It's probably not less than a million tons."[3] While currently concentrating on the rocketry required to get to Mars, he has also considered the destination itself. Mars is cold, there is less sunlight, and the atmosphere is inadequate. But these are all things that can be compensated for, not only in habitats but perhaps, over time, on a planetary scale, with terraforming.[4]

But of course, this raises the ethical question: *Should* Musk—or anyone else—seek to settle Mars? That is the topic of this chapter.

## RATIONALE AND SIGNIFICANCE: SURVIVAL AND FLOURISHING BEYOND EARTH

The idea of living on other worlds seems to call to some humans. Certainly many science-fiction stories have been written about settling other worlds, but the reality is approaching, not only through Musk's work but through organizations like the Mars Society, which is also dedicated to establishing a permanent human presence on Mars.

The ethical problems of surviving in a naturally harsh new place and flourishing with a new governing community are at the heart of these issues. This chapter will pursue two fundamentally ethical questions about space settlement: (1) "Will humans be able to survive the harsh environment of space?" and (2) "Will humans be able to build a government and society which can promote the common good and genuine human flourishing?" In more abstract terms, these questions examine relations with the external (abiotic and/or biotic) environment and relations with the internal (sociopolitical community) environment, or nature and human society, respectively.

## ETHICAL QUESTIONS AND PROBLEMS

### 1. Will humans be able to survive the harsh environment of space?

*A. Obtaining and Consuming Resources*

Human survival relies on obtaining and consuming resources and energy. Both obtaining and consuming resources can be quite difficult, depending on the nature of the resources being used. For example, in space, water can be quite abundant in some places, such as on the Earth or Galilean moons of Jupiter, and quite rare in other places, such as the Earth's Moon or Mercury. Conversely, the Moon and Mercury are good sources of resources related to rocks and metals, while the outer solar system tends to have these resources in asteroid-sized chunks or buried under layers of ice or gas giant atmospheres. And of course, in the vast emptiness between space objects, there are no material resources available at all, though if near to stars there is light energy that can be captured with solar photovoltaic cells—if you already have them.

Obtaining resources could then be quite difficult in space (though, economically speaking, it could be quite a boon for interplanetary trade). Consuming resources is also difficult, not because humans lack the technology to consume the resources but because we would require the starting resources and continuing energy to run the technology that converts the resources into usable forms for us. As Musk noted in his tweets, starting a new civilization takes a lot of stuff, millions of tons of stuff. And moving that much material takes energy. Luckily Earth is awash in energy, but Mars is less so, and the emptiness of space has no energy sources except the Sun—so the solar panels on a vessel or space station better be reliable, or the vessel needs to carry its own source of power, or both.

Of course, obtaining and consuming resources can also have negative impacts on sustainability and the local environment. Obtaining resources involves chewing away with machinery at natural materials (likely ejecting particles in every direction, if on a low-gravity body, thus creating a cloud of space debris, or on higher gravity bodies, burying the local area in tailings) that are then transported (taking energy) to be refined (ejecting waste products of unwanted elements and compounds), then transported again (taking more energy) to factories where they become usable products (using more energy, using more materials, and creating more waste) that are then transported again (taking more energy) to the people who will finally use them (and turn the product, sooner or later, into waste that must be recycled or disposed of). This process is destructive of the natural environment and consumes a lot of energy, yet all of the associated waste products could be potentially recycled into a circular economy.

A **circular economy** is one in which waste products become the inputs for further production, unlike the more common **linear economy**, characterized by "take-make-waste." The linear economy produces destruction at both ends as well as in the middle steps with the use of energy and "making" materials. A circular economy on the other hand closes the loop, reducing environmental impact and increasing efficiency. Unfortunately, the circular economy can also potentially require more energy than a linear economy; however, in space, where everything is more difficult to move around, circularity might be relatively less costly than on Earth. Additionally, in space much is already presumed to be reduced, reused, and recycled, such as water—which on a dry space object (or in space itself) must be recycled and reused over and over again.

Note that on Earth the environment is often responsible for the recycling of used materials, such as the water cycle, which humans merely siphon from. This is what results in environmental damage: a natural system is drawn from (damaging it) and polluted waste is returned to it (further damaging it). Natural systems can typically take a certain amount of this damage; however, humans tend to overload the cycles by moving too fast. Environmental damage that takes humans days or weeks or years to make could take centuries, millennia, or millions of years to repair, if the damage is reparable at all (after all, extinct species, for example, are lost forever). In space we must do better.

We can do better by minimizing the environmental damage through a circular economy, as well as by moving beyond a sustainability model and toward a generative model—a "superenvironmental ethic," as James Schwartz once described it.[5] While on Earth this superenvironmental ethic would focus on restoration and regeneration of the Earth's damaged environment, in space it would focus on **ecopoiesis**: the creation of an ecosystem, or in terms of habitability, the generation of a habitable environment from an uninhabitable one.[6] This will be discussed further in the next chapter on terraforming. Suffice it to say for now that settlers in space will have good reasons to protect their environment and be thrifty with the resource use, while still likely harnessing energies, materials, and technologies beyond those typical of Earth.

### *B. Tensions between Creating a "Backup Earth" and Environmental Ethics*

There is a tension between extraterrestrial environmental ethics and the desire or need for a "backup Earth" or "second biosphere." A backup Earth plan for space exploration emphasizes speed at all costs, likely to the detriment of the environment that is being moved into. Because a backup Earth has an anthropocentric element to it, seeking to preserve human civilization (though likely bringing much other Earth life along with it), the Martian environment, for example, would have a lower priority. Very likely a rushed human settlement would leak Earth life onto the Martian surface while mining resources and developing an industrial base that may well *not* be circular, at least in some ways.

A more Mars-centric environmental ethics would seek to avoid the environmental impact of humans on Mars as much as possible, even to the extent of excluding humans completely. Completely excluding humans would, obviously, completely ruin Mars as a backup Earth, thus demonstrating the incompatibility of the two ethical values, at least in their extreme form. But there are ways to meet partway and perhaps satisfy both sides, at least to some extent.

As a first point, Mars need not be the focus of settlement: there are other places that humans could settle as a backup Earth as well, such as the Moon. The Moon is not ideal because unlike Mars it lacks water and the more volatile elements and compounds. However, on the plus side there is little risk of damaging a living environment (it being a Category II object in planetary protection guidelines), so environmental ethics would be more centered on the nonliving environment and the living environment within habitats. Being only a few days from Earth makes the Moon easily accessible, which is good for getting people and supplies there, but possibly bad if a disaster on Earth somehow spreads to the Moon (for example, a pandemic or war).

As a second point, even settling a potentially sensitive environment like Mars need not necessarily be seriously environmentally destructive, provided it is done right. For example, human activity can be limited to one or a few areas: perhaps the crater of one of Mars's giant volcanoes, which are at very high altitude and are well delineated by their walls. Environmental destruction and pollution can be minimized through careful resource utilization and recycling. Biocontamination can be reduced through careful control over the escape of any living material from the settlement. Human exploration can be limited to areas immediately surrounding settlements, and clean robots can explore farther.

Of note is that a backup Earth is in some ways not anthropocentric but rather Earth-biocentric. That is, it pits Earth life against other potential environments. It would be Earth life, after all, going to other space bodies, even if it is humans, and humans, of course, have an entire microbiome of Earth life with them. We would likely bring many other Earth species with us too, such as crops, larger animals (once the settlement was sufficiently large), and perhaps the preserved genetic material for many more living things, in case disaster befell the Earth. If it was at some point determined that Mars had no native life, then the process of terraformation might begin, thus allowing for the eventual release of life onto Mars's surface, although the ethics of this is highly debatable. If there is native life, then terraformation might still be morally defensible, but the case would be harder to make. After all, perhaps we need not pit Earth life versus other life (if other life even exists) but instead try to determine if all life can get along together. Life of different origins may be intrinsically competitive or dangerous to each other, but they also might not be. Only time and experience will tell.

Whichever path is chosen between the rush to back up Earth and environmental ethics, there will be trade-offs.

### *C. Settlement Environmental Ethics*

Extraterrestrial settlement is likely to occur this century, and perhaps in the next two decades. Because of this, the environmental ethics of settlement should be considered now to prepare. The first question is to consider *where* to place a settlement. There are numerous factors to contemplate, many of which involve safety and access to resources. Some general issues of site selection include: Is the area geologically stable? Is it sheltered from radiation? Is it subject to a chronically unhealthy environment, for example, chemical or physical particle contamination? Is it subject to planetary protection issues with significant science to protect or local organisms? Is it subject to periodic disasters (and if so, what kinds)? Does it have the right natural resources present, such as water, organic chemicals, metals, energy, etc.? How hard is it to access local resources and how hard to import the resources that may be absent? And so on.

Land use is a problem on Earth, and it may also be a problem in space. Should settlements in space be allowed to "sprawl" across a landscape, or should settlements remain small and dense? Should settlement not even occur on the surface of a space object in order to avoid radiation and not damage the surface environment? Should settlers be forced to rely on imported resources or should they be allowed to use local resources? (Note that for any self-sustaining settlement, this second question can only be answered "yes.")

It makes sense to customize settlements to their environment, so, for example, in a high-radiation environment it would make sense to live underground, or at least under a layer of shielding. Likewise, in a place with potential life and planetary protection issues, it would make sense to live in a smaller, denser habitat, carefully designed to limit impacts and contamination. And for a place like the Moon, where there is little issue with PP or biology but there are issues with physical particle danger and radiation, it might make sense to live underground, with careful measures taken against physical particle entry (cleaning anything from outside, filtering air so any contamination that does get inside does not spread, etc.), but with no issues about sprawl.

Last, what does it mean to practice "sustainability" in space? Sustainability, in a very basic sense, means just being able to keep on doing what one is already doing. Humans keep on living, and nature keeps on living. But the problem comes when humans consume resources faster than nature can replace or restore them. Space has resources beyond human imagination, so running out of them is only a relative problem, one of economics, not absolutes. However, sustainability in an artificial environment, as noted earlier, is quite different from sustainability in a natural environment. In an artificial environment, everything depends on human choices, and so the first subject of "sustainability" is keeping the human decision makers alive because without them the entire environment will die (unless some sort of long-surviving AI can operate without people present). In a natural environment, by contrast, sustainability is the natural state, and humans must work with nature in order to keep human activities and the environment in balance.

### D. In Situ Resource Utilization

As one part of settlement, *in situ* resource utilization (ISRU) and local manufacturing will be important for sustainability. Rather than bringing everything with them, settlers will need to extract local resources. Luckily, space bodies are full of useful elements and compounds such as aluminum, silicon, oxygen, iron, and so on. What is less common and yet extremely important, at least in the inner solar system other than Earth and Mars, is water, and more generally elemental hydrogen, which can be combined with elemental oxygen to produce water. A Moon settlement would have good access to many of the heavier elements, but from all present appearances the Moon is sorely lacking in water and hydrogen. These will either have to be mined in extensive strip mines from very poor polar deposits or the water or hydrogen will have to come from Earth or other space bodies such as asteroids or comets.

Which raises another question: With regard to ISRU, how much is allowed? Could an entire asteroid be consumed? Could the entire Moon be strip mined? On lifeless space bodies there is no living environment to destroy, after all, only a dead one. Does the nonliving world have any ethical value to protect?

There are good reasons not to destroy pristine environments wantonly. There are many places on Earth with very little life, for example, Antarctica, which has a "magnificent desolation" like that of the Moon, and we consider it to be aesthetically important and worth protecting. Aesthetics is an instrumental approach to value, however. Might there be an intrinsic value? This has been discussed previously in chapter 2, especially in the context of environmental virtue ethics. There can be arguments made for the protection of lifeless places, but perhaps the easier case to be made is that those humans who choose to wantonly destroy even lifeless places are habituating themselves toward vice and therefore, in order to protect their character, ought not to act in such a way.

### E. How much destruction is permissible?

These limits—or lack of limits—to ISRU raise an even bigger question: How much are humans allowed to destroy for the sake of development? May humans consume, repurpose, and/or destroy small asteroids? Large asteroids? Moons? Entire planets or even stars?

While consuming the larger space bodies is firmly in the realm of science fiction, small asteroids are already the focus of "asteroid miners" who are seeking to mine *in situ* or redirect asteroids toward Earth orbit for consumption. And the Moon has been proposed for mining Helium 3.[7] What should we consider when trying to determine whether these nonliving objects are expendable? Or in other words, what might be some considerations for the instrumental value and intrinsic value of lifeless places?

First, safety is worth taking into account. Any asteroids or other space objects that threaten the Earth or another human settlement should be moved to safer orbits or consumed not only because they have resources but even more because they are dangerous. And like near Earth objects, other bodies in the solar system also have asteroids that may cross their paths, and if humans are in those places, those asteroids might also need to be moved or consumed.

Second, objects with scientific value ought to be protected so that they can be better understood. This is already a core purpose of planetary protection, and so it merely needs to be extended into the future.

Third, any object visible from Earth would be of aesthetic concern. For example, if the Moon began to be covered by unsightly strip mines (or strip mines shaped into large advertisements), people on Earth would justifiably complain. Likewise with other visible bodies such as comets: if they were all consumed and comets were no longer visible from Earth, this would impoverish the human aesthetic environment. What if objects are not visible? Then the human need for resources might make us ask if we can completely consume Jupiter's Trojan asteroids (orbiting 60 degrees ahead and behind Jupiter, in Jupiter's orbit), some of which are enormous yet not visible from Earth.

Fourth is the usefulness of the object itself, as a raw material. If the object is quite valuable because it would help space development, then this would need to be weighed against the other values presented here.

Fifth, after safety, science, aesthetics, and usefulness, it makes sense to balance human needs with the simple size of an object. Larger objects have more resources in them, and so ought to be used after smaller objects have been used first. In this prioritization, a 10-meter asteroid would have less standing than a 1,000-meter asteroid, which would have less than a 10-kilometer asteroid. But why? The simple reason is that as civilization grows, resource needs grow, and so it makes sense to leave bigger resources for later, when bigger resources might be needed. Practically speaking, smaller objects are also easier to consume than larger ones, so as civilization grows in strength, it will be more able to mine larger objects. But, of course, just because humankind might one day be able to consume large asteroids does not mean that it should—that is the entire point of ethical discussion and, more generally, of this book. On that point, if recycling and a circular economy are done well, and resource demands are not too extravagant, smaller asteroids might be all that we need. But under this schema, if resource demands do demand further extraction, there will still be resources to consume.

As a side note on intrinsic value, if nonliving matter has any intrinsic value at all, and that value scales with size, then small objects are not as ethically valuable as larger ones. This is not a strong argument and there are many possible arguments against it, but I think it is plausible. Given this assumption, *ceteris paribus*, larger objects have more value than small ones, therefore smaller objects should be consumed first. This is easy to see with objects of the same type, for example, if one had two diamonds of equivalent quality and had to destroy one, then one ought to destroy the least amount possible, that is, the smaller one. But for comparison between types it fails: a diamond and another kind of rock are hard to compare, and a diamond and an asteroid are harder still; and asteroids come in many types. But as a general idea, and for several reasons, size could be relevant for ethical comparison.

Sixth comes the question of the intrinsic moral value of the nonliving world in itself. We could construe the natural world as concentrating intrinsic value from nonliving things, into living things, into living things with minds. If that is the case, then humans have more value than microbes, and microbes more value than dead rocks. In this schema nonliving things still have value. They should not be dismissed as unimportant. However, they can be made more morally valuable if they are made alive: if they can grow and support life and intelligence (and this intuition underlies some space ethics ideas, such as those of Randolph and McKay[8]). And furthermore, from a virtue ethics perspective, the choices that we make about nonliving things *live in us* and become who we are as human beings. Wanton destruction is not a character trait to encourage, even against targets of very low moral standing. If there are reasons for destruction, then it is no longer wanton and might be justifiable by other ethical considerations. But destruction for the sake of nothing is a vice and should be controlled.

### *F. Space Pollution*

Human activities in space, as on Earth, will create pollution, whether it is space trash, flying rocks, harmful chemicals, biological contamination, or otherwise. What should be done about waste disposal and "pollution" in space and its short- and long-term impacts?

The first step in combating pollution is to prevent it. Earth orbital debris is a prime example: rather than producing debris and cleaning it up, it would be a lot easier to just not make it in the first place. But once it is made, it does need to be cleaned up. Orbital trash collectors can do this job, slowly and laboriously, but will never be perfect, only capturing the worst and riskiest pieces. Space trash could be space treasure if this collected material is recycled; after all, it is already in space, which is hard to get to. People might as well use it up there rather than launching more from Earth.

When mining asteroids, the creation of debris clouds should be strenuously avoided because the orbital sizes are so vast that cleanup becomes nearly impossible. Objects whizzing around in Earth orbit are bad enough, and those in solar orbit, spreading outward, are even worse in terms of speed, detectability, and more. If we want space to be safe, we want it to be empty, not full of stray rocks moving at kilometer-per-second velocities. There should not be mining debris, or any other debris sources, polluting the solar system's orbits.

As for chemicals and biological pollution in the void of space, this is a bit of a different problem. Insofar as those chemicals and biological materials are objects that could be crashed into, they should be covered by future regulations of debris. Insofar as they are dangerous for planetary protection, they ought to be covered by those rules. And if these materials are too small and not dangerous for planetary protection or other concerns, then they might be permissible, for example, if the "pollution" is vented gas.

On the surfaces of objects, these same questions arise, but with stronger concern for human impact and planetary protection impact. If the chemicals discarded are risky to human other life or scientific research, they ought to be controlled. If the biological contamination violates planetary protection standards, then they ought to be controlled. Mine tailings, rubbish, and other merely physical dangers are luckily not so dangerous on surfaces as they are in orbit, but still, for the sake of aesthetics and general cleanliness, they ought to be contained and disposed of properly, including by recycling.

## 2. Will humans be able to build a government and society that can promote the common good and genuine human flourishing?

### *A. The Differences between "Settlement" and "Colonization" (and the Importance of the Words We Choose)*

For decades, future space settlements have been referred to as "colonies." These particular ideas of colonies go back into the middle of the second millennium CE, when European nations began colonizing the world for the sake of territorial expansion, power, and resources. Colonized nations were often treated as mere means to the ends of outsiders, and immeasurable suffering came from this system of exploitation. Finally, most of these colonies either fought for freedom or were freed from their colonial overlords, but the bitter memories still remain. When contemporary people speak of "space colonization," people from formerly colonized nations may naturally bristle at the word: Are we really intending on repeating the horrors of the past at a cosmic scale? And the truth is that, no, that is not the intention. Space settlement is intended to be an ethical endeavor that creates a brighter future for all of humanity, not a dark future of bondage, exploitation, death, and destruction.

In other words, "colonization" is not the right word. Space settlements are not going to be like the colonies of the last millennium, not economically, politically, or ethically. Instead they are precisely meant to be "settlements": places where humans gather and live together.

This is not to say that the word "settlement" does not also come with political and historical baggage; indeed, settlers have long been a tool used by powerful peoples to control weaker ones. But in space there are no "weaker peoples" to oppress, at least not as far as we know. If there were, these weaker entities—even if not "people" at all but only microbes—would have a tremendous ethical standing, not to mention some standing in the Outer Space Treaty (OST), in the forward and backward contamination rules. These ethical and legal conditions could halt human settlement in its tracks.

In any case, settlements are the correct word to use for human gatherings in space. Colonization is a word that should be removed from the space lexicon, not only because it harkens to barbarities of the past but because it is also simply an inaccurate description of what people want to do in space.

*B. What will settlers do about food?*

A self-sustaining settlement needs to be able to grow its own food. Past speculative stories and drawings of space settlements often had greenhouses growing plants, but in some places on Earth farming has started to leave this sort of agriculture behind, with plants being grown in large, enclosed warehouses, in stacked trays up to the ceiling, under LED lights. These new **vertical farming** techniques can be vastly more effective for growing crops in space. As long as energy is available, lights can shine on plants to help them grow, and the density of plants would be many times greater than on a conventional flat farm.

While on Earth these warehouse farms mostly concentrate on small, fast, and valuable crops like greens and herbs, in space they would need to grow staple crops as well, such as grains and root crops. Grains should be no more difficult than greens or herbs, but root crops obviously need a lot of dirt and so might not be practical for vertical farming, at least not without some adjustments.

Animal agriculture might be more of a problem in space. Animals tend to take up a lot of space and are messier than plants, not to mention eating plants that could go to feed humans. In a very tight ecology, humans would need to eat as low on the food chain as possible, while animals would either not be involved at all or would only eat food that humans did not want, such as table waste and agricultural waste (nonfood roots, inedible stems, etc., though these might be processible for human consumption as well). This agricultural waste might be able to support fish (if there is enough water) or other small creatures such as insects, however, and then these small creatures turned into food themselves, thus actually making for a more efficient food chain than if they were not involved. Larger land animals like cattle, pigs, and sheep are going to be a nonstarter for space exploration for a long time, not until there is ample extra land to allow their feed to be grown. Poultry will likely reside in an in-between place, less practical than plants and fish but more practical than domesticated large mammals.

Differences in gravity and radiation will be important concerns for agriculture in space, not to mention a sufficient supply of water, usable soil, and minerals and nutrients. Contaminants like too much salt could easily damage extraterrestrial agriculture, and shortages of basic nutrients like bioavailable nitrogen could also be highly detrimental. Nitrogen, phosphorus, and potassium will be vital for agriculture and can hopefully be locally sourced. If only nitrogen gas is available, settlements will need to build a small facility conducting the **Haber-Bosch process**, fixing nitrogen into a form that is usable by plants.

*C. What of the availability of medical care and pharmaceuticals?*

Among the most necessary and complex items needed by a space settlement would be those involved in medicine, and in particular, medical devices and pharmaceuticals. Current additive manufacturing (3D printing) allows for the production of complex physical items such as tools and structures, including possibly the manufacturing of most devices needed in a hospital or medical settling, given the right equipment and material feedstocks. While this is no simple task, it could be possible. However, many of the chemicals involved in medicine are a level of complexity beyond mere additive manufacturing.

Pharmaceutical and medicine production is a complicated endeavor. Pharmaceuticals must be exceedingly pure and extremely specific: an exact molecule in an exact dose, with no harmful contaminants. This is hard enough to do on Earth, at scale. In a small space settlement, most pharmaceuticals might be beyond their capacity to produce, until their economy has developed to quite a sophisticated level. There are innumerable feedstocks and chemicals necessary as well as reaction vessels, refining and manufacturing equipment, and so on. At the beginning, it might be a few pharmacists working in a laboratory creating specific batches of drugs by hand, and this operation might grow over time. At this scale, drugs will be expensive and only the most valuable ones will be produced until the process can be enlarged to allow for economies of scale. The sheer number of pharmaceuticals also should not be underestimated: there are tens of thousands of medically significant chemicals in the world, and the World Health Organization lists about 450 as "essential" medicines.[9] Some of these are extracted from plants rather than being synthesized, which means that a certain amount of agricultural production will also need to go toward medicine production. As one point of hope, bacteria, fungi, and plants might be engineered to produce some of the pharmaceuticals, chemicals, and feedstocks for the medical industry, which might help to simplify the process of manufacturing for a space settlement.

The chemical and pharmaceutical industries on Earth are immense and often underappreciated sectors of the economy. While they can be polluting and dangerous, contemporary society cannot exist without them. Likewise, any settlements of humans beyond the Earth will need to have these industries as well.

*D. How should distant humans relate to their home planet in terms of governance?*

Article VI of the OST puts all space activities, including settlements, under the jurisdiction of their sending states. And Article II—the shortest article in the entire treaty—makes space "including the Moon and other celestial bodies . . . not subject to national appropriation by claim of sovereignty, by means of use or occupation, or by any other means."[10] In terms of governance, therefore, the OST puts settlements under Earth national jurisdiction, but without any claims of sovereignty. This is an interesting location in terms of governance.

For example, if, say, Elon Musk's Mars settlement were founded by spacecraft launched from the United States of America, that settlement would be under U.S. federal law but not be U.S. territory. If many nations sent settlers to Mars, all of them would be under the jurisdictions of their sending nations, and yet none of them would be territory of their sending states. No amount of "use or occupation" would change that, at least not according to the OST.

Elon Musk has his own ideas, however. For example, those who have signed the licensing agreement for SpaceX's Starlink satellite internet service have agreed with SpaceX that

> for Services provided on Mars, or in transit to Mars via Starship or other colonization spacecraft, the parties recognize Mars as a free planet and that no Earth-based government has authority or sovereignty over Martian activities. Accordingly, Disputes will be settled through self-governing principles, established in good faith, at the time of Martian settlement.[11]

While this legal language might be meant as something of a joke (as some legal experts have wondered[12]), it certainly serves as a provocation in its blatant disregard for the OST. Legal scholars agree that the statement is unenforceable and even "naïve."[13] However, as a "marketing" stunt or otherwise, it has shifted the conversation for some who are thinking about space. By doing this, SpaceX has either made fun of or declared their intent to violate an old but steady treaty, thus weakening the treaty but perhaps also opening the future to new ideas.

Ethics is not the same as the law, though certainly the law (if good) holds ethical force. But whether the law is good or not is precisely what SpaceX is challenging. So we should ask: What ought to be the governing structures for a settlement on Mars or elsewhere in the solar system and beyond? Certainly, throwing out all human law to date and starting over would likely be a counterproductive move; after all, every law can be considered an experiment, and over the course of history many of the bad experiments have been thrown out (though certainly not all). It might be reasonable to start with international law, as described by United Nations treaties, for example. That would at least defend human rights and other basic tenets of international relations, maintaining a unity among humans on Earth and in space. Certainly, choosing to reject international law would open a settlement to possible abuses of human rights that could be very unappealing to potential immigrants.

What about sovereignty? Should a settlement always be under the national jurisdiction of their sending state, or might they at some point become self-governing? This is a reasonable question. Just as colonies on Earth declared independence and are no longer ruled by their colonizers, should not future settlers someday be able to declare independence as well?

There is a sociopolitical principle known as **subsidiarity** that says that governing should take place at the most local level of government capable of solving the problem. In other words, if a city can solve a problem without national or international interference, then let it do so. If instead that city cannot solve that problem without the aid of higher levels, then let those higher levels help, as necessary. Governing a settlement on Mars, then, ought to start with the people of the settlement itself and then involve higher levels of government only as necessary.

However, there are exceptions to subsidiarity. If, for example, a city decided to enact oppressive and racist policies that the rulers of the city were happy with but that outsiders (not to mention the local oppressed people) declared were unethical, this would create a problem in which outsiders might feel justified in interfering with the city, as it had violated international law, for example, the UN Universal Declaration of Human Rights. In other words, according to subsidiarity, the lowest level of governance capable of dealing with the issue—the city—chose wrongly, and therefore the larger community—whatever the next step up in government is—has to step in and solve the problem of local racial oppression. The Reconstruction era after the U.S. Civil War illustrates this point: for about twelve years the U.S. federal government controlled the formerly Confederate states in order to end slavery (which mostly worked) and reduce racism (which mostly did not work). Sometimes larger levels of society should step in and correct what lower levels are doing wrong.

Returning to Mars, and space more generally, this means that it would be wonderful if Mars settlements could rule themselves—as long as they ruled themselves in a way agreeable to all humanity and made the right choices. Following international human rights law should not be difficult; it is, in fact, a rather "low bar" to simply not oppress your own residents and citizens. However, we know from Earth history, and even now, that this "low bar" is regularly not met.

As humans move into space, this "low bar" of protecting basic human rights, as well as other ethical standards, ought to be forcefully protected, preferably at the local level, but if not, then at higher levels as needed. Governance is always an experiment on human lives, and therefore, as with any human subjects test, if the experiment ever begins to clearly go wrong, ethically, it must be stopped.

In sum, all humans should be included in contemporary international law as a bare minimum to protect human survival and flourishing. Ideally, these laws will be followed with no need for higher levels of government to interfere. But because we do not live in an ideal world but rather the real world, we should be prepared to step in in order to correct abuses as they appear.

*E. What should be done or prepared for with regard to emergencies and reliance on Earth support? Is there an obligation for the humans of Earth to support human settlers if they encounter trouble?*

Article V of the OST specifically considers the problem of humans in space in distress. Astronauts are "envoys of mankind" and "States Parties to the treaty" shall "render all possible assistance" to them whether on Earth or in space. Relatedly, if one nation discovers a danger in space, they shall "immediately inform" the other nations.

The law is clear, then, but what about the ethics? In this case, the OST summarizes the ethics of aiding others quite clearly: people who need help should be helped. There is not much complexity to it. Even in war, one side is supposed to help the injured or captured of the other side if possible, and much more so in peacetime.

Legally speaking, again, complexity arises if a state is not party to the OST. If, for example, a Mars settlement decided to declare independence and reject international law, including the OST, it would no longer be part of a state covered by the OST and would both be free of the legal obligation to help others as well as losing the reciprocal benefit of being helped by others should the need arise. As a new settlement would be a weaker party, and therefore more likely to need help than to render help, it would seem to be more prudent to remain in the OST.

Ethically speaking, however, with respect to nonparties to the OST, the obligation to help those in need would still remain. Within reason, people in distress should be rescued, even if they have rejected the very systems that are designed to help them. Helping nonparties would not be a legal act of **reciprocity** as the rescued party is not part of the OST (reciprocity is also one of the most basic ethical rules, repaying good for good) but rather an act of legal **altruism** that is legally **supererogatory**, that is, more than what is required. Ethically speaking, however, it is not supererogatory, it is expected, unless it is supremely difficult to actually help the party in need, that is, very far out of the way, extremely costly, etc. (and many things in space are this difficult). If it is easy to help someone, for example, extending a hand to a drowning person who is right next to you, then the help should be rendered, regardless of legal obligation to do so. Part of this ethical consideration is a simple cost to benefit ratio: if it costs next to nothing to give someone a huge benefit, then it is the right thing to do. (This is another way of describing the reason that the powerful should help the weak: a little help from a powerful person might be near nothing for them but be a lot to someone who is powerless; this is related to social justice.)

*F. What if settlements disagree with Earth or other settlements on proper norms and policies in space? Should space be a heterogeneous libertarian free for all where very different behaviors are accepted (even dangerous cults, racism, sexism, exploitation, slavery, mass murder, and torture), or should the entire human community, perhaps through the UN, enforce homogenizing universal norms and policies on all human settlements? How pluralist should space be?*

Disagreements over ethics are one of the most intractable types of disagreement because they involve not only actions but habits, and ultimately identity. Declaring that *X behavior is wrong*, when another declares that *X behavior is who they are*, creates obstacles that cannot be resolved by any simple process of compromise but rather only by extremes of tolerance or intolerance. For example, if one declares that racial slavery is wrong and then meets a person who defines their identity as being a racist slaveholder, this is not a disagreement that can be solved in any simple way. One person will have to accept the other's assertion, and very often that contest of acceptance will involve force—whether physical, legal, or otherwise.

Diversity on Earth has always been a source of contention. Human groups coalesce around ideas, both theoretical (beliefs) and practical (behaviors), and differentiate themselves from other groups based on differences in these beliefs and behaviors. Note that this is rather odd: humans do not need to do this, but we do. It seems to be a part of our nature: we are bound together by shared ideas and activities. And while in many ways the Earth has less cultural diversity than it did in the past, with languages being lost and smaller groups assimilating to larger ones, in other ways new diversity appears all of the time, in the forms of subcultures, new political and religious movements, etc. Cultures that extoll individualism and free expression will likely diversify rapidly, while those that extoll community focus and shared identity will be likely to diversify less quickly.

So we can expect the future to both be more and less diverse at the same time. Earth will converge around several languages, for example, while others will be spoken less and less. At the same time, within those languages, absurd ideological movements like geocentrism and flat Earth may flourish as people search for communities that give them the feeling of belonging that they so desperately desire. Space will certainly allow for more diversity (though hopefully ideas such as geocentrism and flat Earth will not get far into space) as distance lends itself to cultural divergence.

But how much divergence is acceptable? If a settlement decides, for example, that women are not allowed to have careers but instead must stay at home and raise children, is this acceptable cultural-ethical diversity? Or on the contrary, if a settlement bans women from being homemakers and raising children, would this be acceptable? Why or why not?

There are obvious ethical responses to this involving freedom, nondiscrimination, autonomy, justice, fairness, and more. But at what point does one group of humans get to tell another group of humans what to do? Certainly in the colonial era on Earth, European powers were quite pleased to impose their own ethical rules onto colonized peoples. While in space we speak of settlement, would there not also be some sort of "moral colonialism" if settlers were not also allowed to have their own divergent ethical standards?

As a first approximation, if "ethical diversity" violates international human rights law, which was founded upon ethical principles that were hard won in the twentieth century after World War II, then it is not a legitimate form of diversity but rather one that is a form of oppression. However, if a group refuses to follow international human rights, then how should the rest of humanity respond? There are two general possibilities: do something or do nothing. We will discuss these options next.

*G. There are historical analogs to various religious groups seeking freedom to exercise their religion. Will space finally provide the separation that such groups desire, and should the rest of humanity allow it, or should conformity to international law be enforced by humanitarian military intervention when necessary?*

Certainly on Earth there have been numerous cases of groups seeking the freedom to live out their beliefs, away from what they perceive as the oppressive morals of others. For example, just in the history of the United States, the Puritan Pilgrims who settled in Massachusetts, the Mormons in Utah, and dangerous cults like the Branch Davidians of David Koresh and the People's Temple of Jim Jones all sought out the space and freedom they desired in order to live as they pleased. The second two examples turned into mass murders, which between them resulted in nearly one thousand dead.

Within a country, the government is within its rights to investigate and prosecute crimes, and hopefully the laws of those lands correspond to ethical principles. So when these cults were investigated, this was an appropriate government activity. The cult's responses with violence then demonstrated that they were indeed dangerous groups that were worthy of that investigation. This is all within the typical activities of a nation: maintaining the rule of law and orderly human activities.

If space settlements remain under the national laws of Earth, then jurisdictional questions are solved through those national governments. However, even now on Earth there are certainly some nations that might be following their own laws or breaking their own laws and the government or other organizations violating human rights and ethical principles with impunity within their own borders. In these cases, as with subsidiarity discussed earlier, other states then have to ask if this is something that they will stand by and allow or whether international action should take place to correct these abuses at lower levels.

**Humanitarian intervention** is the name for military action used by outsiders in order to correct unethical activities in another state. The ethics of humanitarian intervention is complicated because it sits astride an ethical tension: first, states should not use violence to impose their will on other states, and second, bystanders should not allow unethical behavior to continue if they could stop it. Humanitarian intervention will be discussed in more detail shortly. But in the context of space it is worth noting that, as discussed in chapter 6 on military activity in space, space is a more vulnerable and dangerous environment for human life. Everything moving at speed is a potential weapon, and human settlements will often be delicate and easily destroyed (though settlements under thick radiation shielding or underground might be less delicate, they will still have significant vulnerabilities, especially considering the energies typically found in space). However, space does offer one thing that Earth is running out of, as our technologies "shrink" the size of the world: distance. The farther away people are, the harder it is to assert power, and so, for those who feel threatened by humanitarian intervention, space may be a final frontier, indeed.

*H. Last, will space allow for human flourishing? This involves all the previous concerns plus the ethical and political questions of what makes for a good life and a good community.*

What makes for a good human life? Certainly, when the basics are available—food, water, air, shelter, health care, etc.—that allows for wellness but does not assure it. These basics are necessary but not sufficient for flourishing.

What about political stability, rule of law, and respect for human rights? Are these enough to assure flourishing? Again, no, they are necessary but not sufficient. Many people in the world have all of their basic needs fulfilled, live in stable nations where their rights are respected, and yet do not feel that their lives are meaningful and flourishing. More is necessary.

But what is this “more” that is needed? Certainly having supportive and caring family and friends are one component; many people find great meaning in their relationships. Other people find great meaning in their work. They feel like they are accomplishing important things and take satisfaction in a job well done. Others find meaning in their group identity, whether it is identity by their nationality, town, region, ethnicity, race, sports team, ideology, political affiliation, and more. Note: these sources of meaning are not necessarily good. But these sources do demonstrate something deep about human beings: we are driven by social communities and cultural ideas, including ideas of purpose. Identity almost always also relates to a purpose, such as remembering the past, rooting for a team, or building a better future. We need these things in order to feel alive and flourishing. We naturally seek these out, though of course many things can get in the way and can prevent our flourishing. And ideas, purpose, and community can be found anywhere, whether on space or on Earth.

The good news, then, is that if space settlements can provide these basics for well-being, then humans will likely be able to supply the community, culture, and purpose to make a flourishing life for themselves. One purpose for early settlers will be simple survival: space will be constantly trying to kill them, and they will need to use their wits and technology to resist these forces. They will also have the purpose of pioneers: to build a new civilization where nothing has existed before. They will also no doubt be able to find religious purpose, political purpose, and other purposes, which will hopefully be good purposes and not bad ones. Ethics, as a search for how to do the right thing and live a good life, should be able to help in these pursuits.

## ETHICAL CONCEPTS AND TOOLS

### Environmental Ethics and Sustainability

Any sort of space settlement (at least in the solar system) will have to be vastly more sustainable, and even generative, than any human settlement on Earth. While on Earth we might seek sustainability (seeking to endure indefinitely) or even restoration or regeneration (seeking to improve the local environment), in space, human settlements will need to go one step farther and generate entirely new livable environments from previously dead matter: ecopoiesis. Therefore we can expect space settlers to become quite preoccupied with their environmental conditions. After all, if their house fills with toxic waste, they cannot simply step outside for some air, rather, stepping outside could well be more immediately harmful than dealing with the toxic waste itself.

We might even expect space exploration to produce spinoff technologies and cultural norms that are highly beneficial to life here on Earth. For example, water recycling technologies are highly developed on the International Space Station (at least on the American side), yet on Earth, while some places recycle water, hardly anywhere recycles water up to the standard of drinking water. Space exploration may develop technologies that allow us to transform our Earthly cities into entities more like space stations—living with greatly reduced environmental impact and allowing various Earth environments to regenerate after decades, centuries, or millennia of abuse.

## Pluralism and Humanitarian Intervention: Sovereignty versus Human Rights

Space settlement will allow for increasing pluralism. Not only will diverse human groups move into space, but we can also expect that once in space humans will start to culturally diverge even more so. Groups on Earth who feel that their freedoms are being limited by the constraints of Earth life and culture may be particularly interested in moving to space. These groups might include various sorts of extremists, religious groups, cults, utopians, transhumanists, political and ideological minorities, and so on. Some of these groups might have or develop sensibilities that are unacceptable to the rest of humanity. For example, think of the worst taboos on Earth—incest, cannibalism, child abuse, slavery, racism, oppression, etc.—these might be core beliefs of some of these settlers. Should the rest of humanity allow smaller human groups to practice these divergent beliefs? And if not, what should be done to encourage (or coerce) conformity?

On Earth, we have a strong tension in international law regarding the absolute sovereignty of nations and the ethical requirements of humanitarian action. Humanitarian interventions are when another country or the international community use force in order to protect a subnational group within a country; for example, during the NATO intervention in Kosovo in 1999, or various operations by various nations and supranational groups in Iraq, Syria, Afghanistan, Somalia, etc.

Of note, sometimes humanitarian intervention might have a compelling cause yet no one wants to carry it out, often depending on the power of the nation in question. For example, North Korea has engaged in tremendous systematic human rights abuses, yet humanitarian intervention is never proposed as an option because not only is North Korea protected by China, it also possesses nuclear weapons. No nation would risk that kind of potential cost in order to engage in a humanitarian activity; the deterrence is too great.

In space this will become an even more acute issue. As mentioned numerous times before, the energies involved in space exploration and use are beyond almost any of the energies typically associated with life on Earth. Any object moving at orbital speed is a potential weapon, and in the vastness of space it could be easy to hide and difficult to approach another object without being seen. Were there human groups engaging in ethically abominable activities in the solar system, it might be very difficult to force them to stop. They would be able to defend themselves quite well and even potentially attack the interveners in a way that would be difficult to stop (for example, via lasers, redirected asteroids, doomsday weapons such as self-reproducing nanotechnology, or other difficult to stop weapons).

Under these conditions, the best solution might be to enforce a minimal ethical conformity among space settlers at the outset (for example, agreement to abide by UN treaties) in order to set expectations and prevent conflicts from occurring in the first place. This may be an unlikely scenario given that we cannot yet do this even on our own planet and that many of the spacefaring nations value freedom of expression and political freedoms that make cultural divergence more likely, but it could at least set a clear minimum for behavior.

What, then, will humanity decide when it comes to legitimate versus illegitimate diversity in space? If some sort of moral code is enforced, will it be enforced in a heavy-handed way or through more "soft-power" means (such as cultural influence and desire for prestige)? This is for us to ponder and the future to decide.

## Respecting "Moral Traces," Part II

There will no doubt be occasions where groups of humans do not get their way in space, and in order to preserve political unity, compromises will need to be made. For those groups who cannot get their way, what sort of respect ought to be shown to them in order to encourage unity despite disagreement? In general, this is highly dependent upon exactly how important the situation is perceived to be by the parties involved, what is at stake, whether compromise is possible or impossible, and so on.

The history of space exploration has relevant cases worth considering. During preparations for Moon exploration in the 1960s, some Indigenous Americans as well as others around the world were adamant that nothing should touch the Moon because the Moon was sacred—to touch it would be to desecrate a sacred entity.[14] There was no way to reconcile these opposing viewpoints. While this could have become a rancorous occasion or an opportunity for mocking or derision, instead the disagreements were treated diplomatically. The opposing perspectives were heard and treated with respect, even though they were not heeded.

For these opponents, this was not a satisfactory outcome: they did not get what they wanted. And ultimately, the decision came down to one of decision-making power. The same science and technology that allowed humans to go to the Moon had previously enabled colonizers to take over Indigenous people's land and subject them to the wills of a more powerful culture. Yet the moral traces of these interactions and others have stayed with NASA leadership, and as we go into the future, these moral traces should at least remind us to treat the Moon with respect, even if we do not choose to leave it alone entirely. From the opponents' viewpoint, people might desecrate the Moon by our presence, which we might still choose to do over their protestations, but, remembering their concerns, we should be as careful as we can be by not marring it, actively destroying it, and so on.

The question of "desecration" should not be taken as a purely religious one or as a perspective to be dismissed too easily. Environmental author Wendell Berry once said, "There are no unsacred places; there are only sacred places and desecrated places,"[15] and this perspective on the environment is one that many peoples on Earth can find comprehensible. This is not dependent upon religion: completely secular, nonreligious, and antireligious people can still understand the concept of the sacred when viewing the magnificence of nature, instead construing it as a feeling of wonder or amazement at the sublime beauty of the universe. In this secular, nonsectarian sense of the sacred, we should not think of the Moon and other extraterrestrial objects as being "unsacred." Instead we could view them as sacred to many of us, if not even most of humanity, or even simply "sacred" in themselves, apart from human minds. Objects far above the Earth's surface are literally in the "heavens" and "celestial."[16] These feelings should not be merely tossed aside; we should keep them in mind while we go about exploration and settlement and navigate exactly what sacredness means in every new context that we find ourselves in.

In one place it might mean that human interests allow complete consumption of an asteroid in an Earth-endangering orbit. As a defensive measure, deconstructing and utilizing such a body could make sense, and in the words of sacredness it protects the sacredness of life on Earth. In another place, respect might mean not marring the face of the Moon, for example, by reducing human impact on the visible side of the Moon. Perhaps the far side of the Moon can be much more developed, but the near side ought to remain more aesthetically pleasing. Importantly, this aesthetic is not only important to humanity but to many other life forms on Earth that use the Moon for navigation, noting time, etc.

## The basis of the right to exploit space resources and by what legitimate authority do we have that right (or is there no claim or right to be had)?

We might ask ourselves, by what authority do humans have any legitimate right to do anything in space, whether utilizing resources or even existing there? We might we ask the same question of ourselves on Earth. Much of the time it reduces to a question of power: we cannot use other people's property as if it were our own, or else they will call the police and we will be forced to stop. In the international realm, we cannot simply go into other people's countries and do whatever we want, we must respect their rules, and if we do not, once again we will be forcibly stopped, or else we must have a legitimate reason for a humanitarian intervention.

However, on Earth, against the natural world, we assert our authority by raw power: we can strip mine the land and empty the seas of fish, clear cut the forest and lay waste with nuclear fire if we so decide. Nature cannot stop us, and so we proceed. The legitimacy of this rule is questionable because ethically speaking *might does not make right*. "Might makes right" is the rule of barbarism and violence, ultimately leading to totalitarianism and self-destruction. If sustainability is good, then nonsustainability is not good—and "might makes right" is an unsustainable way of life because it favors the short term (power over others) over the long term (power together). Long-term thinking sees that violence delegitimizes those who practice violent short-termism; over the long term, "might makes right" is actually a weaker strategy than building ethical power together. True ethical legitimacy lies in the opposite formulation: **right makes might**, discussed previously in chapter 2 as the Gandhian idea of *satyagraha* or "truth-force." In other words, ethical action gives legitimacy, legitimacy is oriented toward the good, and orientation toward the good ultimately creates a stronger, more sustainable, more powerful whole.

Legitimate human authority in space must therefore begin with ethics. Legitimacy requires justice, sustainability, and humility. As we step into space we should acknowledge that we do not "own the place." Rather, we are humble visitors, powerful but short-lived compared to the ancientness of the landscapes we will see. It will make sense to judiciously utilize the resources that we find, but we should not greedily assume that everything belongs to us. Indeed, we can imagine if the tables turned how unjust it would be if powerful ETIs existed and they had already staked a claim to our own solar system. We would have no say in the matter and be unfortunate subjects of their will, much like the colonized peoples of Earth's past. Like the Vogons in Douglas Adams's *The Hitchhiker's Guide to the Galaxy*, they might just take the most dramatic of actions—destroying the Earth—without inquiring of us at all.[17] We, in our own actions, should strive not to behave in such a manner.

This might all sound quite unrealistic, and sometimes ethics can sound that way. But the role of ethics is by its very nature one that seeks the good, and not just any good, but the best good: all the goods that can be attained together. If we cannot envision a better world, then we will never have one. Even if this ethical world (perhaps comparable to Immanuel Kant's "Kingdom of Ends"[18]) is very hard to attain, we should still envision it. However, we should not envision impossible worlds to compare ourselves against. Impossibility only breeds frustration and cynicism. Harkening to Kant again, "ought implies can"—we cannot hold ourselves to impossible standards, we can only expect ethics to include that which is possible. Indeed, if we allow ethics to forget reality, it becomes unethical because it becomes impractical, and the whole point of ethics is to *do things*. Actions that are ethical or unethical in a perfect world also do not necessarily align with what is ethical and unethical in this world, for example, in a perfect world people might never have to destroy natural objects in order to obtain necessary resources, but in the real world in some situations people do need to destroy nature to obtain resources.

Our forays into space ought to be impact minimalist, using only what we need and not catering to outlandish desires and wants. And this control of desire is not only a key virtue from the past—temperance or moderation—it is also a key virtue of the future. In a world where technology is continually catering to our every want and increasingly delivering even on wants that we did not even know that we had, controlling our wants becomes ever more important. We, humanity in general and humans as individuals, not only need to desire the right things but also, at an even more basic level, we need the second-order desire to desire the right things. The desire for good cannot be superficial, it must be at our core. If we can have anything, ranging from universal happiness and peace to universal suffering and death, our desires become the absolute most important thing. If technology can give us anything, and with technological power, this is the end humanity is seeking, knowingly or not, then it is our desires that will ultimately make our future good or bad.

This metadesire allows us to seek our ethically better selves and become better people. We do not become more ethical and gain better characters by succumbing to our every gut instinct and acquisitive whim. Instead when we control these impulses, we become ethically stronger: more temperate, more courageous, more just, more prudent.

Having more power does not make us better people, and in fact, power coupled with poor ethics is a recipe for disaster and suffering. What makes us better people is ethical behavior and developing our best character traits. Returning to the matter of legitimate authority, a legitimate authority is one with ethical authority—one that at the deepest level seeks the best ends for not only all human beings but for all living creatures, in their diverse and particular ways. This includes non-Earth life. It may even include nonliving things such as unique celestial objects (though what it means for a nonliving object to flourish is worthy of consideration, perhaps processing it into an environment for living things—indeed turning a dead thing into a living thing—is the best way to do this: but that is for the next chapter). A legitimate authority wields its power in order to avoid unnecessary conflicts and facilitate the survival and flourishing of all.

## BACK IN THE BOX

Because this chapter has a twofold focus, it seems right to consider two Earth-focused questions. First, considering survival and flourishing, we should determine how to enhance efficient and sustainable resource utilization on Earth. Contemporary resource use is highly inefficient, and the facilitation of consumerism and waste damages the environment and unjustly takes resources from future generations. This is not a mere triviality; the future of humanity is at stake. Remarkably, this impact on the environment can be reduced by intelligent use of technology and behavioral and policy choices. While space might inspire us to do better here on Earth, we should not wait for space to do this for us and instead take the initiative on our own.

Second, the pluralism of politico-religio-ethical diversity on Earth certainly raises the question of how to make decisions when many parties have differing views of the meaning of the concept of good—in other words, the ethics that go into decision making when groups fundamentally differ in their ethical allegiances. Establishing and navigating a viable politico-religio-moral pluralism on Earth is a problem we have not yet solved. At the very least there will need to be many compromises in order to keep the majority of people satisfied, and all groups need to keep in mind the common good. Variances in the conception of what "the common good" means precisely will be at stake, however, which means that the process of negotiation, and sometimes applications of power, will sometimes be difficult.

## CLOSING CASE

Humans have yet to build any real settlements in space (space stations in low Earth orbit being "stations" not "settlements"), however, many science-fiction authors have considered life in space settlements, and many engage questions concerning the struggle to survive and flourish. Kim Stanley Robinson's award-winning *Mars Trilogy* stands out as a profound exploration not only of the Red Planet but also of the environment, politics, and culture of Mars. Starting with *Red Mars*, as the first hundred settlers travel to Mars and start digging the series of chambers for "Underhill," their first permanent settlement (which is underground to protect it from radiation), moving through terraformation in *Green Mars*, and ending with the oceans and cities of *Blue Mars*, the series is an epic about human settlement of a new place.[19]

The political aspects of the series are particularly interesting, as various factions struggle to implement their own conception of good onto the entire planet and everyone else living there. In particular, the "Red" and "Green" factions represent those who want to keep Mars less developed and those who want to terraform it, and their hostility is a major driver of the story. We can hope that in the future any settlement of Mars will be less contentious and more orderly than what is depicted in the Mars trilogy. One of the benefits of speculative fiction is that we get to imagine future worlds and how they might be if "X" variable is adjusted in "Y" way. Hopefully thinking more about ethics beforehand and setting clear ethical expectations for how a future will unfold might allow us to adjust one variable in a way that will permit humankind to flourish together in a harsh environment and perhaps set an example for humanity elsewhere too.

## DISCUSSION AND STUDY QUESTIONS

1. What are some of the top ethical issues that settlers on Mars should be concerned with? How would you rank them in terms of priority, and why?
2. Which do you find to be more concerning with respect to the ethical situation of any future settlers in space: the natural environment or the social environment? Why?
3. Which ethical tools in this chapter do you find to be most helpful for thinking about future settlements in space? Which are least useful? Why?

## FURTHER READINGS

Charles S. Cockell, *Extra-Terrestrial Liberty: An Inquiry into the Nature and Causes of Tyrannical Government Beyond the Earth* (Edinburgh: Shoving Leopard, 2013).

Martyn J. Fogg, "The Ethical Dimensions of Space Settlement," *Space Policy* 16 (2000): 205–11.

Patrick Lin, "Viewpoint: Look Before Taking Another Leap for Mankind—Ethical and Social Considerations in Rebuilding Society in Space," *Astropolitics* 4, no. 3 (2006): 281–94.

## NOTES

1. Elon Musk, @elonmusk, Twitter, 5:56 p.m., January 16, 2020, https://twitter.com/elonmusk/status/1217989066181898240, and Elon Musk, @elonmusk, Twitter, 6:01 p.m., January 16, 2020, https://twitter.com/elonmusk/status/1217990326867988480.

2. Amy Thompson, "SpaceX's Elon Musk and His Plans to Send 1 Million People to Mars," Teslarati, January 26, 2020, https://www.teslarati.com/spacex-ceo-elon-musk-plan-colonize-mars-1-million-people/.

3. Eric Berger, "Inside Elon Musk's Plan to Build One Starship a Week—and Settle Mars," *Ars Technica*, March 5, 2020, https://arstechnica.com/science/2020/03/inside-elon-musks-plan-to-build-one-starship-a-week-and-settle-mars/.

4. Tim Childers, "Elon Musk Says We Need to Live in Glass Domes Before We Can Terraform Mars," *Popular Mechanics*, November 20, 2020, https://www.popularmechanics.com/space/moon-mars/a34738932/elon-musk-glass-domes-terraforming-mars/.

5. Smith et al., "The Great Colonization Debate," 9.

6. "*Ecopoiesis*" originates from Robert H. Haynes, "*Ecce Ecopoiesis*: Playing God on Mars," in *Moral Expertise*, ed. MacNiven, 161–83.

7. Tony Milligan, *Nobody Owns the Moon: The Ethics of Space Exploration* (Jefferson, NC: McFarland and Company, Inc., 2015), 94; and Milligan, "Scratching the Surface: The Ethics of Helium-3 Extraction," paper presented to the 8th IAA Symposium on the Future of Space Exploration: Towards the Stars, Torino, Italy, July 3–5, 2013, http://uhra.herts.ac.uk/bitstream/handle/2299/12133/Scratching_the_Surface.pdf?sequence=2.

8. Randolph and McKay, "Protecting and Expanding the Richness and Diversity of Life."

9. World Health Organization, "World Health Organization Model List of Essential Medicines, 21st List, 2019" (Geneva: World Health Organization, 2019).

10. United Nations, Outer Space Treaty.

11. Mike Brown, "SpaceX Mars City: Legal Experts Respond to 'Gibberish' Free Planet Claim," *Inverse*, November 3, 2020, https://www.inverse.com/innovation/spacex-mars-city-legal.

12. Ibid.

13. Ibid.

14. Young, "'Pity the Indians of Outer Space'"; and Pop, "Lunar Exploration and the Social Dimension."

15. Wendell Berry, "How to Be a Poet (to Remind Myself)," in *Given: Poems* (Berkeley, CA: Counterpoint LLC, 2005). It should be noted that Wendell Berry has been extremely opposed to space "colonies" in the past; see Wendell Berry, "Comments on O'Neill's Space Colonies" and "The Debate Sharpens," in *Space Colonies*, ed. Stewart Brand, 36–37 and 82–85.

16. Arnould, "An Urgent Need to Explore Space," 160–61.

17. Douglas Adams, *The Hitchhiker's Guide to the Galaxy* (London: Pan Books, 1979).

18. Kant, *Foundations of the Metaphysics of Morals*, 50.

19. Kim Stanley Robinson, *Red Mars* (New York: Bantam Books, 1993); Robinson, *Green Mars* (New York: Bantam Books, 1994); and Robinson, *Blue Mars* (New York: Bantam Books, 1996).

*Chapter Fourteen*

# Planetary-Scale Interventions on Earth and Afar

**Figure 14.1. Earth and Mars**

## WARMING MARS IN ONE HUNDRED YEARS

In 2001, now NASA senior scientist Christopher McKay and SpaceX senior engineer Margarita Marinova proposed a plan to warm up Mars to habitable temperatures in just one hundred years by using supergreenhouse gases.[1] These super-greenhouse gases, including carbon tetrafluoride and sulfur hexafluoride, are thousands of times more effective at retaining atmospheric heat than the greenhouse gases we are more typically familiar with, such as carbon dioxide. (These fluorine-filled gases would also prevent the formation of an ozone layer to reduce the penetration of ultraviolet light to the Martian surface, but at the beginning of terraformation of Mars this would not be a significant concern, as humans could not expose their skin to direct sunlight in any case.) Carbon, sulfur, and fluorine are common on the Martian surface, so automated roving factories could begin to comb the surface of Mars and refine these ingredients into supergreenhouse gases and release them. This would trigger a global warming chain reaction, first melting carbon dioxide ice caps, then water-soaked permafrost, and ultimately creating a planet more suitable for life.

We have the scientific theory and may soon have the technological power. We could try to begin to implement this plan. So the question is: Should we do it?

## RATIONALE AND SIGNIFICANCE: POWER BEYOND ANYTHING HUMANS HAVE WIELDED BEFORE

This chapter will examine the possible future of extraterrestrial settlements beyond the "habitat" stage, moving into thoughts of and plans for terraforming planets and living in "open-air" or at least very large open spaces in "tents" (tent terraforming or "**paraterraforming**").[2] Terraforming—the intentional engineering of objects in space to become more Earth like—has been a staple of science-fiction literature for decades, but with growing awareness of climate change here on Earth, some humans have become conscious that we are already "terraforming" (or perhaps "unterraforming") our own planet by changing the Earth's atmospheric composition. This, in turn, has raised the issue of intentionally trying to regain control of Earth's climate through technological means, which has been dubbed "climate engineering" or "geoengineering." As we have become aware of our own power here on Earth, terraforming other planets seems less like fiction and more like plans awaiting action. But as with all technical challenges with ethical implications, the question soon becomes: Should we? And if we should, what is the proper way to do it? Terraforming would be one of the ultimate cases of "playing God," and as such, ethical tensions can be expected to be at a commensurate level.

## ETHICAL QUESTIONS AND PROBLEMS

### What is terraforming and how might it be done?

Terraforming seeks to turn other environments in space closer to the environment of Earth. But how might this be done? The answer highly depends on what an environment needs to have done to it in order to make it more Earth like. For example, Mars is cold, dry, receives less sunlight than Earth, and lacks sufficient atmosphere. If Mars were warmed up, its carbon dioxide ice caps would naturally melt, which would thicken the atmosphere and further act to melt the water ice frozen in various places around the planet. Means of warming Mars might include supergreenhouse gases to trap solar heat or giant orbital mirrors or lenses to focus more light onto the planet. Merely exposing the planet to more heat would do much toward thickening the atmosphere and making water more available, however this still might not be enough to be habitable. In this case, the planet might become more Earth like but not enough to allow humans to live outside of habitats, at least not without much further work such as smashing watery comets into the planet's ice caps or bombing the ice caps with nuclear weapons (needless to say, both of these alternatives are rather ethically questionable, with nuclear weapons being worse). Less violent techniques might see Mars "paraterraformed" as a series of giant tents eventually encircled the planet in a bubble held up by air pressure inside. Paraterraforming has the advantage of not losing atmosphere to the erosive effects of solar wind, which has already degraded Mars's atmosphere over geologic time scales. Other solutions to the atmospheric erosion problem might include creating a magnetic shield around Mars to protect its atmosphere, just as Earth has a magnetic field to protect our atmosphere. Mars is a "fixer-upper" of a planet, but it is actually already remarkably Earth like, almost like a "freeze-dried" planet waiting to thaw, and with other desirable traits as well, such as a nearly twenty-four-hour day-night cycle, which is quite rare among planets.

Other planets are harder to terraform than Mars. Venus, for example, would first need to be cooled from its hellish temperature of over 460 degrees Celsius (nearly 900 degrees Fahrenheit), which would need to be done by some sort of orbital Sun shade, thus reducing light to the planet. The tremendously thick atmosphere of mostly carbon dioxide (over ninety Earth atmospheres) would then begin to liquefy or solidify and perhaps react with the surface to form carbonate rocks: though this could take an incredibly long period of time if left to nature. The reaction can be hastened with catalysts and by injecting carbon dioxide deep into the crust to react with subsurface rocks. Venus lacks sufficient hydrogen to convert the oxygen in its carbon dioxide into water, so ocean-like quantities of hydrogen would need to be brought from somewhere else (no mean feat) in order to create water. The nitrogen in its atmosphere would still be three times the amount on Earth and so would also need to be solved, perhaps by converting it into nitrates. The rotational period of Venus is 243 Earth days, so the day-night cycle (approximately 117 Earth days from one sunrise to the next: 58 Earth days of light and 58 Earth days of night) will not be suitable for most normal Earth life, unless it is artificially controlled by mirrors and shades.

The solar system also contains the planet Mercury, which, like Venus, is another terraforming nightmare. The gas giant planets are out of the question for terraforming within any time scales worth talking about, though they might supply hydrogen or water for other terraforming projects. However, there are numerous other moons, asteroids, and cometary bodies that could be turned into habitats if enclosed in domes or bubbles and then given the right terraforming treatments.

In general, terraforming would be incredibly expensive, though this would depend, of course, on the difficulty of the object in question. Mars is easier than Venus, and some places are simply economically unviable, even with foreseeable technology. The difficulty really raises the next question of *why* terraforming might be considered, if it at all.

## Why should we terraform?

Humans are evolved for life on Earth. Everywhere else in the universe that we know of is currently unsuitable for human habitation—even planets in the habitable zones of other stars may have unsuitable atmospheres or other problems. Therefore, if humans are ever to live openly on the surfaces of other objects besides Earth or upon or within artificial constructs large enough to have their own ecosystems, we will need to consider terraforming these surfaces, and by extension we will need to consider the ethics of terraforming.

We also know that the ideal environment for humans can be supportive of many other forms of life. Therefore this choice to terraform might not only benefit humans but also these other forms of Earth life. Depending on the extraterrestrial object, terraforming might also include Indigenous life forms that are currently existing under conditions that would seem harsh to Earth life, such as any possible microbes on Mars. Therefore among the ethical considerations of terraforming we should consider not only human life but also other Earth life forms and any potential Indigenous life forms on natural space objects.[3]

Humans are in some respects "natural terraformers." We have been remaking environments on Earth for thousands of years, as we not only build structures to live in, such as houses, but also convert forests, grasslands, and other ecosystems into farms and pastures and eventually even the cities and megacities that now dot the Earth. There are few places on Earth untouched by human action, and often that action has destructive results rather than constructive ones. When we speak of sustainable development on Earth, we should also speak of restorative and regenerative development, which repairs the damage that humans have done in the past. Similarly, there are environments in space that, while not destroyed by human action, currently lack the conditions appropriate for thriving life. Terraformation can be viewed as an attempt by humanity to generate thriving environments *de novo*, where there have previously been none. The techniques developed to produce these sustainable new environments might help those on Earth to make more sustainable choices as well.

There still remains the deepest question of *why*. *Why* should humans want to live anywhere besides the Earth? At the bare minimum, we have the same questions of *why explore* that we have been considering throughout the entire book. There is survival. There is science. There is technology. There is the purpose and meaning of human existence (however that might be specified). But all of these things could be pursued without altering planets as though we were gods. Is terraformation just a terrible hubris, or might it reflect the human desire to make the universe a better place? Likely it is both. And thus with these ethically bad and good motivations both at play, humans ought to be very careful about how any potential future terraforming might go.

### What is the purpose of terraforming—merely to expand human life, to expand all Earth life, or to assist alien life as well, or even biotechnologically create new life forms never before seen in the universe?

While terraforming has obvious benefits for humans by expanding our potential range within the universe, what about other life forms? There are at least four categories of life forms that are relevant for this discussion: life forms we practically cannot leave behind, life forms we might want to take with us, life forms we might encounter in extraterrestrial environments, and life forms we might create for new environments.

Particular kinds of Earth life would almost certainly accompany us, for example, innumerable species of microbes that would inhabit our microbiomes and would stow away on our ships. We cannot choose to leave our microbiomes behind—we need our symbiotic gut bacteria and other organisms that associate with us (although this might be a good opportunity to prevent bringing along bad microbes and pathogens).

But what about intentionally bringing along a wide variety of Earth life in order to protect and expand upon the biodiversity of Earth life? At the bare minimum, humans would probably like to bring along various types of animals that we bear an affinity toward: pets and other "popular" animals. Dogs, cats, and other domesticated animals will tug on our heart strings and be difficult to leave behind—this despite the fact that they would be another mouth to feed and that domesticated pets gone feral can be extremely environmentally destructive, such as feral cats in numerous places on Earth like Australia and New Zealand. Decorative plants will also have an appeal, as well as food species including animals, plants, fungi, and microbes. Nondomesticated creatures may also be of interest to us, such as the popular creatures found in zoos: birds, large mammals, reptiles, and so on. Each of these animals would require incredible care—to keep them not only alive but also healthy. They also raise questions of animal ethics. Will the animals be forced to live in cages? Will they be able to run and have their own wild territories? Will they have adequate nutrition and all the needs met of their own microbiomes? Will their favorite plants also be there? Certainly some plants are likely to be there, such as food crops, but what of crops for creatures other than humans? To create a functional ecosystem, these sorts of relationships will be necessary and will have to be considered with great care in advance.

What of terraforming for the sake of improving the conditions for the native life of planets that are only borderline habitable? For example, Mars was more habitable in the past but is less habitable now—should we improve Mars's habitability in order to "help" the native life (if there is any) thrive once more? This is a genuinely interesting and difficult question. We cannot restore Mars's past environment, and indeed, if we did, it would likely harm much of the life currently living there anyway because it would need to rapidly adapt to conditions that existed millions of years ago (if there is any life there at all). Thus, while we might think we are creating a long-term benefit for native Mars life, terraformers would certainly be creating short-term chaos, as well as introducing competing organisms from Earth.

What of creating completely new life forms through biotechnology and synthetic biology, whether chimeras of contemporary creatures or creatures completely *de novo*? This is another complicated question. While humans have been selectively breeding plants and animals for millennia, more extreme means of modification, such as synthetic biology, are a qualitatively different approach to the manipulation of organisms. This does not make it intrinsically wrong, but it does make it intrinsically powerful and therefore proportionately more subject to ethical analysis.[4] Given the differences between Earth and other planets and the potential for synthetic biological organisms to help the terraforming effort, it might make sense to create such organisms.

## Is it ethically permissible to terraform places that are lifeless?

It might seem that lifeless places would be without intrinsic ethical value and would instead only have instrumental value. Often when humans discuss ethical value we consider only intelligent creatures, or only creatures with sensation (sentient creatures), or only living things and not dead ones. But the intrinsic ethical value of lifeless places can be accounted for in some ethical systems, for example, some religious systems or secular ecological philosophies. Instrumental value can also often be enough to set aside areas as protected against human interference. For example, many parks and scenic locations around the world have been preserved for their spectacular landforms. Environmental virtue ethics can also serve for thinking about the protection of lifeless places, not out of concerns for the places themselves but rather for the ethical well-being of the decision makers.[5]

An additional distinction can be made between natural and artificial lifeless places. Artificial lifeless places—places made specifically by humans for human purposes—are places that exist because of human will and therefore can be subject to it, within reasonable limits. For example, a space station could be adjusted internally as its human decision makers saw fit, adjusting the temperature, humidity, plant and animal life, etc., but all within limits. It would not be ethical to terminate life support functions while humans were still on board, for example, and likely not ethical to do so even if only animals were on board, except for a good reason (such as carrying disease, etc.). Intuitively, artificial places are places where humans exercise a more legitimate authority, however, this does have some oddness to it because, after all, the resources used to make that artificial place all came from nature, and thus everything artificial is in some sense also natural. All of which returns to the original question of whether it is permissible to terraform lifeless places, which can only be answered that it depends upon the instrumental value of the natural space, as well as the ethical system that might give intrinsic value to such a place. Places that humans deem to be intrinsically worthy as lifeless can reasonably be left as such, and places that humans deem to be instrumentally more valuable either as a resource to consume or as territory for terraformation can be reasonably argued to be treated as such as well.

### What if a place has microbial or more complex forms of life—should humanity then not terraform it, or should we intentionally terraform it in order to create a more hospitable climate for the local life forms, that is, to "help" them?

The question of possible microbial life is likely to be a serious one when considering terraforming in many desirable environments. For example, it is quite possible that Mars has microbial life on it, though we do not yet have strong evidence of this. We do know that Mars was warmer and wetter in the past and had flowing water on its surface and that life on Earth seems to have appeared very quickly once the planet cooled; so if Mars also had these favorable conditions it would seem that it could have evolved life too. This life might still be on Mars, underground or in other refuges. Is our ethical duty toward this life to leave it alone or to "help" it? If we turned Mars into a "planetary park" and peacefully ignored the life there, not interfering, would that be ethically blameworthy or praiseworthy?[6] If we terraformed Mars in order to help the life there, would we actually be helping it or actually harming it by changing its conditions and forcing different forces of natural (really artificially forced) selection to operate on that life?

Among ethical theories, there is sometimes a prioritization that places "do no harm" above "do good." The Latin phrase *primum non nocere*, for example, translates to "first, do no harm." (While this phrase is sometimes attributed to the Hippocratic Oath, the exact wording is not found there.) However, the prioritization of avoiding evil over doing good is debatable because while it is good to avoid doing evil, simply avoiding evil does not itself result in anything good necessarily happening. If good is desired, it must be actively pursued, which is one of the reasons that the first principle of practical reason prioritizes doing good over avoiding evil.

If "do no harm" were prioritized over "do good," in the case of terraformation, then microbial life ought to be left alone, and humans ought to move on to other objects in space. However, if "do good" is prioritized above "do no harm," then terraformation for the sake of benefitting local life forms could seem to be the correct course of action, creating a better place for local life to survive. However, it is nearly inconceivable that humans could terraform a planet and not then proceed to become even more intimately involved with it: permanent human settlement and introduction of innumerable Earth life forms would seem like an irresistible lure. Given this fact, it becomes a trade-off between the ethical value of local life and the ethical value of human and Earth life, not to mention the cost in resources to terraform a planet.

**Does it matter if the planet is an "improving" planet or a "dying" planet, and how would we judge such things? For example, Mars has lost much of its water over history, thus becoming less habitable, but in several more billion years, as the Sun expands, Mars will warm and may become more habitable again.**

If a planet's habitability is degrading, it might seem that terraforming would be a good thing, because it would be preserving the environment toward which its life forms (assuming there are any) are accustomed. On the other hand, if a planet is improving in habitability, then terraforming would speed up that process but perhaps take the planet in a direction it would not otherwise go. For example, the early Earth had little or no oxygen in its atmosphere, and if ETI terraformers had come to Earth at that point and started changing the environment, Earth's entire natural history would have been altered.

As a general rule, I would propose that planets that are young and are becoming more habitable ought to be excluded from terraforming. These planets have their own trajectories toward the future, and humans ought not to tamper with them. Tampering with them would damage the future richness and diversity of life in the universe,[7] even if that time is billions of years away. Such planets should belong to their own evolving future inhabitants, not to curious or acquisitive humans.

Concomitantly, planets that have a poor prognosis for ever naturally forming life or that have their best days behind them are better candidates for terraforming because their trajectory toward this rich and diverse future is closed off. In this case it makes sense (if our end value is expanding life to the universe—but perhaps not if otherwise) to help these planets meet their potential as places where life can flourish.

Notice that this argument is presented in a consequentialist fashion and therefore is subject to the weaknesses of that form of ethical reasoning, most especially our lack of ability to predict the future. Planets can be unpredictable, after all: they can be hit by asteroids and comets, experience ruinous volcanic activity, and even be flung completely out of their solar systems. Humans will make mistakes in predicting these future states of planets, and so we ought to be very cautious in deciding what to do.

But in the case of the solar system we have a fairly reasonable view of the future. The Sun will continue to slowly increase in brightness over hundreds of millions of years, eventually overheating and then finally consuming the inner planets, including the Earth. Mars will become warmer in this phase, but it will still lack sufficient water and time to form a thriving biosphere. It looks like the only way for Mars to thrive (perhaps, as it once did) is if humans make it happen.

Last, terraforming that involves biological organisms from other planets intrinsically violates planetary protection standards for forward contamination. Biological terraformation would therefore also need serious consideration by the proper authorities.

## Should humans actively spread life to the universe (directed panspermia)?

**Panspermia** is the idea that life can move between planets seeding new biospheres as planets become habitable. Directed panspermia would involve humans choosing to do this intentionally. If humanity does begin to terraform and/or paraterraform extraterrestrial objects, should we also attempt to spread life to the far reaches of space?

Certainly science fiction as well as scientists themselves have explored these ideas, but there are great differences between speculation and action. Once again, these ideas involve the meaning of human existence: What are we here for and what are humans supposed to do? Asking about our role in the universe is the only way to answer this question. Perhaps humanity is here for the sake of spreading life beyond the Earth, or perhaps we are here for some other reason (and perhaps there are several reasons, hopefully not exclusive of each other) or no reason at all.

Because purposes and meanings are not empirically verifiable, we can only evaluate them as best we can as theories, as discussed previously, through such means as Ian Barbour's four criteria for evaluating a theory: correspondence, coherence, comprehensiveness, and consequences.[8] Applied to the question of whether humans ought to seed life in the universe we still come to a hard case, however. Given two binary options (for theories can only be relatively explanatory, not absolutely) of "yes, spread life" and "no, do not spread life," they would seem to tie on the first three criteria and the fourth is a matter of opinion: Do people want to do this or not, does it benefit us somehow or otherwise prove to be a fruitful course of action? Those who are risk averse on matters of planetary protection would oppose it then, and those who want to hedge against human and Earth life extinction might be in favor. It is, at least for now, an unsatisfactory conclusion on so significant a matter, to say the least. In the future as we gain more information, these comparisons may become clearer.

### What are the ethics of creating artificial locations for human habitation, ranging from paraterraforming natural objects (such as asteroids) to space stations and proposed megastructures?

Humans have already created space stations such as Skylab, Mir, the International Space Station, and Tiangong-1 and -2. Other than the exorbitant costs and associated risks to life and health, such stations are not controversial; however, in the future space stations could become larger, even growing to massive size, such as the stations proposed by Gerard O'Neill.[9] Such massive constructions would effectively need to be "terraformed" inside to transform them from being a mere construction into being a living system. Some authors have even proposed enclosing asteroids in giant bubbles to house very large numbers of people.[10] Terraforming the inside of an artificial object or tent terraforming smaller space objects such as asteroids or moons can be considered paraterraforming.

Rather than ignore natural objects incapable of holding an atmosphere, they could instead be enclosed in a giant bubble (perhaps two layers of translucent plastic, with meters of water in between to act as radiation shielding[11]) and then be brought up to the correct atmospheric pressure and temperature through various means of terraforming. Because space stations are made completely *de novo*, there would be no native life to endanger (unless it was somehow transported aboard), which could make them ideal from that perspective. Likewise, asteroids and certain moons (with no subsurface oceans) have no possibility to harbor life as we know it (life as we don't know it always being a possible exception), so from that perspective they might seem to be ideal for paraterraformation.

Paraterraforming artificial or small extraterrestrial objects could be ethically simpler than attempting to terraform planetary-sized objects. There would be fewer planetary protection concerns or disruption to native extraterrestrial life, it would mostly be a matter of resource extraction, use, and consumption/destruction, which is certainly not insignificant but is less complicated than with potentially living places. Once again, the matter of the ethical value of lifeless places comes to the fore of our analysis, and the six criteria of safety, scientific value, aesthetics, usefulness, size, and relative concentration of value should be considered. Resources can be used if there is a good reason to use them, but not without that reason. If people and other life forms need places to live or materials to support their life, then it makes sense to reasonably accommodate these needs.

Artificial objects will require massive resources to build. Natural objects in space would be destroyed in order to produce these artificial environments, therefore these artificial places better be "worth it," not only economically but ethically. Unlike natural places that often have no relationship to human needs (such as asteroids with extremely weak gravity and rapid rotational periods), artificial objects have the benefit of being technology that we design to suit humans, complete with whatever gravity, atmospheric pressure and composition, and other traits we might desire. These artificial objects will create an easier environment for other Earth life as well.

### What governance and policy considerations are necessary before, during, and after such massive projects? How will we assign responsibility when things go right or wrong?

Constructing immense objects or terraforming planets will take institutional governance strategies with vastly much longer time horizons than anything that humans have ever embarked upon. While some of the oldest institutions on Earth include various religions and small businesses, and previously humans have spent decades or even centuries constructing monuments and cathedrals, in today's high-speed world, long-term thinking is relatively rare. How will we govern projects that will outlive any individual human working on them? How can we prepare to embark on a project where the end has never before been attained? Will people be able to pay attention to a project where there won't be any expected payoffs for decades or centuries? Governance of such a project will be a complicated matter in every sense: ethically, politically, economically, scientifically, and technologically.

The good news is that some cases will be easier than others. If Mars settlement goes as well as some hope (like Musk), then the locals will be tasked with terraforming and those not on Mars will have little say in the matter. No life will be found there (they hope) and so planetary protection will not be a concern. Supergreenhouse gases will warm the planet and carbon dioxide and water ice will sublimate or melt, respectively, and fill the atmosphere. The locals will benefit from their efforts, and their descendants will know that their ancestors worked hard and made a biosphere from nonliving matter.

However, hopefully from reading everything else in this book, the reader will recognize that this scenario is not necessarily realistic. If Mars has any signs of past or present life, then the judgments change dramatically. The Outer Space Treaty will raise questions about planetary protection and jurisdiction. And any terraforming enterprise will require money in order to work. It will need investors who will demand to be paid, likely in land, goods, or economic rights.

All these things complicate matters greatly, and the entire scenario is one of unimaginable stakes: the future of the surface habitability of an entire planet. Billions of people could live on Mars, where none live now; trillions of other life forms could live there too. The stakes raise the question of what happens if something goes wrong.

Inevitably, because terraforming will be such a complex and long-term process, things will go wrong. And the causes of those wrongs may be choices that were made long ago, by people who are long dead. Those responsible parties might even include those living today, by our actions now, as we consider this subject matter and shape future thinking. In such a high-stakes matter it would be tempting to hold people tightly accountable for their choices that can have such momentous impacts on the future of not only humanity but entire planets. But in the midst of so much uncertainty, not only ethical but technological and scientific, a measure of humility is warranted both on the part of those making decisions and of those judging those decisions.

Not to concentrate only on what might go wrong, there are also occasions where things will go right. Those people who make right choices that lead to helping millions of future people ought to be honored for their insights by those beneficiaries. Of course, in some sense we are all deeply in debt to our ancestors, who have made our current lives possible, and yet we do not necessarily honor them very much, perhaps not as much as we should. One ethical universal among humans is the show of gratitude for good acts, and it is beneficial ethically, socially, and psychologically to show such gratitude. So if things go well in the future, future humans should be thankful.

## ETHICAL CONCEPTS AND TOOLS

### Metaethics, Part II

Metaethics is the branch of ethics that deals with fundamental questions such as "what is good?" and "what is the difference between good and bad?" We have discussed metaethics previously, but as is visible from these questions, finding ethical guidance becomes a key concern when dealing with the enormously important questions posed by terraforming.

When dealing with the ethical value of life forms and even nonlife, fundamental ethical assumptions come into play. For example, do natural objects have any intrinsic value at all, or are all ethical values something that humans assign to nature? Metaethics considers ethical first principles and, for example, the ethical value of life-filled versus lifeless places. It also gets into deeper existential questions about our purpose in the universe and what, ultimately, we should be aiming for, not only with terraforming or space exploration but also as the meaning of our lives and the purpose of our species in existence.

This is the domain that would have previously been held by religion and spirituality, but given the declines and shifts in affiliations and worldviews in recent decades, the answers of the past are at least in some cases no longer popular. While popularity is not necessarily correlated to truth or goodness, it is an important consideration in any sort of democratic decision-making process because democracies are majority ruled. And because the world is pluralistic, and these decisions concern all of humankind, particular care is warranted.

Luckily, there are good precedents for cooperation on deciding ethical issues despite metaethical uncertainty. The United Nation's Universal Declaration of Human Rights is one example, as well as many more human rights treaties that have been added since. Metaethical uncertainty is a serious theoretical problem, but there are solutions in the realm of the practical. Humans can dream up many more theories than we can practice actual realities, and because reality is the natural realm of ethics, it at least has that going for it, to limit its incorrect ideas. In other words, ethical theories that do not work in reality are bad theories that should be set aside in favor of other theories that do work: this is Barbour's fourth criteria, again.[12] Soviet Marxist Communism as a theoretical-practical system failed in practice, and so has been discarded. Other varieties of Marxist Communism are still being tried, not to mention myriad other theories and practices. The space for experimentation is large, but it is not at all infinite—it is probably not anywhere near as large as we might imagine.[13] In the words of author Nassim Nicholas Taleb, we should be skeptical of anyone who advocates a theory yet remains disconnected from that theory's impact: trustworthy people should have "skin in the game," that is, people who feel the impacts of their beliefs are likely to have beliefs with practically better effects than ones who do not.[14]

Interestingly, this project of searching for metaethical truth can be approached from the other side as well, from practice rather than theory. We know this because some practices have lacked accurate theories and have worked just fine. For example, for thousands of years Indigenous Americans knew to nixtamalize maize with alkaline compounds in order to make niacin bioavailable and thus prevent the disease pellagra (niacin deficiency). It was a custom that was not to be broken; there was no accurate theory required. The true theory lay in the science of nutrition, only to be understood in the twentieth century.[15] These practices without theories are functional, although it might be nice to know why they work as that would provide useful insights. Anthropologists and sociologists love to collect these stories, and there are many of them.[16]

In relation to terraforming, then, we might ask what relevance these ideas hold. Perhaps the key point here is that practice will make clearer that which theory cannot foresee. One becomes good at terraforming by terraforming, not by just talking about terraforming. With such a powerful technology, this practice is essential and yet also terrifying if it fails or is misused. Metaethics is a complicated subject, but it is connected to reality. Reality ultimately has veto power over bad metaethical theories if humans choose to learn these lessons from our experience.

While we should not always just go with "whatever works," we should keep it in mind. In Barbour's fourfold structure for evaluating truth content, if everything else is equal, choose the theory that works in practice. By working backward from that which produces human flourishing and yet remains equally valid in terms of correspondence, coherence, and comprehensiveness, we can generate theories about reality that allow for full consistency with reality and at the same time promote human flourishing. Of course, all this is easier said than done, which is why governance is also a key issue.

## The Ethics of Governance

Terraforming raises questions of the governance of massive projects extending over space and time. Among the first questions that need answers is: Who gets to make these decisions? At first pass, the answer is obvious: human beings get to make these decisions, but of course it is not actually all of us but just a few who have that decision-making power. Even if many humans did get to have a say, it would be because a small group of experts decided to make it that way. These experts will reside in only a few nations, and, like with development of the Outer Space Treaty or the Apollo missions that touched the Moon despite the objections of some people in the world, the vast majority of humanity will likely have no say at all.

Because the majority will likely not have a say, what then can we say about the minority who do? As one point, should those living closer to or on the object in question have more say about the decision than those who are far away? As with the previously mentioned principle of subsidiarity, there are good reasons to try to localize governance as much as possible, but no more so than guarantees the well-being of the people and environment in that locale. If, for example, the locals on Mars voted to start destroying unique environments and life forms regardless of their scientific and moral value and in clear opposition to the protestations of the rest of humanity, this would be a significant problem. The rest of humanity would have a strong case for intervening on the people of Mars to stop their destructive ways, and yet the people of Mars would also have a case to support their actions because they are the people closest to the impacts of these decisions. This is one reason that it is absolutely necessary to establish clear political, ethical, and behavioral expectations before settlements occur. In the absence of clear expectations and enforcement mechanisms, governance disputes could arise and lead to dangerously volatile disagreements.

As another point, very often small groups hold very strong ethical or political positions, while the majority either do not care much, slightly oppose them, or oppose them relatively strongly. If there are small groups of people vehemently in favor or opposed to terraformation, should their small but strong opinions weigh more than the tepid opinions of vastly more people? This consideration of minority rights in majority-ruled societies is a classical one in political theory, and in the absence of the particulars of the case it lacks a clear answer (and even with those particulars, it might still lack a clear answer). In general, ethics is not decided by majority vote, but neither is it decided by minority. Ethics is decided by reasoning, practical experience, and time. Ideally those in power would be able to evaluate the best arguments and theories, most relevant experience, and longest histories of experience to determine the best choices, but, especially when it comes to technology ethics, there may not be relevant experience—that experience may lie in the future.

We are left in a situation then in which we still have to decide, or at least determine the way in which to decide who gets to decide. Reasoning matters. Experience matters. Proximity matters. And numbers and vehemence matter, hence the importance of democracy as well as respect for moral traces. Subsidiarity helps with the moral issue of proximity. Experience can be gathered through the study of history, anthropology, sociology, psychology, political science, ethics (including cases), natural sciences, technology, and so on. And reasoning is developed by education, character development, logic, math, philosophy, the scientific method, long-term thinking, and so on. Somehow all of these difficult areas of human endeavor must be integrated into one whole, working for the sake of the common good of humanity and the environment: everything that we may affect. This is doable, but it will take much more work and more successful use of ethics than humanity has ever managed to have before. On the path to good governance we have to start now.

## Space Ethics Frameworks

The reasoning and experience components of governance are of particular importance, and here I will summarize a few resources for future decision makers. There have been many proposals for space ethics, and there will be, no doubt, many more.

One proposed ethical framework for how humanity ought to engage with the universe, which is of particular relevance for terraforming, is that of Richard Randolph and Christopher McKay.[17] Their proposal to "protect and expand the richness and diversity of life" gives humankind a clear purpose in the universe, and it is one that should aid not only in human flourishing but in the flourishing of all Earth life and any other life that we might find in the universe. This is an optimistic view of the future, with opportunity for meaning and integration with other worldviews and human psychology.

Kelly Smith takes a different perspective and argues that "manifest complexity" of sociality, reason, and culture—three traits that make ethics possible—gives moral value to life in the universe.[18] Like Hans Jonas, Smith argues for the ethical value of ethics, a position that cannot be denied without contradicting itself (that is, if one argues that being able to ethically value is of no ethical value, then one argues that one's own ethical argument is of no value and can therefore be legitimately ignored[19]). Here Smith prioritizes intelligent life with these traits over other types of life or nonliving objects. His would be an ethic of terraforming that is more permissive for human desires and intentions.

There are innumerable more options, all with relevance for terraforming as well as many other questions in space ethics. For example, Mark Lupisella and John Logsdon argue for a cosmocentric ethic in which reciprocity is a key value, including reciprocity toward the weak, such as microorganisms.[20] Jacques Arnould advocates for a "space humanism," in which human choice is defended against both utilitarianism and moral relativism.[21] Robert Zubrin and James D. Heiser of the Mars Society tirelessly argue for the settlement of Mars.[22] And many more thinkers have written on these topics, including Cockell, McLean, Milligan, Schwartz, etc.[23] These scholars have provided some of the first explorations into these deep questions of what is most valuable about space and what thinking about space means for humankind and our contemporary ethical systems.[24] There are also many ideas from associated fields such as environmental ethics, technology ethics, ethics of war and peace, and so on that should be put into further conversations with space ethics.

These contemporary theories are also worth comparing to older ethical theories such as those offered by Aristotle, Kant, Mill, and others.[25] For example, in relation to terraforming there is Aristotle's useful distinction between cultivation and construction. In Aristotle's thought, cultivation takes natural objects with their own entelechies and leads them toward their proper ends, their own flourishing, like a farmer cultivating crops or a physician cultivating bodily health. On the other hand, construction takes objects without entelechies, nonliving things, and imposes upon them external purposes: the purposes of humans. However, as humans have grown in power, this line between cultivation and construction has become progressively more blurred, and terraforming will continue this trend as life forms are put at the service of constructing a biosphere and machines are put in the service of cultivating life.

Additionally, non-Western perspectives should not be ignored when we are considering an endeavor like space exploration or terraformation, which is relevant to all of humanity; there is certainly much more work to be done by people of all cultures.[26] Certainly the field of AI has highlighted the need for considering non-Western perspectives as well, particularly in relation to China (which is a world power both in AI and space exploration) and hopefully much of the work of this conversation between cultures will continue fruitfully over the coming decades.[27]

Terraformation is a power unlike anything humans have ever intentionally wielded. While we have begun to change the Earth's climate, that has been by accident, and our intentional efforts to reverse this accident have been halting at best. This raises the issue of whether humans ought to be playing with these god-like powers at all or if we are intrinsically unsuited to such activities.

## The Ethics of "Playing God," Part II

Perhaps nothing could call to mind the worry of "playing God" more so that having the fate of a planet—or multiple planets—completely in human hands.[28] As noted in chapter 9, Mizrahi summarized the "playing God" argument as ultimately nontheological and nonreligious, relying only on the existence of the *concept* of God and the idea that a God would be all powerful, all knowing, and all good.[29] Humans, in contrast, are too weak, too unintelligent, and too unethical to properly control powerful new technologies. The "playing God" argument, then, is really a critique of humanity and asks us to remember the virtue of humility and avoid *hubris* or face inevitable humiliation.

Among the most important considerations of terraformation must also come the questions of power, and the role of humanity in taking unto itself god-like power. Changing planets from deserts into gardens or repurposing materials from asteroids into colossal space stations were beyond the imaginations of most of humanity for most of human history. In the past, stories like these could only have involved gods, but now we propose that we humans do them. We might well ask: "Who are we to make such choices and think we can do such things?" Perhaps these choices do not or should not belong to us—perhaps they properly belong only to nature, or deities, or intelligent aliens. Or perhaps they only belong to us because we are the only choice makers here that we know about.

Whatever the case, the playing God objection will almost certainly be thrown at terraforming because it is perhaps an archetypal example of humans doing something that requires enormous power, incredible intelligence, and the highest ethics. Because humans are not perfect, there will be mistakes: we will fall short of perfection. If humankind chooses to embark on such a difficult course of action, then we need to be prepared to approach it with humility and more care than humans have perhaps ever applied before. We need to remember that we are humans and not gods.

## Anthropotence: The Ability of Humankind to Get Things Done

Some have begun to call the current period of Earth's natural history "the Anthropocene," calling to mind human power over the natural order of the planet. Following on the previous section, aligned with this word is another idea, derived from the idea of the omnipotence of a monotheistic God but applicable to humanity—"anthropotence"—that captures the idea of the power of humanity as a whole.[30] This word was unnecessary in the past when humankind was weaker, but now, as Hans Jonas notes, the nature and scope of human power has changed.[31] Anthropotence is a measure that can be applied both to whole-species power and to the power of individuals, but here I will mainly concentrate on whole-species power because presumably it will be an act of many people to terraform space objects and not generally the acts of individuals, at least not for a long time. Of course, individuals within humanity will still be the ones exercising power, through their wealth, charisma, influence, intelligence, labor, etc.[32]

At this point we also need to ask exactly what we mean by "power." Power comes in many types, as just noted. On a more collective level, it can also be political, military, economic, cultural, and so on. Given the varieties of power, perhaps power in the most abstract form can be thought of as *the ability to get things done* or, more scientifically, as energy in the form of watts of power or joules of energy at humankind's disposal. This quantification of power as energy provides a direct point of connection to the Soviet astronomer Nikolai Kardashev's scale of civilizations, which classifies Type I civilizations as those that control planetary scales of energy, Type II civilizations as those that control stellar scales of energy, and Type III civilizations as those that control galactic scales of energy.[33]

Carl Sagan reinterpreted Kardashev's scale logarithmically, in increments of $10^{10}$, so that Type I civilizations have at their disposal $10^{16}$W, Type II have $10^{26}$W, and Type III have $10^{36}$W. Carl Sagan estimated that in 1973 human civilization was Type 0.7.[34] On a logarithmic scale, humankind has moved up only a few hundredths of a point since then, despite roughly doubling energy use and significantly advancing in many areas of technology.[35] This, then, raises an important concern: perhaps energy use is not a good measure of a civilization's level. Following this, Sagan, John Barrow, and Robert Zubrin have offered alternative scales for civilization based on information content,[36] the sizes of the objects being manipulated,[37] and the range of a civilization's settlement,[38] respectively. Certainly each are relevant for power in their own ways.

What do these have to do with terraforming and ethics? Certainly, energy is needed for terraforming, for example, whether heating or cooling a planet or other space body. Information is needed too, specifically information on how to conduct terraforming. Mastery over scales of intervention from atomic to planetary are needed, and a wide range of civilization too, for a place cannot be terraformed without first being contacted by human activity. Which then brings us again to ethics: Should we choose to affect these places? The answer is that we should, if it is good, and we should not, if it is bad. So which is it?

This entire book has hopefully started to answer that question, or at least started to give us tools to think about how to answer these questions. The answer is always, with ethics, that right and wrong, good and evil depend on the particulars of the situation. Ethics cannot be done in the abstract, only in the concrete circumstances of a specific situation. This is what makes ethics so hard: we can know the general ideas of ethics, but their application to actual reality is the hard part, and this is not a new discovery: Aristotle stated as much twenty-three centuries ago. Moral theories and practices do not eliminate complexity, but they do help us to manage complexity and work within it.[39]

As we go into space we will still have our general ethical rules, but the contexts for application will expand enormously not only in spatial scale but also in temporal scale, on the scale of information, and on the scale of power. In all of these ways, we should then consider what is best for the common good of all humanity and life.

Ultimately, when it comes to power, we sometimes do not seem to have much choice when it comes to wielding it. When nature threatens to impose destruction and death upon us unless we act, then we should use our power to act and protect human life and well-being, as well as the life and well-being of others. This is a good use of power, and having more power to do good things is in itself a good thing. And if we currently lack the power to do good, then in general we should develop those powers unless there are good reasons not to, such as dual-use dangers. Along with the power to do good often comes the power to do evil, and this maleficent side of power needs to be controlled. It will someday be possible to terraform planets, but the question is whether or not we can do it well enough to manage the powers we are reaching for. "With great power comes great responsibility," and we are gaining power consistently.[40] Can we handle terraforming, or will the attempt merely reveal to us our own hubris? Only time will tell.

## BACK IN THE BOX

Thinking of modifying other planets should of course make us consider the modifications we are currently making to our own. For several centuries humans have been burning fossil fuels at an increasingly furious pace, releasing enormous amounts of carbon dioxide into the air, and for millennia our agricultural systems and land use have been slowly contributing to changes in the Earth's atmospheric composition, releasing, for example, methane and nitrogen oxides.[41] These changes long went unnoticed, but now we can no longer claim to be ignorant—overwhelming evidence indicates that humans are causing climate change on Earth through inadvertent climate engineering and geoengineering (via changing atmospheric gas levels, land use changes, deforestation, draining swamps, ocean acidification, etc.). While these planetary changes began in ignorance, they should culminate in wisdom. Our actions are damaging the environment that we and the rest of Earth's life depend on. These actions are bad and should be changed. How, then, can we restore our planet to the conditions prior to this devastation?

Climate engineering considers both **solar radiation management** (controlling the amount of sunlight warming the Earth) and atmospheric **carbon dioxide removal** (managing carbon dioxide levels and levels of other greenhouse gases). These are similar tools to those needed for terraforming, though on other planets the conditions, of course, vary considerably. Even seeking to reduce $CO_2$ output or to plant trees to uptake $CO_2$ are simple forms of terraforming—all we need to do is think of our fossil fuel–burning power plants, factories, and vehicles as small pieces of terraforming machinery (for warming the Earth) rather than as what we typically consider to be their primary use.

In the past humankind was involuntarily limited by its own weakness. Now we must learn to be voluntarily constrained by our own good judgment, our ethics.[42] Our weakness can no longer decide for us if we want to change the composition of the Earth's atmosphere—in our strength we have already begun that process and now we need to get it right. Our weakness can no longer decide for us not to drive humanity extinct through nuclear war or other technological mishaps—in our strength, the decision is now ours and we must choose wisely.

## CLOSING CASE

As with the previous chapter on settlement, this chapter on terraforming moves far beyond human experience and into the realm of fiction. In 1992 Marshall T. Savage wrote a book called *The Millennial Project: Colonizing the Galaxy in Eight Easy Steps*, which has a following among enthusiasts of both seasteading and extraterrestrial settlement.[43] While presented as a plan for the future and extensively referenced, the book should be viewed more as a vision for what the role of humanity might be in the universe. In Savage's vision, the role of humanity is to extend life to everywhere that we can. In this way his vision is similar to that of Randolph and McKay and others, who would like to expand "the richness and diversity of life" in the universe.[44]

Savage's ideas include paraterraforming domed craters on the Moon, terraforming Mars, and settling the asteroid belt by enclosing asteroids in thick water-filled bubbles as clear radiation shields and then paraterraforming the asteroids. Because there are about one million asteroids larger than 1 kilometer in diameter, this is an almost inconceivably large amount of living space.[45] But this vision raises a question: Should we make these nonliving places alive? Or is there intrinsic value in nonliving places that should make us leave them alone (for example, Richard York's version of Leopold, or Robinson's "Red" faction in his Mars trilogy[46])? How could we even come to a clear decision on such a high-stakes question? This decision will be momentous, no matter which side is chosen. But hopefully the ideas in this book might help to at least provide some tools for thinking about these topics.

## DISCUSSION AND STUDY QUESTIONS

1. Should humans seek to terraform other worlds? Why or why not?
2. What would justify the terraformation of another world? Why would this be something that humans should seek? What gives humankind legitimacy to carry out such powerful actions?
3. Do you think it might be appropriate to try to expand "the richness and diversity of life" in the universe through such means as terraforming planets? Why or why not?
4. If humankind found a planet with deteriorating habitability and extraterrestrial lifeforms struggling to survive on this planet, would it be appropriate to terraform that world for the sake of "helping" those ETLs gain a better living situation? Why or why not?
5. If a planet has any native life at all, should that make it off-limits to terraforming and human habitation? Or should humans seek to settle or terraform the planet anyway? Why or why not?
6. Of the ethical tools and concepts listed in this chapter, which do you find to be the most useful and why? Which do you find to be the least useful and why?

## FURTHER READINGS

M. M. Averner and R. D. MacElroy, eds., *On the Habitability of Mars: An Approach to Planetary Ecosynthesis* (Washington, DC: NASA Scientific and Technical Information Office, 1976).
Martin Beech, *Terraforming: The Creating of Habitable Worlds* (New York: Springer, 2009).

Arthur C. Clarke, *The Snows of Olympus: A Garden on Mars* (London: Victor Gollancz, 1995).
Martin J. Fogg, *Terraforming: Engineering Planetary Environments* (Warrendale, PA: SAE International, 1995).
Kim Stanley Robinson, *Blue Mars* (New York: Bantam Books, 1996).
Kim Stanley Robinson, *Green Mars* (New York: Bantam Books, 1994).
Kim Stanley Robinson, *Red Mars* (New York: Bantam Books, 1993).
Carl Sagan and Ann Druyan, *Pale Blue Dot: A Vision of the Human Future in Space* (New York: Ballantine Books, 1997).
Robert Sparrow, "The Ethics of Terraforming," *Environmental Ethics* 21 (1999): 227–45.
Marshall T. Savage, *The Millennial Project: Colonizing the Galaxy in Eight Easy Steps* (Boston: Little, Brown and Company, 1992).

## NOTES

1. Christopher P. McKay and Margarita M. Marinova, "The Physics, Biology, and Environmental Ethics of Making Mars Habitable," *Astrobiology* 1, no. 1 (2001): 89–109.
2. Richard L. S. Taylor, "Paraterraforming—The Worldhouse Concept," *Journal of the British Interplanetary Society* 45, no. 8 (August 1992): 341–52.
3. McKay, "Does Mars Have Rights?" 193–95.
4. Race et al., "Synthetic Biology in Space."
5. James, "For the Sake of a Stone?"
6. Cockell and Horneck, "A Planetary Park System for Mars"; and Cockell and Horneck, "Planetary Parks."
7. Randolph and McKay, "Protecting and Expanding the Richness and Diversity of Life."
8. Barbour, *Religion in an Age of Science*, 34–39.
9. Brand, ed., *Space Colonies*.
10. Marshall T. Savage, *The Millennial Project: Colonizing the Galaxy in Eight Easy Steps* (Boston: Little, Brown and Company, 1992).
11. Ibid.
12. Barbour, *Religion in an Age of Science*, 34–39.
13. James C. Scott, *Seeing Like a State: How Certain Schemes to Improve the Human Condition Have Failed* (New Haven, CT: Yale University Press, 1998)
14. Nassim Nicholas Taleb, *Skin in the Game: Hidden Asymmetries in Daily Life* (New York: Random House, 2020).
15. Joseph Henrich, *The Secret of Our Success: How Culture Is Driving Human Evolution, Domesticating Our Species, and Making Us Smarter* (Princeton, NJ: Princeton University Press, 2016), 102–4.
16. For example, Emile Durkheim, *The Elementary Forms of Religious Life* (New York: The Free Press, 1995); Durkheim, *On Suicide* (New York: Penguin Books, 2006); Roy Rappaport, *Ecology, Meaning, and Religion* (Berkeley: North Atlantic Books, 1979); Rappaport, *Ritual and Religion in the Making of Humanity* (Cambridge: Cambridge University Press, 1999); J. Stephen Lansing, *Priests and Programmers: Technologies of Power in the Engineered Landscape of Bali* (Princeton, NJ: Princeton University Press, 1991); and David Sloan Wilson, *Darwin's Cathedral: Evolution, Religion, and the Nature of Society* (Chicago: University of Chicago Press, 2002).
17. Randolph and McKay, "Protecting and Expanding the Richness and Diversity of Life."
18. Smith, "Manifest Complexity."
19. Green, "Self-Preservation Should Be Humankind's First Ethical Priority and Therefore Rapid Space Settlement Is Necessary."
20. Lupisella and Logsdon, "Do We Need a Cosmocentric Ethic?"
21. Arnould, *Icarus' Second Chance*, 183–86.
22. Robert Zubrin with Richard Wagner, *The Case for Mars: The Plan to Settle the Red Planet and Why We Must* (New York: Touchstone, 2011); and James D. Heiser, *The Myth of Mars: Imagining the Course of Human Destiny* (Malone, TX: Repristination Press, 2015).
23. Cockell, *Extra-Terrestrial Liberty*; McLean, "Reaching Out from Earth to the Stars"; Milligan, *Nobody Owns the Moon*; Schwartz, *The Value of Science in Space Exploration*; Schwartz, "The Accessible Universe"; etc.
24. Green, "Convergences in the Ethics of Space Exploration."
25. Green, "Ethical Approaches to Astrobiology and Space Exploration."
26. Francisca Cho, "Comparing Stories about the Origin, Extent, and Future of Life: An Asian Religious Perspective," in *Exploring the Origin, Extent, and Future of Life: Philosophical, Ethical, and Theological Perspectives*, ed. Constance M. Bertka (Cambridge: Cambridge University Press, 2009); James A. Dator, *Social Foundations of Human Space Exploration* (New York: Springer Science and Business Media, 2012); and John W. Traphagan, "Religion, Science, and Space Exploration from a Non-Western Perspective," *Religions* 11, no. 8 (2020): 397.
27. Vallor, *Technology and the Virtues*; Takeshi Kimura, "Masahiro Mori's Buddhist Philosophy of Robot," *Paladyn, Journal of Behavioral Robotics* 9, no. 1 (May 15, 2018); James Hughes, "Compassionate AI and Selfless Robots: A Buddhist Approach," in *Robot Ethics: The Ethical and Social Implications of Robotics*, ed. Patrick Lin, Keith Abney, and George A. Bekey (Cambridge, MA: MIT Press, 2011).

28. Haynes, "*Ecce Ecopoiesis*."
29. Mizrahi, "How to Play the 'Playing God' Card," 1445–61.
30. Brian Patrick Green, "Transhumanism and Roman Catholicism: Imagined and Real Tensions," *Theology and Science* 13, no. 2 (May 2015): 194.
31. Jonas, *The Imperative of Responsibility*, 1–12.
32. Green, "Transhumanism and Roman Catholicism," 194.
33. Nikolai Kardashev, "Transmission of Information by Extraterrestrial Civilizations," *Soviet Astronomy* 8 (1964): 217.
34. Carl Sagan, *The Cosmic Connection: An Extraterrestrial Perspective* (Cambridge: Cambridge University Press, 2000 [1973]), 234.
35. Green, "Transhumanism and Roman Catholicism," 195.
36. Sagan, *The Cosmic Connection*, 234–38.
37. John D. Barrow, *Impossibility: The Limits of Science and the Science of Limits* (Oxford: Oxford University Press, 1998), 132–33.
38. Robert Zubrin, *Entering Space: Creating a Spacefaring Civilization* (New York: Jeremy P. Tarcher/Putnam, 1999).
39. Vallor, Green, Raicu, "Ethics in Tech Practice," slide 35.
40. Lee and Ditko, "Spider-Man," exact phrase from Straczynski, *Amazing Spider-Man*.
41. Fagan, *The Long Summer*.
42. Green, "The Catholic Church and Technological Progress," 10.
43. Savage, *The Millennial Project*.
44. Randolph and McKay, "Protecting and Expanding the Richness and Diversity of Life."
45. Edward F. Tedesco and François-Xavier Desert, "The Infrared Space Observatory Deep Asteroid Search," *The Astronomical Journal* 123, no. 4 (April 2002): 2070–82.
46. York, "Toward a Martian Land Ethic"; and Robinson, *Mars Trilogy*.

*Chapter Fifteen*

# Conclusion

**Figure 15.1. "Earthrise"**

Space is the next step for ethics, and ethics is the next step for space. Ethical issues are all around us and add complexity to human life everywhere and for every culture and generation. It should therefore not be surprising that current and proposed exploration and activities in space would offer a wealth of case examples that encourage us to think outside the box in developing notions of what decisions and actions are ethically supportable. As humanity grows in power, our ethics must grow with us. Everywhere that humans go, ethics accompanies us because we are decision-making and habit-forming creatures who can either act well or badly. And we must act—we have no choice not to: the choice not to choose is a choice in itself.

As we approach the "high frontier," we should be prepared for ethical situations—some of which will be similar to those we have encountered on Earth or in our current explorations and activities, others that we can anticipate but have not yet experienced, and others that may be unpredictable, novel, and/or unusually challenging. Just as the connection of the Old World and New World in 1492 transformed so many things about our world, so too will settlement of space unleash momentous choices and risks upon humanity and broaden and change human horizons and thinking. As we go forth in discovery, whether physically or mentally, new topics and questions will inevitably arise. In the future we may come to see the various ethical systems that we practice now on Earth are just a few particular cases of a more universal ethic that we will develop in space.

Pondering these questions may not only help us plan for the future of human space exploration and use but also help our imaginations and moral deliberations here and now. Ethics is not only about actions but also about the people who perform those actions—whether they are people in the future, in the past, or living right now. We judge those who have gone before us, and those in the future will judge us as well. We should do our best to be, in the words of famed virologist Jonas Salk, "good ancestors," for though we will make mistakes, we should try to leave the best legacy to the future that we can.[1] Ethics is always under review, and so this discussion will continue.

Additionally, while space is the next step for ethics, it is not the only next step. Because of advances in technology, human power is growing and our scope of action with it. We are facing ever more new and complex ethical decisions all the time, not just in space but in many fields of emerging technology. Very often these ethical questions are not even recognized and are instead perceived as mere business decisions, choices for the sake of "national security," or other such abstract or ill-defined goals. The problem with such goals is not that they exist, for surely national security is a paramount concern for all nations on Earth, but that in their abstract ill definition they are hard to achieve, and therefore more concrete—and sometimes evil—goals tend to be substituted in their place, goals such as obtaining more destructive weapons or hoarding more wealth. And, of course, such concrete goals may actually be inimical to national security because possessing weapons makes one feared, and therefore hated, and being wealthy makes one a target for thieves.

It is precisely in the interest of achieving our wiser aims that we should not allow the self-defeating, easy to conceptualize, concrete, and unethical goals substitute for better but more general goals such as national security. Because, in fact, national security might be better achieved through international diplomacy than through arms races, and in fact overall security for the human species might be better attained that way as well.

We should not allow our narrowly scoped minds and goals to obstruct us from achieving the ethical greatness of which humanity may be capable. I say "may be" because while that choice of greatness is within our reach, we could choose otherwise, or choose it and fail to achieve it.

Ethics, ultimately, is the theory and practice of how to make better choices and thereby live a good life and be a good person. But we should not delineate the boundaries of ethics to be too small and only involve individuals. If there are enough good people living good lives, they might be able to produce an ethically focused culture and even an ethically focused government. Space exploration and settlement may give humanity this opportunity, just as previous opportunities have been made over the course of human history on Earth. In order to achieve this morally good life, not only as individuals but as an entire species, humanity—we—will need to work as hard as we ever have. We will need to coordinate across all cultures, nationalities, classes, and races in order to create a better future together.

This, then, might be our provisional goal as a species at this time in history: we need to find our purpose in the universe and, once found, live that purpose. This is also the purpose of any individual human: we need to find our purpose and live it. Ethics, as a field of study and practice, is here to help us determine what goals, among the many possible for us, are actually good. No matter the field of study—whether relationships, politics, technology, exploration, or any combination of those fields and more—the purpose of ethics is to help us make better decisions. As we go forward into the future and technology empowers us to make ever more significant choices, ethics will become ever more important. Those who understand this reality now will have an advantage in helping to make our future a better place.

## DISCUSSION AND STUDY QUESTIONS

1. Thinking of all the topics in space ethics discussed in this book, which do you think are the most important and why? Which are the least important and why?
2. Thinking of all of the ethical tools and concepts discussed in this book, which do you find to be the most useful and why? Which do you find to be the least useful and why?
3. Take a case from the book and apply an ethical tool or concept to it. What can you learn by this application?
4. Take a space ethics case from outside of this book and apply an ethical tool or concept from the book to it. What can you learn by this application?
5. This question is about human teleology: Does human existence have a purpose or goal and, if so, what? How does this relate to space? Why do you think this? How do you know?
6. This question is about universal teleology: Does the universe have a purpose or goal? What is the universe about, if anything? Is the universe doing something and, if so, what? Should humankind help the universe in its activities or not? Why do you think this? How do you know?
7. Last, this question is about personal teleology: What is your purpose in life? What gives your life meaning? Does this purpose or meaning relate to space? Why do you think this? How do you know?

## NOTE

1. Jonas Salk, "Are We Being Good Ancestors?" *World Affairs: The Journal of International Issues* 1, no. 2 (December 1992): 16–18.

# Acknowledgments

This book is the product of a lot of help, and I am deeply thankful to those who gave it. I would like to thank my editors and everyone at Rowman & Littlefield International for their incredible patience and steadfast encouragement and assistance.

I would like to thank Keith Abney, Margaret Race, and Jim Schwartz for invaluable help improving the text. A text is never perfect, always just "good enough," but their help certainly helped to raise the text toward perfection. In particular Margaret Race was there for years helping to refine the book, and her input was invaluable. As acknowledgments always say (because it is true), the remaining errors are mine.

I would be remiss to exclude all the other wonderful colleagues who work on space-related issues who helped to inspire this work, including Seth Baum, Mark Lupisella, Christopher McKay, Carlos Mariscal, Lucas Mix, Joshua Moritz, Ted Peters, Adam Pryor, Bob Russell, Kelly Smith, and many more. Margaret Race deserves special thanks, again, because she, along with Mary Ashley, first got me into space ethics, so this book would not exist without their efforts.

I would like to thank Santa Clara University for being the kind of institution that makes it possible to write a book on space ethics. I also want to thank my colleagues at the university who encouraged my interests through conversations and events, including Ahmed Amer, Erin Bradfield, Chris Kitts, and Erick Ramirez. And I certainly must thank my colleagues at the Markkula Center for Applied Ethics, especially Margaret McLean, who originally proposed writing this book. Without her this book would never have been attempted. I also want to thank Kirk Hanson, Don Heider, and Thor Wasbotten for their encouragement even when there was so much else to be done.

This book was a long labor over seven years. I would like to thank my wife and children for their gracious and graceful patience during the writing of this book through more than a few long days, even during "vacation." They also helped me talk over many ideas and found a good number of issues in the text to remedy. My daughter deserves special appreciation for reading the entire manuscript and finding numerous ways to improve it. I would also like to thank my mother and father for their steadfast encouragement of my interests ever since I was a child. Only with that kind of support would I ever think that I could write a book with a title like *Space Ethics*.

A philosopher once said that it is never possible to thank one's parents, one's teachers, or the gods enough, and that is true. So I would like to thank my teachers from grade school through graduate school. By educating future generations, teachers make every future endeavor possible. I would also like to thank the people of the Marshall Islands, who set me on the course of studying the ethics of technology. For two years I was their teacher, but they were also my teachers, and they taught me to truly appreciate the differences and experiences of others, regardless of expectations, with wisdom and forgiveness. And to complete the philosophic trifecta, and echo the Marshallese who said so at every special occasion, I would like to thank God for everything. AMDG.

Last, I would like to thank you, the reader, for spending time with these ideas and thinking about space ethics. Whether you agree with the ideas presented here or not, I invite your feedback at bpgreen@scu.edu. Philosophy is a conversation that is never done, so your comments and questions—whether directed to me or elsewhere—can help the conversation and, with respect to ethics, hopefully improve decisions and actions.

This book is dedicated to hope for the future flourishing of Earth life, and all life, in the universe and to the human choices that will get us there. While it might not seem like the ethical issues of space are that close, the future approaches every day, and therefore every day we need to do the right things that will make for a better tomorrow. Starting later is not good enough; we always have to start now.

# Glossary/Keywords

**acute technosocial opacity:** a situation in which it is increasingly difficult to predict even the near-term future due to rapid technosocial change

**additive manufacturing:** also known as 3D printing, creates objects by adding materials together rather than by subtraction (as in milling or sculpting) or by assembling parts

**altruism:** doing good to others merely for the sake of it, not for selfish motives

**artificial general intelligence (AGI):** an artificial intelligence with general applicability; typically thought of as being able to do anything that a human can do

**artificial intelligence (AI):** attempts to reproduce various aspects of natural intelligence in computerized form

**asteroid:** a small rocky or metallic nonplanetary body orbiting a star

**asteroid, carbonaceous chondrite:** silicate asteroids that are sometimes relatively high in carbon and water

**asteroid, undifferentiated, metamorphosed:** asteroids heated enough to dehydrate them and likely kill any life forms yet not heated so much that minerals separate at large scale

**astronomical unit:** the distance from the Earth to the Sun, 150 million kilometers

**autonomy:** when a person can govern their own actions, "self-rule"

**backup Earth:** the idea of duplicating the Earth's biosphere and human cultural information in another location, preferably a self-sustaining settlement on another planet, so that humanity can make itself and Earth life more likely to survive existential risks

**Belmont Principles:** the three bioethical principles of respect for persons, beneficence, and justice

**best interests:** choosing what is best on behalf of another person, regardless of how it might affect oneself or others

**bias, IIDDISEW:** "if I don't do it, someone else will," a rationalization intended to justify or excuse one's own bad behavior because it may happen in any case, so one might as well be the one to profit from it

**bias, ISEP:** "it's someone else's problem," a rationalization that excuses action because hopefully someone else will do the work

**bias, TINA:** "there is no alternative," an appeal to lack of choices, that this course of action must continue, may include a sense of disempowered fatalism

**biases, cognitive:** biases in the way human think

**biases, irrational:** do not make sense because they benefit no one

**biases, rational:** make sense to a limited extent because they benefit at least one person for some period of time

**bioethics:** expands upon medical ethics to additionally include biological research, biotechnology, genetics, the role of animals in research, and so on

**carbon dioxide removal:** managing carbon dioxide levels and levels of other greenhouse gases

**case-based analysis/casuistry:** that approach to ethics that involves comparing cases and reasoning by analogy

**centrism:** depending on the context either asks where the focus of ethical value lies or asks from what perspective ethical value should be viewed

**chemical toxicity:** toxicity caused by chemical compounds

**circular economy:** one in which waste products become the inputs for further production

**climate change:** any change in climate, though typically referring to Earth's climate due to human activity

**climate engineering and geoengineering:** technological means directed toward changing the climate or other geosystem on Earth

**collaboration:** working together with more intent and trust

**Columbian Exchange:** the process of transporting trade items, technologies, crops, diseases, peoples, etc., between the Old World and New World beginning in 1492

**comet:** an icy nonplanetary body orbiting a star

**commission:** to do an action that could have been not done; in contrast to omission that is a failure to do something that one ought to have done

**common good:** seeks to protect the community and society through protecting such shared resources and benefits as clean air and water, quality education and health care, safe and rapid transportation, and quality relationships and social institutions

**concentrated solar:** focusing or reflecting the energy of the Sun in order to concentrate that energy on one spot; with regard to asteroid or comet redirection, to try to induce material ablation on one side of the object and thereby nudge it in a new direction

**conflict of interest:** occurs when one is tasked with judging fairly but may benefit from choosing in a biased way

**consequentialism:** the approach to ethics that makes judgments based on consequences, results, outcomes, effects, etc.; utilitarianism is a type of consequentialism

**contamination, backward:** contamination from extraterrestrial objects to the Earth

**contamination, forward:** contamination from Earth to extraterrestrial objects

**cooperation:** working together for a common cause; can be collaborative (with more intent and trust) or coordinative (not necessarily with intent or trust)

**cooperation and noncooperation with evil:** a process for thinking about whether or not it is right to cooperate with evil

**cooperation, formal:** the cooperator has the same intent as the primary actor/evildoer; always wrong

**cooperation, material:** the cooperator has a different intent from the primary actor/evildoer; may be justified or unjustified but requires further investigation; there are two types: immediate and mediate

**cooperation, material, immediate:** the cooperator cooperates on the same evil act as the primary but for a different reason

**cooperation, material, mediate:** the cooperator cooperates on a different evil act from the primary and for a different reason; can be proximate or remote

**cooperation, material, mediate, proximate:** the cooperator cooperates on a different act, with a different intent, but is more morally close, and can sometimes raise ethical questions: Should you have known better?

**cooperation, material, mediate, remote:** the cooperator cooperates on a different act, with a different intent, and is more morally distant; often required otherwise society will not function

**coordination:** working in a cooperative fashion, even if not directly intentionally or with trust

**coronal mass ejection (CME):** a large release of plasma from the Sun directed into space

**cosmic rays:** atomic nuclei traveling at a significant fraction of the speed of light, which can be dangerous to human health

**cost-benefit analysis:** a form of consequentialist reasoning often employed in business that compares costs and benefits, typically in monetary terms

**culpable:** being responsible for an action of commission or omission

**cyber-physical infrastructure:** physical infrastructure that is controlled by computers such that controlling those computers might destroy the infrastructure; for example, power plants, water treatment plants, etc.

**deontology:** "the study of duty," an ethical system traced to Immanuel Kant and his prioritization of goodwill and following the categorical imperative

**deterrence:** aiming to prevent a conflict before it starts by inspiring fear in the opponent

**double-effect reasoning:** an ethical tool that helps to think through whether to act when that action has both a good effect and a bad effect

**Drake Equation:** developed by Frank Drake as a thought experiment to consider how many extraterrestrial intelligent species might exist in the galaxy

**dual-use and dual-use technology:** a power or technology that can have both good, beneficial uses and bad, harmful uses

**Earth-crossing object:** an orbital object that crosses Earth's orbit and therefore might impact with the Earth at some point

**eco-nihilists:** adherents of an ideology that says that humanity is evil and deserves to go extinct

**ecopoiesis:** in the language of terraforming, the creation of an ecosystem, or in terms of habitability, the generation of a habitable environment from an uninhabitable one

**effective altruism:** a social movement to maximize the positive impact of charities and charitable work, etc.

**electromagnetic pulse (EMP):** a wave of electromagnetic energy that can cause power surges and damages electrical infrastructure and electronics

**ends-justify-the-means reasoning:** rationalizes doing a bad action now so that a good effect may occur later

**entelechy:** a teleology (purpose or goal seeking behavior) built in to an organism or entity

**environmental control:** or "life support," consists of everything environmental that humans need in order to stay alive

**environmental ethics:** considers forms of ethical thinking and behavior that promote good human interactions with the natural environment

**environmental toxicity:** the presence of substances in the environment that may harm living things

**epigenomic:** studying the complete set of epigenetic (heritable changes in gene expression that do not involve changes to the DNA sequence) changes in an organism

**ethical intuitionism:** a form of ethics that relies on impressions, feelings, and intuitions, sometimes in opposition to more theoretical ethical approaches

**ethics:** the theory and practice of how to make good decisions, perform good actions, and thereby become a good person

**existential:** having to do with existence; in the context of philosophy, related to existentialism, a philosophical approach that emphasizes the realness of personal experience, and in the context of risk, related to risks that could drive humankind extinct

**existential risks:** risks that threaten human extinction

**extraterrestrial life (ETL):** life that originates from somewhere other than Earth

**extraterrestrial intelligence (ETI):** intelligence that originates from somewhere other than Earth

**ET artificial intelligence (ETAI)** and **ET artificial general intelligence (AGI/ETAGI):** artificial intelligences of extraterrestrial origin, including general purpose AI (AGI), which is of human level or greater

**feminist ethics:** ethics that takes the experience of female humans seriously and works toward equality and equity of the sexes

**Fermi Paradox:** the seeming paradox that with such a vast universe with so many planets, extraterrestrial intelligent life seems like it ought to be common, and yet we see no evidence of it

**final cause:** in Aristotelian terms, the purpose for the sake of which something acts or exists

**flag of convenience:** a nation without much regulatory oversight, used for registering ships, etc.

**free rider:** someone who gets a benefit without paying for it

**gain-of-function research:** research that adds functions to an organism or technology, sometimes dangerous and dual-use functions

**Galilean moons:** the four largest moons of Jupiter, visible to Galileo Galilei when he first discovered them: Io, Europa, Ganymede, and Callisto

**game theory:** investigates how cooperative and competitive interactions develop between agents under differing situations

**gamma ray burst:** an extremely energetic pulse of gamma rays from a distant galaxy

**gene drive:** the use of biased inheritance to force a gene through a population or drive it extinct

**geomagnetic storm:** waves of electromagnetic energy moving through the Earth's magnetosphere, which may damage electrical and electronic equipment on Earth or in near Earth space

**global catastrophic risks:** risks that threaten to kill large portions of the human population or devastate large areas of a planet

**gravity tractor:** the use of one or more satellites to pull an object through space via gravitational interactions

**Great Filter:** related to the Fermi Paradox, asks at what point intelligent life gets "filtered out" of the universe: before life evolves, before intelligent-like evolves, before intelligent life becomes spacefaring, etc.

**guns versus butter problem:** the question in government spending of whether to spend money on military power or helping more direct civilian interests

**Haber-Bosch process:** takes atmospheric nitrogen and transforms it into nitrogen compounds that are usable by organisms such as plants

**hermeneutic of suspicion:** a skeptical way of interpreting communication that looks for nonobvious or negative valences

**heuristic of fear:** a way of making decisions that prioritizes stopping worse outcomes rather than seizing beneficial ones

***hostis humani generis:*** enemies of humankind

**human dignity:** a concept of the human as being sacred, inviolable, and intrinsically worthy of respect and protection

**humanitarian intervention:** defending the vulnerable in another country against their oppressive government

**humility, epistemological:** recognizing that one cannot know everything and therefore must have some uncertainty about knowledge

**humility, ethical:** recognizing that ethical judgment is difficult and ethical values hard to know, and therefore allowing for some uncertainty in ethics

**ignorance, culpable:** ignorance that is one's fault (one should have known better) and for which one is therefore responsible

**ignorance, nonculpable:** ignorance that is beyond anyone's fault and for which one is therefore not responsible

**imperative of responsibility:** from Hans Jonas, the idea that humans, as the only morally capable organism that we know of, and therefore the precondition for ethical evaluation, ought to exist

***in situ* resource utilization (ISRU):** utilization of resources on site rather than relying upon importation of goods from elsewhere

**induced seismicity:** seismicity that is caused by human activities, such as filling a reservoir, pumping water underground, hydraulic fracturing of rocks for petroleum exploration, etc.

**informed consent:** the ethical idea that under normal circumstances people (1) ought to fully know the risks of participating in an activity and (2) ought to fully consent to the activity before engaging in it or being subjected to it

**intergenerational justice:** the concept that justice not only extends spatially between all humans currently alive but also extends temporally in how we treat the past and the future

**instrumental value or good:** values or goods that are useful for attaining other objectives
**internalization of economic externalities:** making users pay the full cost for their use of a resource, including long-term environmental impacts and costs that others might otherwise bear
**intersectional ethics:** ethics that takes seriously the experience of those at the intersection of multiple identities, such as sex, gender, ethnicity, race, sexual orientation, and so on
**intrinsic value or good:** values or goods that are ethically valuable in themselves
**ion beam:** an energetic beam of charged particles that delivers energy at a distance that may thereby act as a weapon or act to adjust the vectors of moving objects by ablation
**ionizing radiation:** intense forms of radiation that are capable of stripping electrons from atoms thus causing significant damage to materials including biological organisms and their DNA
***jus ad bellum:*** refers to justice in the conditions necessary for going to war
***jus in bello:*** refers to justice in the conduct of the war itself
***jus post bellum:*** refers to justice in the conditions after a war has occurred
**just war theory:** seeks to limit the barbarity of war for the sake of protecting justice and preserving human dignity and the common good
**just war theory, *jus ad bellum*, just cause:** a cause that makes a war morally worth it, for example, protecting the innocent
**just war theory, *jus ad bellum*, last resort:** because war is such an extreme deterioration of the international order, all other options short of war should be utilized prior to entering war
**just war theory, *jus ad bellum*, legitimate and competent authority:** a government that has been rightfully put in place, for example, by democratic processes, and has in mind the common good
**just war theory, *jus ad bellum*, probability of success:** the war needs to be winnable; if a war can only be lost, then such a futile action will merely cost lives rather than preserve the common good
**just war theory, *jus ad bellum*, right intention:** the intention of a war should be to promote the common good, including what is good for one's opponent, not merely selfish political, military, or personal gain
**just war theory, *jus in bello*, discrimination between combatants and noncombatants:** also known as "noncombatant immunity," states only enemy combatants who pose a threat should be intentionally targeted and civilians should never be intentionally targeted
**just war theory, *jus in bello*, exercise proportionality:** a military should not use excessive force; weapons or weapon effects that linger such as land mines, radiation, and toxins may violate not only proportionality but also discrimination, *jus post bellum*, and no *malum in se* criteria
**just war theory, *jus in bello*, military necessity:** military targets are legitimate targets, civilian targets such as power plants, bridges, and communications systems may only be targeted if the harms they do to civilians are not disproportionate
**just war theory, *jus in bello*, no means that are *malum in se:*** there should be no sexual assault, torture, mutilation, mass starvation, razing cities, genocide, or otherwise killing in an intrinsically disproportionate or indiscriminate manner (such as using nuclear and biological weapons)
**just war theory, *jus post bellum*, restore environmental peace:** seeks restoration of natural areas damaged or devastated during the war
**just war theory, *jus post bellum*, restore political peace:** seeks an international peace treaty to be mutually agreed upon by legitimate authorities on all sides of the conflict
**just war theory, *jus post bellum*, restore social peace:** this may include war crimes trials, reconstruction of civil infrastructure, and reconciliation commissions
**justice and fairness:** the idea that people should get what they are due, whether in terms of benefits or burdens, based on their actions, behaviors, and capabilities
**kinetic impactor:** high-speed and/or heavy impactor used as a weapon or to adjust the vector of another object, for example, to nudge the asteroid or comet into another orbit
**Kessler Syndrome:** a debris cascade in Earth orbit that potentially cuts off Earth from space
**Lagrange points:** relatively stationary orbital places in space where gravity balances between objects
**libertarianism:** the idea that people ought to be as free as possible, the limit of which is harm to others
**linear economy:** produces destruction at both ends as well as in the middle steps with the use of energy and transformation of materials
**locus of ethical value:** the reason for which something has ethical value, for example, intrinsic value or extrinsic (instrumental) value
**long-termism:** making ethical decisions with a longer time horizon
**low Earth orbit (LEO):** orbits around Earth that are below 2,000 kilometers
**machine learning (ML):** a method for processing data in which algorithms analyze a data set and find patterns
**magnetosphere:** a magnetic field around the Earth that diverts ionizing radiation away from the Earth and into specific areas such as the atmosphere near the poles and the Van Allen radiation belts
**Malthusian:** the idea that overpopulation will lead to starvation and disaster
***malum in se:*** meaning "evil in itself" or intrinsically evil; actions that cannot be done in a good way, only badly
**manifest destiny:** an idea from the history of the United States that asserted that the United States was destined and entitled to rule a band of territory across North America from the Atlantic to the Pacific

**metabolomic:** studying metabolic changes in an organism, including metabolites (molecules involves in metabolism)
**metaethics:** the study of fundamental questions underlying ethics, such as the definition of good
**metanarratives:** overarching concepts that organize and interpret subordinate narratives of groups, peoples, and cultures and draw ethical lessons from those narratives
**metaphysics:** the study of fundamental questions underlying the nature of reality that are not subject to scientific inquiry, such as whether purpose exists in the universe
**microbiome:** the microorganisms that live on another organism, for example, in humans, all the microbes and other life forms that live on and in us
**microgravity:** colloquially called "weightlessness," caused by being in a state of free fall while in orbit
**micrometeoroids:** tiny high-velocity meteoroids that can cause damage to equipment and personnel in space
**moral agent:** a person who has agency and is capable of making ethical decisions
**moral patient:** a person who is subject to the ethical decisions of others
**nanotechnology:** technology built or designed to operate at the molecular level
**narrative ethics:** relates stories and identity to ethics; seeks suitable courses of action for a particular life narrative
**naturalistic fallacy:** a poorly defined and widely misunderstood and misapplied philosophical dictum that says that "good" cannot be defined in natural or nonnatural terms
**nature and scope of human action and power:** the qualitative and quantitative side of human capabilities to act in the universe
**near Earth object (NEO):** asteroids and comets with perihelion distances less than 1.3 astronomical units (AU)
**noninterference:** the ethical idea of leaving something alone, neither helping nor harming it
**normal accident** or **system accident:** an accident based purely on the probabilities of failures of parts within complex systems
**nutrition:** the study of health as related to food and liquid intake
**ocean acidification:** a lowering of the pH of the ocean into a more acidic state, with concomitant negative effects on oceanic chemistry, biology, and ecology
**oceanic anoxia:** loss of oxygen in the oceans, harming oceanic life and altering planetary geochemistry
**omission, ethical:** failing to do what ought to have been done
**open-question argument:** in G. E. Moore, goodness is an indefinable intuition because it is always open to the retort "but why is *that* good?"
**Outer Space Treaty (OST):** the UN treaty governing human use of space since 1967
**panpsychism:** a worldview that believes that consciousness in some form is core to reality and pervades the universe
**panspermia:** the idea that life can move between planets seeding new biospheres as planets become habitable
**parentalism:** a gender-neutral version of such concepts as "paternalism" and "nanny state" wherein outside decision makers intervene in order to protect others, including limiting their freedom
**parentalism versus autonomy:** a tension in ethics that considers when legitimate authorities should intervene to make decisions on behalf of others, for example, by limiting freedom for safety
**physical particle danger:** health risks posed by physical particles themselves, such as being sharp, damaging mucous membranes, lodging in the lungs, etc.
**planetary protection:** the practice of controlling forward and backward contamination between objects in space
**principle of proportionate care:** states that the riskier a situation is, the more care that ought to be taken in it
**principlism:** a quick applied ethical system for medical ethics, developed by Beauchamp and Childress, including the principles of autonomy, beneficence, nonmaleficence, and justice
**probabilism:** an early modern approach to moral uncertainty that asserts that if a few expert opinions affirm that an action is acceptable, then the action may be done
**probabilism, aequiprobabilism:** asserts that if there are equal expert opinions on each side, then the action may be done
**probabilism, minus probabilissimus:** asserts that if one expert opinion affirms the course of action, then the action may be done
**probabilism, probabiliorism:** asserts that if most expert opinions affirm a course of action, then it may be done
**probabilism, tutiorism:** asserts that one ought to follow the safest (most morally conservative) option, regardless of the number of expert opinions
**proteomics:** the study of the proteome, the expressed protein production of an entire organism
**proxy consent:** giving consent for another who is unable; typically found in medical ethics literature
**psychological stress:** stress due to isolation, lack of social diversity, lack of freedom of movement, and other psychological stressors
**radiation:** the transmission of energy in the form of waves (typically electromagnetic) or particles (for example, alpha and beta particles)
**radiation exposure:** because radiation is a form of energy, it can change objects that are hit by it; in living organisms this can include genetic and tissue damage, which can lead to illness and death, therefore exposure should be measured and limited if possible
**radio telescope:** a telescope that operates in the radio frequency of the electromagnetic spectrum

**reasonable person/reasonability standard:** a standard based on what reasonable people would or should be expected to do in a situation; it places both expectations and limits on responsibility based on what a human being or group of humans are reasonably capable of foreseeing; not always a clear standard, but suitable for use in many situations

**reciprocity:** one of the most basic ethical rules, that individuals and groups ought to repay good for good; conversely, descriptively, negative might also be repaid for negative

**right, objective:** summarized by the phrase "that is right," something empirically or logically right (in the sense of being true) or objectively ethically right (that is, affirming fundamental ethical principles and values)

**right makes might:** the idea that power comes from moral legitimacy; connected with the Gandhian idea of *satyagraha*, "truth force"

**rights, civil-political:** rights such as voting, freedom of speech, exercise of religion, etc.

**rights, human:** ethical standards for behavior owed to persons simply because they are human

**rights, negative:** those needing no actions on the part of others in order to protect and enable them, the rights holder simply needs to not be interfered with

**rights, positive:** rights that require outside inputs in order to be exercised, for example, rights to safe food, drinking water, housing, health care, education, etc.

**rights, social-economic:** rights to food, drinking water, housing, health care, education, etc.

**rights, subjective:** summarized by the phrase "that is *my* right," a right that belongs to a person, such as human and/or civil rights, or rights gained by privilege of social role

**risk:** as described by Martin and Schinzinger, "the probability that something unwanted or harmful may occur"

**risk adaptation:** summarized by the phrase "managing the unavoidable," coping with those risks that are inescapable; adaptation adjusts human behavior and infrastructure to cope with problems that cannot be mitigated

**risk aversion:** the level of unwillingness to take a risk or risks

**risk-benefit analysis:** a form of cost-benefit analysis that understands that costs may be not only determinate but also probabilistic

**risk compensation** or **risk homeostasis:** the tendency to return to a set level of risk tolerance, in other words, if an activity becomes safer in one way, the person engaged in the activity might take more risks in other ways

**risk equation:** risk = harm × probability

**risk ethics:** conducting ethical analysis under conditions of uncertainty; combines risk analysis with ethical decision making

**risk mitigation:** summarized by the phrase "avoiding the unmanageable," seeks to reduce the probability that something bad will happen by reducing the likelihood of scenarios that are beyond control

**risk principle, gambler's:** an approach to risk that counsels "don't bet more than you can afford to lose"; in other words, for any elective scenario with a finite probability of an intolerable outcome, the scenario should not be chosen

**risk principle, polluter pays:** an approach to risk that permits risk-taking behavior and, if something goes wrong, requires restitution from those who created the problem

**risk principle, precautionary:** an approach to risk that prioritizes caution: if a policy or action might cause harm, even if there is still no scientific consensus, the policy or action in question should not be pursued

**risk principle, prevention:** an approach to risk that follows the general rule that "prevention is better than cure"; generally uncontroversial because cause and effect are understood

**risk principle, proactionary:** an approach to risk that seeks to advance technology as quickly as possible because the current situation is intolerable

**risk principle, prudent vigilance:** an approach to risk that does not demand an extreme aversion to all risks but continuously assesses benefits and risks

**risk tolerance:** the amount of risk that individuals or organizations are willing to accept in a certain situation

**safe/safety:** an individual and social judgment regarding what is not dangerous and what is an acceptable behavior or condition from the perspective of health and harm; individuals can have their own conceptions of safety, but no individual person can legitimately decide for everyone what safety means

**safe exits:** ways to escape when technology fails or disaster strikes; should be considered in the design process

***satyagraha:*** "truth force," a concept promoted by Mohandas Gandhi that truth has its own power and that "right makes might"

**Search for Extraterrestrial Intelligence (SETI):** using radio astronomy and other tools to search for intelligent life in the universe

**second genesis:** an origin of life not common to Earth

**second-order intentions:** *what we should intend to intend*, in other words, we should *desire to desire* the good, and we should want to want good goals

**serum:** the part of blood not including blood cells or blood components involved in clotting

**seven generation sustainability rule:** a rule attributed to the Constitution of the Iroquois Confederacy stating that one ought to be concerned for future generations, in some interpretations, to seven generations

**short- versus long-term consequences:** in consequentialist thinking, balancing short-term needs and long-term needs

**short-termism:** tends to make decisions mostly with a short time horizon

**should and ought:** have two senses, one relating to behavioral prescription—an agent ought to do something because it is the right or good thing to do, or is at least not wrong or evil—and one relating to prediction—that something ought to happen, in a future descriptive sense

**siderophilic:** elements that "love iron" and therefore over the Earth's natural history have tended to migrate toward the core; tend to be metallic, dense, and near platinum on the periodic table

**sleep disturbance:** the absence of gravity as well as lack of airflow, noise, exposure to cosmic rays (causing light flashes in closed eyes), and other phenomena generally reduce the amount of sleep that astronauts can get

**social justice:** expresses the idea that there is a responsibility of the strong to care for the weak

**societal informed consent:** the idea that, like individual informed consent, society ought to consent to risks by both having full knowledge of the situation and full consent to participate

**solar flare:** a large burst of energy from the Sun

**solar photovoltaic (PV):** panels that collect solar energy and turn it into electricity

**solar radiation management:** controlling the amount of sunlight warming the Earth

**space adaptation syndrome** or **"space sickness":** a condition related the effects of microgravity upon the vestibular system, causing the upward redistribution of fluids within the body, motion sickness, nausea, "moon face," increased eye pressure, etc., typically associated with the first few days of transition into space

**space weather:** the conditions in space including solar wind, electromagnetic conditions, radiation, particle flux, temperature, etc.

**Strategic Defense Initiative:** an initiative under U.S. president Reagan's administration to consider putting weapons or dual-use devices into Earth orbit

**subsidiarity:** a principle of governance that states that governance should take place at the most local level capable of solving the problem

**supererogatory:** doing more than what is ethically required in a situation

**supernova:** the explosion of a star

**supervolcano or flood basalt eruption:** enormous releases of lava and volcanic ash from under the surface of the earth, more than 1,000 cubic kilometers

**synthetic biology:** using powerful engineering techniques to operate directly on the genetic material of an organism and create new functions and controls, interchangeable parts, easily transferred modules, etc.

**techno-conservative** and/or **techno-pessimistic:** a perspective on technology that sees our current situation as fairly good and future changes as risking that good for something worse; optimistic about the present and pessimistic about the future

**techno-optimistic** and/or **techno-progressive:** a perspective on technology that sees our current situation as inadequate or poor, while the future is very bright, so haste is warranted; pessimistic about the present but optimistic about the future

**telomeric:** pertaining to the telomeres at the ends of chromosomes; shortened telomeres can indicate age or damage

**teleology:** from the Greek word *telos* meaning end or goal, philosophically meaning "that for the sake of which" something is done; teleology is the study of goals, purposes, ends, aiming points, meanings, significance, etc.

**teleonomic:** apparent purpose-seeking behavior that is merely programmed and not truly purposive

**terraforming:** changing another planet or extraterrestrial body into something more like the Earth

**theistic:** religions with a belief in one of more gods

**trade-off:** occurs when two or more ethical values come into conflict with each other and one or more must be compromised

**tragedy of the commons:** a situation in which an unregulated shared resource is unsustainably exploited

**triage:** under conditions of scarcity, triage allocates resources in this manner: (1) no resources are devoted to those patients who cannot be saved or problems that are not solvable, (2) no resources are devoted to those patients who will survive without intervention or problems that do not need immediate attention, and (3) all resources and attention are devoted toward those patients who can be saved with immediate attention or problems that can be solved with immediate attention

**trust:** believing that another entity will work in good faith to carry out agreed upon duties

**universalizable:** ethical rules that everyone can obey and yield a flourishing society

**utilitarianism:** a variety of consequentialism that seeks "the greatest happiness for the greatest number"

**utilitarianism, hedonistic:** utilitarianism that tries to maximize pleasure

**utilitarianism, preference:** utilitarianism that tries to maximize personal preference satisfaction

**Van Allen radiation belts:** belts of intense radiation around the Earth that make certain Earth orbits dangerous for humans and electronics

**vertical farming:** farms with plants grown in large warehouses, under LED lights, in trays stacked up to the ceiling to maximize efficiency

**virtue ethics:** the approach to ethics that concentrates more on developing good habits and dispositions of character rather than focusing on discrete actions

**weakening of the immune system:** an effect of space travel where the human immune system becomes less effective; especially noticeable upon return to Earth

**Whipple shielding:** layered and spaced shielding to mitigate the effects of hypervelocity impacts in orbit

**winter, volcanic, nuclear,** or **asteroid induced:** a cooling of the climate due to dust, ash, or soot from a disastrous event

# Bibliography

80,000 Hours. https://80000hours.org/.
Abney, Keith. "Ethics of Colonization: Arguments from Existential Risk." *Futures* 110 (2019): 60–63.
Abney, Keith A. Personal correspondence, February 10, 2021.
Abney, Keith A. "Space War and AI." *Artificial Intelligence and Global Security* (July 15, 2020).
Abney, Keith, and Patrick Lin. "Enhancing Astronauts: The Ethical, Legal and Social Implications." In *Commercial Space Exploration: Ethics, Policy and Governance*, edited by Jai Galliott, 245–57. London and New York: Routledge, 2015.
Adams, Douglas. *The Hitchhiker's Guide to the Galaxy*. London: Pan Books, 1979.
Aldrin, Buzz. "Cyclic Trajectory Concepts." SAIC presentation to the Interplanetary Rapid Transit (IRT) Study Meeting, Jet Propulsion Laboratory, Pasadena, California, October 28, 1985.
Allwood, A., D. Beaty, D. Bass, C. Conley, G. Kminek, M. Race, S. Vance, and F. Westall. "Conference Summary: Life Detection in Extraterrestrial Samples." *Astrobiology* 13, no. 2 (2013): 203–16.
Almar, I., and M. S. Race. "Discovery of Extraterrestrial Life: Development of Scales Indicative of Scientific Importance & Associated Risks." *Philosophical Transactions of the Royal Society A* 369 (2011): 679–92. http://rsta.royalsocietypublishing.org/content/369/1936/679.full.pdf+htm.
Amer, Ahmed. Personal correspondence, June 28, 2013.
Amos, Jonathan. "Hayabusa Asteroid-Sample Capsule Recovered in Outback." *BBC News*, June 14, 2010. https://www.bbc.co.uk/news/10307048.
Anderson, Douglas S. "A Military Look into Space: The Ultimate High Ground." *Army Lawyer* (November 1995).
Anscombe, G. E. M. "Modern Moral Philosophy." *Philosophy* 33 (1958).
Aquinas, Thomas. *Quaestiones Disputate de Veritate*. Translated by Robert W. Mulligan, James V. McGlynn, and Robert W. Schmidt. Html edition edited by Joseph Kenny. Chicago: Henry Regnery Company, 1952–54. Accessed December 29, 2020. https://isidore.co/aquinas/english/QDdeVer25.htm.
Aquinas, Thomas. *Summa Theologiae*. Translated by Fathers of the English Dominican Province, Complete English Edition in 5 Volumes. New York: Benziger Bros., 1947, republished by Notre Dame, IN: Ave Maria Press, 1981.
Aristotle. *History of Animals*. Translated by d'A. W. Thompson. In *The Complete Works of Aristotle: The Revised Oxford Translation*, edited by Jonathan Barnes, 774–993. Princeton, NJ: Princeton University Press, 1984.
Aristotle. *Metaphysics*. Translated by W. D. Ross. In *The Basic Works of Aristotle*, edited by Richard McKeon. New York: Random House, 1941.
Aristotle. *Nicomachean Ethics*. Translated by Terence Irwin. Second edition. Indianapolis, IN: Hackett Publishing Co., 1999.
Aristotle. *Nicomachean Ethics.* Translated by W. D. Ross. In *The Basic Works of Aristotle*, edited by Richard McKeon. New York: Random House, 1941.
Aristotle. *Politics*. Translated by Benjamin Jowett. In *The Basic Works of Aristotle*, edited by Richard McKeon. New York: Random House, 1941.
Arnould, Jacques. "The Emergence of the Ethics of Space: The Case of the French Space Agency." *Futures* 37 (2005): 245–54.
Arnould, Jacques. *Icarus' Second Chance: The Basis and Perspectives of Space Ethics.* New York: SpringerWeinNewYork, 2011.
Arnould, Jacques. "An Urgent Need to Explore Space." In *The Ethics of Space Exploration*, edited by James S. J. Schwartz and Tony Milligan. Switzerland: Springer International, 2016.
Arnould, Jacques, and Andre Debus. "An Ethical Approach to Planetary Protection." *Advances in Space Research* 42 (2008): 1089–95.

Arrhenius, Gustaf. "Astrobiology at NASA: Life in the Universe." Astrobiology at NASA, 2021. Accessed January 6, 2021. https://astrobiology.nasa.gov/.

Arrhenius, Gustaf. "An Impossibility Theorem for Welfarist Axiologies." *Economics and Philosophy* 16 (2000): 247–66.

Aven, Terje. "On the Ethical Justification for the Use of Risk Acceptance Criteria." *Risk Analysis* 27, no. 2 (2007): 303–12.

Averner, M. M., and R. D. MacElroy, eds. *On the Habitability of Mars: An Approach to Planetary Ecosynthesis.* Washington, DC: NASA Scientific and Technical Information Office, 1976.

Baker, Joanne. "The Falcon Has Landed." *Science* 312, no. 5778 (June 2, 2006): 1327. https://science.sciencemag.org/content/312/5778/1327.

Barbour, Ian. *Religion in an Age of Science: The Gifford Lectures*, Volume I. New York: Harper Collins, 1990.

Barbree, Jay. "Chapter 5: An Eternity of Descent: Evidence Hints That Astronauts Were Alive During Fall." *NBC News*, January 1997. Accessed April 16, 2020. http://www.nbcnews.com/id/3078062#.XpiDw_1Kjcs.

Barrow, John D. *Impossibility: The Limits of Science and the Science of Limits.* Oxford: Oxford University Press, 1998.

Baum, Seth. "The Ethics of Outer Space: A Consequentialist Perspective." In *The Ethics of Space Exploration*, edited by James S. J. Schwartz and Tony Milligan, 109–23. Switzerland: Springer, 2016.

Baum, Seth D. "Viewpoint: Cost–Benefit Analysis of Space Exploration: Some Ethical Considerations." *Space Policy* 25 (2009): 75–80.

Baum, Seth D., Jacob D. Haqq-Misra, and Shawn D. Domagal-Goldman. "Would Contact with Extraterrestrials Benefit or Harm Humanity? A Scenario Analysis." *Acta Astronautica* 68, nos. 11–12 (2011): 2114–29.

Beauchamp, Tom L., and James F. Childress. *Principles of Biomedical Ethics.* Fifth edition. New York: Oxford University Press, 2001.

Beech, Martin. *Terraforming: The Creating of Habitable Worlds.* New York: Springer, 2009.

Bekoff, Mark, and Jessica Pierce. *Wild Justice: The Moral Lives of Animals*. Chicago: University of Chicago Press, 2009.

Bell, Trudy E. "Preventing 'Sick' Spaceships." *NASA Science: Science News*, May 11, 2007. http://science.nasa.gov/science-news/science-at-nasa/2007/11may_locad3/.

Benatar, David. *Better Never to Have Been: The Harm of Coming into Existence.* Oxford: Oxford University Press, 2006.

Bentham, Jeremy. *A Fragment on Government.* London, 1776. Accessed August 27, 2014. http://www.constitution.org/jb/frag_gov.htm.

Berger, Eric. "Inside Elon Musk's Plan to Build One Starship a Week—and Settle Mars." *Ars Technica*, March 5, 2020. https://arstechnica.com/science/2020/03/inside-elon-musks-plan-to-build-one-starship-a-week-and-settle-mars/.

Berkman, John, and Brian Green. "Lecture 19: Cooperation with and Appropriation of Evil." Fundamental Moral Theology, Dominican School of Philosophy and Theology, Spring 2008.

Berry, Wendell. "Comments on O'Neill's Space Colonies." In *Space Colonies*, edited by Stewart Brand, 36–37. San Francisco, CA: Waller Press, 1977.

Berry, Wendell. "The Debate Sharpens." In *Space Colonies*, edited by Stewart Brand, 82–85. San Francisco, CA: Waller Press, 1977.

Berry, Wendell. "How to Be a Poet (to Remind Myself)." In *Given: Poems.* Berkeley, CA: Counterpoint LLC, 2005.

Bertka, Constance M. *Exploring the Origin, Extent, and Future of Life: Philosophical, Ethical, and Theological Perspectives.* Cambridge: Cambridge University Press, 2009.

Billings, Linda. "Overview: Ideology, Advocacy, and Spaceflight—Evolution of a Cultural Narrative." In *Societal Impact of Spaceflight*, edited by Stephen J. Dick and Roger D. Launius, 483–99. Washington, DC: NASA, 2007.

Bostrom, Nick. "Astronomical Waste: The Opportunity Cost of Delayed Technological Development." *Utilitas* 15, no. 3 (2003): 308–14.

Bostrom, Nick. "Existential Risk Prevention as Global Priority." *Global Policy* 4 (February 2013): 15.

Bostrom, Nick. "Existential Risks: Analyzing Human Extinction Scenarios and Related Hazards." *Journal of Evolution and Technology* 9 (March 2002).

Bostrom, Nick. "Infinite Ethics." *Analysis and Metaphysics* 10 (2011): 9–59.

Bostrom, Nick. *Superintelligence: Paths, Dangers, Strategies.* Oxford: Oxford University Press, 2014.

Bostrom, Nick, and Milan M. Circovic. "Introduction." In *Global Catastrophic Risks*, edited by Nick Bostrom and Milan M. Circovic. Oxford: Oxford University Press, 2008.

Boteler, D. H. "A 21st Century View of the March 1989 Magnetic Storm." *Space Weather* 17, no. 10 (2019): 1427–41. https://agupubs.onlinelibrary.wiley.com/doi/full/10.1029/2019SW002278.

Braithwaite, Richard Bevan. *Theory of Games as a Tool for the Moral Philosopher.* Cambridge: Cambridge University Press, 1955.

Brand, Stewart, ed. *Space Colonies.* San Francisco, CA: Waller Press, 1977.

Braverman, Irus. "Robotic Life in the Deep Blue Sea." In *Blue Legalities: The Life and Laws of the Sea*, edited by Irus Braverman and Elizabeth R. Johnson, 147–64. Durham, NC: Duke University Press, 2020.

Breakthrough Starshot. Accessed January 2, 2021. https://breakthroughinitiatives.org/initiative/3.

Brecht, Bertholt. "The Threepenny Opera." 1928.

Brennan, Pat. "Exoplanet Exploration: Planets Beyond Our Solar System." Exoplanet Exploration Program and the Jet Propulsion Laboratory for NASA's Astrophysics Division. January 2, 2021. https://exoplanets.nasa.gov/.

Bridges, James. *Colossus: The Forbin Project*. Universal City, CA: Universal Pictures, 1970.

Brin, David. "The 'Barn Door' Argument, The Precautionary Principle, and METI as 'Prayer'—An Appraisal of the Top Three Rationalizations for 'Active SETI.'" *Theology and Science* 17, no. 1 (February 2019): 16–28.

Brown, Mike. "SpaceX Mars City: Legal Experts Respond to 'Gibberish' Free Planet Claim." *Inverse*, November 3, 2020. https://www.inverse.com/innovation/spacex-mars-city-legal.

Bruger, Steven J. "Not Ready for the First Space War: What about the Second?" *Naval War College Review* 48, no. 1 (Winter 1995): 82.

Cameron, James. *Avatar*. Century City, Los Angeles, CA: 20th Century Fox, 2009.

Cameron, James, David Giler, and Walter Hill. *Aliens*. Century City, Los Angeles, CA: 20th Century Fox, 1986.

Carrigan, Richard A., Jr. "Do Potential SETI Signals Need to Be Decontaminated?" *Acta Astronautica* 58, no. 2 (January 2006): 112–17.

Carrigan, Richard A. "The Ultimate Hacker: SETI Signals May Need to Be Decontaminated." In *Symposium—International Astronomical Union* 213 (2004): 519–22.

Casadevall, Arturo, and Michael J. Imperiale. "Risks and Benefits of Gain-of-Function Experiments with Pathogens of Pandemic Potential, Such as Influenza Virus: A Call for a Science-Based Discussion." *American Society for Microbiology* 5, no. 4 (July/August 2014): e01730–14.

Casler, Krista, and Deborah Kelemen. "Developmental Continuity in Teleo-Functional Explanation: Reasoning about Nature Among Romanian Romani Adults." *Journal of Cognition and Development* 9 (2008): 340–62.

Cassidy, William A. *Meteorites, Ice, and Antarctica: A Personal Account.* Cambridge: Cambridge University Press, 2003.

Center for Applied Rationality (CFAR). https://www.rationality.org/.

Centre for the Study of Existential Risk (CSER). https://www.cser.ac.uk/.

Chamberlin, John R. "Ethics and Game Theory." *Ethics and International Affairs* 3 (1989): 261–76.

Chayes, Abram, and Antonia Handler Chayes. "On Compliance." *International Organization* 47, no. 2 (Spring 1993): 175–205.

Chayes, Abram, and Antonia Handler Chayes. "Compliance Without Enforcement: State Behavior under Regulatory Treaties." *Negotiation Journal* 7 (1991): 311–30.

Chayes, Abram, Antonia Handler Chayes, and Ronald B. Mitchell. "Managing Compliance: A Comparative Perspective." In *Engaging Countries: Strengthening Compliance with International Environmental Accords*, edited by Edith Brown Weiss and Harold Jacobson, 39–62. Cambridge, MA: MIT Press, 1998.

Chesterton, G. K. *Illustrated London News*, April 19, 1924.

Childers, Tim. "Elon Musk Says We Need to Live in Glass Domes Before We Can Terraform Mars." *Popular Mechanics*, November 20, 2020. https://www.popularmechanics.com/space/moon-mars/a34738932/elon-musk-glass-domes-terraforming-mars/.

Childress, James F. "Moral Norms in Practical Ethical Reflection." In *Christian Ethics: Problems and Prospects*, edited by Lisa Sowle Cahill and James F. Childress, 196–217. Cleveland: Pilgrim Press, 1996.

"Chinese ASAT Test." *Center for Space Standards & Innovation*, December 5, 2007. http://www.centerforspace.com/asat/.

Cho, Francisca. "Comparing Stories about the Origin, Extent, and Future of Life: An Asian Religious Perspective." In *Exploring the Origin, Extent, and Future of Life: Philosophical, Ethical, and Theological Perspectives*, edited by Constance M. Bertka, 303–20. Cambridge: Cambridge University Press, 2009.

Chodas, Paul. "Hypothetic Impact Scenarios." Center for Near Earth Object Studies (CNEOS), NASA Jet Propulsion Laboratory. https://cneos.jpl.nasa.gov/pd/cs/.

Chodas, Paul. "Overview of the 2019 Planetary Defense Conference Asteroid Impact Exercise." *EPSC Abstracts* 13, EPSC-DPS Joint Meeting, Geneva, Switzerland, September 15–20, 2019.

Chodas, Paul. "Planetary Defense Conference Exercise—2019." Center for Near Earth Object Studies (CNEOS), NASA Jet Propulsion Laboratory. https://cneos.jpl.nasa.gov/pd/cs/pdc19/.

Chui, Michael, Martin Harryson, James Manyika, Roger Roberts, Rita Chung, Ashley van Heteren, and Pieter Nel. "Notes from the AI Frontier: Applying AI for Social Good." McKinsey and Company, December 2018. https://www.mckinsey.com/~/media/mckinsey/featured%20insights/artificial%20intelligence/applying%20artificial%20intelligence%20for%20social%20good/mgi-applying-ai-for-social-good-discussion-paper-dec-2018.pdf.

Clarke, Arthur C. *The Snows of Olympus: A Garden on Mars.* London: Victor Gollancz, 1995.

Clinton, William J. "President Clinton Statement Regarding Mars Meteorite Discovery." The White House: Office of the Press Secretary, August 7, 1996. https://www2.jpl.nasa.gov/snc/clinton.html.

Cocconi, Giuseppe, and Philip Morrison. "Searching for Interstellar Communications." *Nature* 184 (September 19, 1959): 844–46.

Cockell, Charles S. "Liberty and the Limits to the Extraterrestrial State." *Journal of the British Interplanetary Society* 62 (2009): 139–57.

Cockell, Charles S. "Essay on the Causes and Consequences of Extraterrestrial Tyranny." *Journal of the British Interplanetary Society* 63 (2010): 15–37.

Cockell, Charles S. "The Ethical Status of Microbial Life on Earth and Elsewhere: In Defense of Intrinsic Value." In *The Ethics of Space Exploration*, edited by James S. J. Schwartz and Tony Milligan, 167–80. Switzerland: Springer International, 2016.

Cockell, Charles S. "Ethics and Extraterrestrial Life." In *Humans in Outer Space—Interdisciplinary Perspectives*, edited by Ulrike Landfester, Nina-Louisa Remuss, Kai-Uwe Schrogl, and Jean-Claude Worms, 80–101. New York: SpringerWienNewYork, 2011.

Cockell, Charles S. *Extra-Terrestrial Liberty: An Inquiry into the Nature and Causes of Tyrannical Government Beyond the Earth.* Edinburgh: Shoving Leopard, 2013.

Cockell, Charles, and Gerda Horneck. "A Planetary Park System for Mars." *Space Policy* 20, no. 4 (November 2004): 291–95.

Cockell, Charles S., and Gerda Horneck. "Planetary Parks—Formulating a Wilderness Policy for Planetary Bodies." *Space Policy* 22, no. 4 (November 2006): 256–61.

Consolmagno, Guy. *Brother Astronomer: Adventures of a Vatican Scientist.* New York: McGraw Hill, 2000.

"The Constitution of the Iroquois Nations." IndigenousPeople.net, August 5, 2016. http://www.indigenouspeople.net/iroqcon.htm.

Covault, Craig. "Chinese Test Anti-Satellite Weapon." *Aviation Week & Space Technology*, January 17, 2007. https://web.archive.org/web/20070128075259/http://www.aviationweek.com/aw/generic/story_channel.jsp?channel=space&id=news%2FCHI01177.xml.

Conway Morris, Simon. *Life's Solution: Inevitable Humans in a Lonely Universe.* Cambridge: Cambridge University Press, 2003.

Craig, Paul P., and John A. Jungerman. *Nuclear Arms Race: Technology and Society*. Second edition. New York: McGraw-Hill, 1990.

Crenshaw, Kimberlé. "Demarginalizing the Intersection of Race and Sex: A Black Feminist Critique of Antidiscrimination Doctrine, Feminist Theory and Antiracist Politics." *University of Chicago Legal Forum* (1989): 139–67.

Crenson, Matt. "After 10 Years, Few Believe Life on Mars." *USA Today*, August 6, 2006. https://usatoday30.usatoday.com/tech/science/space/2006-08-06-mars-life_x.htm.

Criswell, M. E., M. S. Race, J. D. Rummel, and A. Baker, eds. "Planetary Protection Issues in the Human Exploration of Mars, Pingree Park Final Workshop Report." NASA/CP-2005-213461. Mountain View, CA: NASA Ames Research Center, 2005.

Crosby, Alfred W. *The Columbian Exchange: Biological and Cultural Consequences of 1492.* Westport, CT: Greenwood Press, 1972.

Crowl, Adam, John Hunt, and Andreas Hein. "Embryo Space Colonisation to Overcome the Interstellar Time Distance Bottleneck." *Journal of the British Interplanetary Society* 65 (2012): 283–85.

Cucinotta, Francis A., Nobuyuki Hamada, and Mark P. Little. "No Evidence for an Increase in Circulatory Disease Mortality in Astronauts following Space Radiation Exposures." *Life Sciences in Space Research* 10 (August 2016): 53–56.

Dator, James A. *Social Foundations of Human Space Exploration.* New York: Springer Science and Business Media, 2012.

David, Leonard. "Russian Fireball Explosion Shows Meteor Risk Greater Than Thought." Space.com, November 1, 2013. https://www.space.com/23423-russian-fireball-meteor-airburst-risk.html.

Davies, Richard J., Maria Brumm, Michael Manga, Rudi Rubiandini, Richard Swarbrick, and Mark Tingay. "The East Java Mud Volcano (2006 to Present): An Earthquake or Drilling Trigger?" *Earth and Planetary Science Letters* 272, no. 3–4 (August 15, 2008): 627–38.

Davis, Michael. "Three Nuclear Disasters and a Hurricane." *Journal of Applied Ethics and Philosophy* 4 (August 2012): 8.

de Lazari-Radek, Katarzyna, and Peter Singer. *Utilitarianism: A Very Short Introduction.* Oxford: Oxford University Press, 2017.

Delp, Michael D., Jacqueline M. Charvat, Charles L. Limoli, Ruth K. Globus, and Payal Ghosh. "Apollo Lunar Astronauts Show Higher Cardiovascular Disease Mortality: Possible Deep Space Radiation Effects on the Vascular Endothelium." *Scientific Reports* 6 (July 28, 2016).

Denkenberger, David, and Joshua M. Pearce. *Feeding Everyone No Matter What: Managing Food Security after Global Catastrophe.* London: Elsevier, 2015.

Denkenberger, David C., and Robert W. Blair Jr. "Interventions That May Prevent or Mollify Supervolcanic Eruptions." *Futures* 102 (September 2018): 51–62.

Deudney, Dan. *Dark Skies: Space Expansionism, Planetary Geopolitics, and the Ends of Humanity.* New York: Oxford University Press, 2020.

de Vera, Jean-Pierre, Diedrich Mohlmann, Frederike Butina, Andreas Lorek, Roland Wernecke, and Sieglinde Ott. "Survival Potential and Photosynthetic Activity of Lichens Under Mars-Like Conditions: A Laboratory Study." *Astrobiology* 10, no. 2 (March 2010): 215–27.

Diamond, Jared. "Ch. 14: Why Do Some Societies Make Disastrous Decisions?" In *Collapse: How Societies Choose to Fail or Succeed*, 419–40. New York: Penguin Books, 2005.

Dick, Steven J. "Back to the Future: SETI Before the Space Age." *The Planetary Report* 15, no. 1 (1995): 4–7.

Donnelly, Jack. *Universal Human Rights in Theory and Practice*. Second edition. Ithaca, NY: Cornell University Press, 2003.

Doorn, Neelke. “The Blind Spot in Risk Ethics: Managing Natural Hazards.” *Risk Analysis* 35, no. 3 (2015).
“The Drake Equation.” SETI Institute, 2020. Accessed January 2, 2020. https://www.seti.org/drake-equation-index.
Dreier, Casey. “Reconstructing the Cost of the One Giant Leap: How Much Did Apollo Cost?” The Planetary Society, June 16, 2019. https://www.planetary.org/blogs/casey-dreier/2019/reconstructing-the-price-of-apollo.html.
Drengson, Alan. *The Deep Ecology Movement.* New York: North Atlantic Books, 1995.
Drexler, K. Eric. *The Engines of Creation: The Coming Era of Nanotechnology*. New York: Anchor Press Doubleday, 1986.
Durante, Marco. “Space Radiation Protection: Destination Mars.” *Life Sciences in Space Research* 1 (April 2014): 2–9. https://www.sciencedirect.com/science/article/abs/pii/S2214552414000042.
Durkheim, Emile. *The Elementary Forms of Religious Life.* New York: The Free Press, 1995.
Durkheim, Emile. *On Suicide.* New York: Penguin Books, 2006.
Eckersley, Peter. “Impossibility and Uncertainty Theorems in AI Value Alignment or Why Your AGI Should Not Have a Utility Function.” arXiv.org, 2019. Accessed December 29, 2020. https://arxiv.org/abs/1901.00064.
Edelstein, Karen S. “Orbital Impacts and the Space Shuttle Windshield.” *Proceedings of SPIE 2483 Space Environmental, Legal, and Safety Issues*, Symposium on OE/Aerospace Sensing and Dual Use Photonics, Orlando, Florida, June 23, 1995.
Editors. “Our Manifest Destiny Is to Move Beyond Earth.” *Financial Times*, December 23, 2014. https://www.ft.com/content/56e28fda-8447-11e4-bae9-00144feabdc0.
Effective Altruism. https://www.effectivealtruism.org/.
Ehrenfreund, P., M. S. Race, and D. Labdon. “Responsible Space Exploration and Use: Balancing Stakeholders Interests.” *New Space Journal* 1, no. 2 (2013): 60–72.
Eisenhower, Dwight D. “Chance for Peace.” April 16, 1953. https://millercenter.org/the-presidency/presidential-speeches/april-16-1953-chance-peace.
Ellsberg, Daniel. *The Doomsday Machine: Confessions of a Nuclear War Planner.* New York: Bloomsbury, 2017.
Ersdal, G., and T. Aven. “Risk Informed Decision-Making and Its Ethical Basis.” *Reliability Engineering and System Safety* 93, no. 2 (2008): 197–205.
Evans, J. H. *Playing God? Human Genetic Engineering and the Rationalization of Public Bioethical Debate.* Chicago: University of Chicago Press, 2002.
Faden, Ruth R., and Tom L. Beauchamp. *A History and Theory of Informed Consent.* New York: Oxford University Press, 1986.
Fagan, Brian. *The Long Summer: How Climate Changed Civilization.* New York: Basic Books, 2004.
Flannery, Kevin L., SJ. *Cooperation with Evil: Thomistic Tools of Analysis.* Washington, DC: Catholic University of America Press, 2019.
Fogg, Martyn J. “The Ethical Dimensions of Space Settlement.” *Space Policy* 16 (2000): 205–11.
Fogg, Martin J. *Terraforming: Engineering Planetary Environments.* Warrendale, PA: SAE International, 1995.
Foley, Duncan K. “Recent Developments in the Labor Theory of Value.” *Review of Radical Political Economics* 32, no. 1 (March 2000): 1–39.
Foot, Philippa. *Natural Goodness.* Oxford: Clarendon, 2001.
Foot, Philippa. “The Problem of Abortion and the Doctrine of the Double Effect.” *Virtues and Vices and Other Essays in Moral Philosophy* 19 (1978).
Forden, Geoffrey. “How China Loses the Coming Space War (Pt. 1).” *Wired*, January 10, 2008. https://www.wired.com/2008/01/inside-the-chin/.
Frankena, William. “The Naturalistic Fallacy.” In *Perspectives on Morality: Essays of William Frankena*, edited by Kenneth E. Goodpaster. Notre Dame, IN: University of Notre Dame Press, 1976.
Frasz, Geoffrey B. “Environmental Virtue Ethics: A New Direction for Environmental Ethics.” *Environmental Ethics* 15, no. 3 (Fall 1993): 259–74.
Freud, Sigmund. *The Future of an Illusion*. Edited by Todd Dufresne, translated by Gregory C. Richter. Peterborough, Canada: Broadview Press, 2012.
Friedman, David D. *The Machinery of Freedom: Guide to a Radical Capitalism*. Third edition. Chicago: Open Court, 2014.
Future of Humanity Institute (FHI). https://www.fhi.ox.ac.uk/.
Future of Life Institute (FLI). https://futureoflife.org/.
Galliott, Jai, ed. *Commercial Space Exploration: Ethics, Policy and Governance.* Emerging Technologies, Ethics and International Affairs. London and New York: Routledge, 2015.
Galton, David J. “Greek Theories on Eugenics.” *Journal of Medical Ethics* 24 (1998): 263–67.
Garrett-Bakelman, Francine E., et al. “The NASA Twins Study: A Multidimensional Analysis of a Year-Long Human Spaceflight.” *Science* 364 (April 12, 2019): 1. https://science.sciencemag.org/content/364/6436/eaau8650.full.
Gladwell, Malcolm. “Blowup.” *The New Yorker*, January 22, 1996. http://gladwell.com/blowup/.
Global Catastrophic Risk Institute (GCRI). https://gcrinstitute.org/.
Green, Brian Patrick. “Are Science, Technology, and Engineering Now the Most Important Subjects for Ethics? Our Need to Respond.” Paper presented at the 2014 IEEE International Symposium on Ethics in Engineering, Science, and Technology, Chicago, Illinois, *IEEE Xplore*, May 23–24, 2014.

Green, Brian Patrick. "Artificial Intelligence, Decision-Making, and Moral Deskilling." Markkula Center for Applied Ethics, March 15, 2019. https://www.scu.edu/ethics/focus-areas/technology-ethics/resources/artificial-intelligence-decision-making-and-moral-deskilling/.

Green, Brian Patrick. "Artificial Intelligence and Ethics: Sixteen Challenges and Opportunities." Markkula Center for Applied Ethics, August 18, 2020. https://www.scu.edu/ethics/all-about-ethics/artificial-intelligence-and-ethics-sixteen-challenges-and-opportunities/.

Green, Brian Patrick. "Astrobiology, Theology, and Ethics." In *Anticipating God's New Creation: Essays in Honor of Ted Peters*, edited by Carol R. Jacobson and Adam W. Pryor, 339–50. Minneapolis, MN: Lutheran University Press, 2015.

Green, Brian Patrick. "The Catholic Church and Technological Progress: Past, Present, and Future." *Religions* 8 (June 2017): 10. https://www.mdpi.com/2077-1444/8/6/106.

Green, Brian Patrick. "Constructing a Space Ethics upon Natural Law Ethics." In *Astrobiology: Science, Ethics, and Public Policy*, edited by Octavio A. Chon Torres, Ted Peters, Joseph Seckbach, and Richard Gordon. Hoboken, NJ: Wiley, 2021.

Green, Brian Patrick. "Convergences in the Ethics of Space Exploration." In *Astrobiology: The Social and Conceptual Issues*, edited by Kelly C. Smith and Carlos Mariscal, 179–96. Oxford: Oxford University Press, 2020.

Green, Brian Patrick. "Emerging Technologies, Catastrophic Risks, and Ethics: Three Strategies for Reducing Risk." IEEE International Symposium on Ethics in Engineering, Science, and Technology, *ETHICS 2016 Symposium Record*, *IEEE Xplore*, Vancouver, British Columbia, May 13–14, 2016. http://ieeexplore.ieee.org/abstract/document/7560046/.

Green, Brian Patrick. "Ethical Approaches to Astrobiology and Space Exploration: Comparing Kant, Mill, and Aristotle." Special Issue "Space Exploration and ET: Who Goes There?" *Ethics: Contemporary Issues*, 2, no. 1 (2014): 29–44.

Green, Brian Patrick. "Ethical Reflections on Artificial Intelligence." *Scientia et Fides* 6, no. 2 (2018). https://apcz.umk.pl/czasopisma/index.php/SetF/article/view/SetF.2018.015.

Green, Brian Patrick. "Ethics Is More Important Than Technology." Markkula Center for Applied Ethics, August 10, 2020. https://www.scu.edu/ethics/all-about-ethics/ethics-is-more-important-than-technology/.

Green, Brian Patrick. *The Is-Ought Problem and Catholic Natural Law*. Dissertation, Graduate Theological Union, Berkeley, California, 2013.

Green, Brian Patrick. "Little Prevention, Less Cure: Synthetic Biology, Existential Risk, and Ethics." Workshop on the Research Agendas in the Societal Aspects of Synthetic Biology, Arizona State University. https://cns.asu.edu/sites/default/files/greenp_synbiopaper_2014.pdf.

Green, Brian Patrick. "Self-Preservation Should Be Humankind's First Ethical Priority and Therefore Rapid Space Settlement Is Necessary." *Futures* 110 (June 2019): 35–37.

Green, Brian Patrick. "Should Christians Explore Space?" The Moral Mindfield, 2016. https://moralmindfield.wordpress.com/2012/01/05/should-christians-care-about-space-exploration/.

Green, Brian Patrick. "Six Approaches to Making Ethical Decisions in Cases of Uncertainty and Risk: The Principles of Prevention, Precaution, Prudent Vigilance, Polluter Pays, Gambler's, and Proaction." Markkula Center for Applied Ethics, November 14, 2019. https://www.scu.edu/ethics/focus-areas/technology-ethics/resources/six-approaches-to-making-ethical-decisions-in-cases-of-uncertainty-and-risk/.

Green, Brian Patrick. "The Technology of Holiness: A Response to Hava Tirosh-Samuelson." *Theology and Science* 16, no. 2 (2018): 223–28.

Green, Brian Patrick. "Transhumanism and Catholic Natural Law: Changing Human Nature and Changing Moral Norms." In *Religion and Transhumanism: The Unknown Future of Human Enhancement*, edited by Calvin Mercer and Tracy Trothen, 202–15. Santa Barbara, CA: Praeger, 2015.

Green, Brian Patrick. "Transhumanism and Roman Catholicism: Imagined and Real Tensions." *Theology and Science* 13, no. 2 (May 2015): 187–201.

Green, Brian Patrick, with Irina Raicu. "A Template for Technology Ethics Case Studies." Markkula Center for Applied Ethics, March 5, 2019. https://www.scu.edu/ethics/focus-areas/technology-ethics/a-template-for-technology-ethics-case-studies/.

Grier, Peter. "The Flying Tomato Can." *Air Force Magazine*, February 2009, 66–68.

Grossman, Lisa. "Ambitious Mars Joy-Ride Cannot Succeed without NASA." *New Scientist*, November 21, 2013. https://www.newscientist.com/article/dn24633-ambitious-mars-joy-ride-cannot-succeed-without-nasa/.

Gustafson, James M. *Ethics from a Theocentric Perspective: Theology and Ethics*. Chicago: University of Chicago Press, 1983.

Haines, Michael R., and Richard H. Steckel, eds. *A Population History of North America*. Cambridge: Cambridge University Press, 2000.

Hanson, Robin. "The Great Filter—Are We Almost Past It?" Personal website, September 15, 1998. http://mason.gmu.edu/~rhanson/greatfilter.html.

Hardin, Garrett. "The Tragedy of the Commons." *Science* 162, no. 3859 (December 13, 1968): 1243–48.

Hargrove, Eugene C., ed. *Beyond Spaceship Earth: Environmental Ethics and the Solar System*. San Francisco: Sierra Club Books, 1986.

Harsanyi, John C. "Morality and the Theory of Rational Behavior." *Social Research* 44, no. 4 (Winter 1977): 623–56.

Hasan, Zubair. "Labour as a Source of Value and Capital Formation: Ibn Khaldun, Ricardo, and Marx—A Comparison." *Journal of King Abdulaziz University: Islamic Economics* 20, no. 2 (2007): 39–50.

Haynes, Robert H. "*Ecce Ecopoiesis*: Playing God on Mars." In *Moral Expertise*, edited by D. MacNiven, 161–83. New York: Routledge, 1990.

Heiser, James D. *The Myth of Mars: Imagining the Course of Human Destiny.* Malone, TX: Repristination Press, 2015.

Henrich, Joseph. *The Secret of Our Success: How Culture Is Driving Human Evolution, Domesticating Our Species, and Making Us Smarter.* Princeton, NJ: Princeton University Press, 2016.

"Historical Information 'Satellite Fighter'—Program." Army.lv, October 9, 2016 (in Russian). https://web.archive.org/web/20161009144831/http://www.army.lv/ru/istrebitel-sputnikov/istorija/894/342.

Hoffman, Ross N. "Controlling Hurricanes." *Scientific American* 291, no. 4 (October 2004): 68–75.

Hogan, J. A., M. S. Race, J. W. Fisher, J. A. Joshi, and J. D. Rummel. "Life Support and Habitation and Planetary Protection Workshop, Final Report." NASA/TM-2006-213485. Moffett Field, CA: NASA Ames Res. Ctr. 2006.

Horneck, Gerda, Ralf Moeller, Jean Cadet, et al. "Resistance of Bacterial Endospores to Outer Space for Planetary Protection Purposes—Experiment PROTECT of the EXPOSE-E Mission." *Astrobiology* 12, no. 5 (May 2012): 445–56.

Hughes, James. "Compassionate AI and Selfless Robots: A Buddhist Approach." In *Robot Ethics: The Ethical and Social Implications of Robotics*, edited by Patrick Lin, Keith Abney, and George A. Bekey. Cambridge, MA: MIT Press, 2011.

Hunt, Gavin R. "Manufacture and Use of Hook-Tools by New Caledonian Crows." *Nature* 379 (January 18, 1996): 249–51.

Hunt, John. "The EGR Mission—Rationale and Design of the First True Interstellar Mission: A Crazy Presentation for the PI Club, Appendix 1: Minimizing the Ethical Concerns of an EGR Mission." Peregrinus Intersteller: PI Club. Accessed August 5, 2015. http://www.peregrinus-interstellar.net/images/Files/Crazy_Ideas/EGR/egr_appendix_1_ethics.pdf.

Hursthouse, Rosalind. "Environmental Virtue Ethics." In *Working Virtue: Virtue Ethics and Contemporary Moral Problems*, edited by Rebecca L. Walker and Philip J. Ivanhoe, 155–72. Oxford: Clarendon, 2007.

Inspiration Mars. Accessed January 3, 2021. https://web.archive.org/web/20151013012955/http://inspirationmars.org/.

Inter-Agency Space Debris Coordination Committee. "Report of the IADC Activities on Space Debris Mitigation Measures." Presented to the 41st Session of the Scientific and Technical Subcommittee of the United Nations Committee on the Peaceful Uses of Outer Space, 2004. https://web.archive.org/web/20150402103645/http://www.iadc-online.org/Documents/IADC-UNCOPUOS-final.pdf.

Inter-Agency Space Debris Coordination Committee, Steering Group and Working Group 4. "IADC Space Debris Mitigation Guidelines." IADC Action Item number 22.4, September 2007.

Ipsos. "Majority of Americans Believe There Is Intelligent Life and Civilizations on Other Planets." Ipsos Press Release, Washington, DC, January 28, 2020. https://www.ipsos.com/sites/default/files/ct/news/documents/2020-01/topline-medium-aliens-012820.pdf.

"ISS Crew Take to Escape Capsules in Space Junk Alert." *BBC News*, March 24, 2012. https://www.bbc.com/news/science-environment-17497766.

James, Simon P. "For the Sake of a Stone? Inanimate Things and the Demands of Morality." *Inquiry* 54, no. 4 (2011): 384–97.

JAXA. "Asteroid Explorer, Hayabusa2, Reporter Briefing." *JAXA Hayabusa2 Project*, September 15, 2020. http://www.hayabusa2.jaxa.jp/enjoy/material/press/Hayabusa2_Press_20200915_ver9_en2.pdf.

Johnson, Monte Ransome. *Aristotle on Teleology.* Oxford: Oxford University Press, 2005.

Jonas, Hans. "The Heuristics of Fear." In *Ethics in an Age of Pervasive Technology*, edited by Melvin Kranzberg, 215. Boulder, CO: Westview Press, 1980.

Jonas, Hans. *The Imperative of Responsibility.* Chicago: University of Chicago Press, 1984.

Jonsen, Albert R., and Stephen Toulmin. *The Abuse of Casuistry: A History of Moral Reasoning.* Berkeley: University of California Press, 1988.

Joy, Bill. "Why the Future Doesn't Need Us." *Wired*, April 2000. http://archive.wired.com/wired/archive/8.04/joy_pr.html.

Kalbian, Aline H. "Moral Traces and Relational Autonomy." *Soundings: An Interdisciplinary Journal* 96, no. 3 (2013): 280–96.

Kant, Immanuel. *Critique of Judgement*. Translated by J. H. Bernard. New York: Hafner Press, 1951.

Kant, Immanuel. *Critique of Pure Reason*. Translated by Norman Kemp Smith. London, 1933.

Kant, Immanuel. *Foundations of the Metaphysics of Morals and What Is Enlightenment?* Translated by Lewis White Beck. Second edition. Upper Saddle River, NJ: Prentice Hall, Inc., 1997.

Kant, Immanuel. *Religion Within the Boundaries of Mere Reason*. Translated by George di Giovanni. In *Religion and Rational Theology*, translated and edited by Allen W. Wood and George di Giovanni. Cambridge: Cambridge University Press, 1996.

Kant, Immanuel. *Universal Natural History and Theory of the Heavens*. Translated by Ian Johnston. Arlington, VA: Richer Resources Publications, 2008.

Kardashev, Nikolai. "Transmission of Information by Extraterrestrial Civilizations." *Soviet Astronomy* 8 (1964): 217.

Kass, Leon R. "Forbidding Science: Some Beginning Reflections." *Science and Engineering Ethics* 15 (2009): 271–82.

Kelemen, Deborah. "Function, Goals and Intention: Children's Teleological Reasoning about Objects." *Trends in Cognitive Sciences* 3, no. 12 (December 1999): 461–68.

Kelemen, Deborah, and Cara DiYanni. "Intuitions About Origins: Purpose and Intelligent Design in Children's Reasoning About Nature." *Journal of Cognition and Development* 6 (2005): 3–31.

Kelemen, Deborah, and Evelyn Rosset. "The Human Function Compunction: Teleological Explanation in Adults." *Cognition* 111 (April 2009): 138–43.

Kennedy, Ann R. "Biological Effects of Space Radiation and Development of Effective Countermeasures." *Life Sciences in Space Research* 1 (April 2014): 10–43. https://www.sciencedirect.com/science/article/abs/pii/S2214552414000108.

Kennedy, John F. "We Choose to Go to the Moon/Address at Rice University on the Nation's Space Effort." Rice Stadium, Houston, Texas, September 12, 1962. https://er.jsc.nasa.gov/seh/ricetalk.htm.

Kerr, Richard A., and Richard Stone. "A Human Trigger for the Great Quake of Sichuan?" *Science* 323 (January 16, 2009): 322.

Kerwin, Joseph P. "Letter, Joseph P. Kerwin to Richard H. Truly." July 28, 1986. https://history.nasa.gov/kerwin.html.

Kessler, Donald J. "Collisional Cascading: The Limits of Population Growth in Low Earth Orbit." *Advances in Space Research* 11, no. 12 (1991): 63–66.

Kessler, Donald J. "Earth Orbital Pollution." In *Beyond Spaceship Earth: Environmental Ethics and the Solar System*, edited by Eugene C. Hargrove. San Francisco: Sierra Club Books, 1986.

Kessler, Donald J., and Burton G. Cour-Palais. "Collision Frequency of Artificial Satellites: The Creation of a Debris Belt." *Journal of Geophysical Research: Space Physics* 83, no. A6 (June 1, 1978): 2637–46.

Kimura, Takeshi. "Masahiro Mori's Buddhist Philosophy of Robot." *Paladyn, Journal of Behavioral Robotics* 9, no. 1 (May 15, 2018).

King, Martin Luther, Jr. "The Man Who Was a Fool." In *Strength to Love.* Minneapolis, MN: Fortress Press, 2010 [1963].

Kminek, G., C. Conley, V. Hipkin, H. Yano, and COSPAR. "COSPAR Planetary Protection Policy." COSPAR Panel on Planetary Protection, March 2017. https://cosparhq.cnes.fr/assets/uploads/2019/12/PPPolicyDecember-2017.pdf.

Knight, L. U. "The Voluntary Human Extinction Movement." 2019. www.vhemt.org.

Konopinski, Emil, Cloyd Margin, and Edward Teller. "Ignition of the Atmosphere with Nuclear Bombs." *Classified US Government Report*, declassified 1979, August 14, 1946. https://fas.org/sgp/othergov/doe/lanl/docs1/00329010.pdf.

Kramer, W. R. "To Humbly Go: Guarding Against Perpetuating Models of Colonization in the 100-Year Starship Study." *Journal of the British Interplanetary Society* 67 (2014): 180–86.

Kraus, John. "We Wait and Wonder." *Cosmic Search* 1, no. 3 (Summer 1979): 31. http://www.bigear.org/CSMO/PDF/CS03/cs03p31.pdf.

Krulwich, Robert. "Aliens Found in Ohio? The 'Wow!' Signal." *Weekend Edition Saturday, National Public Radio*, May 28, 2010. https://www.npr.org/sections/krulwich/2010/05/28/126510251/aliens-found-in-ohio-the-wow-signal.

Kubrick, Stanley, and Arthur C. Clarke. *2001: A Space Odyssey*. Beverly Hills, CA: Metro-Goldwyn-Mayer, 1968.

Kubrick, Stanley, Terry Southern, and Peter George. *Dr. Strangelove or: How I Learned to Stop Worrying and Love the Bomb.* Culver City, CA: Columbia Pictures, 1964.

Kumar, Chethan. "India Shows Off Tech to 'Kill' Satellites, Will Also Help Tackle High-Altitude Missiles." *The Times of India*, March 27, 2019. https://timesofindia.indiatimes.com/india/india-shows-off-tech-to-kill-satellites-will-also-help-tackle-high-altitude-missiles/articleshow/68602482.cms.

Lambeth, Benjamin S. *Mastering the Ultimate High Ground: Next Steps in the Military Uses of Space.* Santa Monica, CA: RAND, 2003.

Lane, Mira, Josh Lovejoy, Arathi Sethumadhavan, Katherine Pratt, Neil Coles, Karen Chappell, and Harmony Mabrey. "Foundations of Assessing Harm." Microsoft Azure, May 18, 2020. https://docs.microsoft.com/en-us/azure/architecture/guide/responsible-innovation/harms-modeling/.

Lane, Mira, Josh Lovejoy, Arathi Sethumadhavan, Katherine Pratt, Neil Coles, Karen Chappell, and Harmony Mabrey. "Types of Harm." Microsoft Azure, May 18, 2020. https://docs.microsoft.com/en-us/azure/architecture/guide/responsible-innovation/harms-modeling/type-of-harm.

Lansing, J. Stephen. *Priests and Programmers: Technologies of Power in the Engineered Landscape of Bali.* Princeton, NJ: Princeton University Press, 1991.

Lee, Stan, and Steve Ditko. "Spider-Man." *Amazing Fantasy* 1, no. 15 (August 1962).

Levinson, Paul, and Michael Waltemathe, eds. *Touching the Face of the Cosmos: On the Intersection of Space Travel and Religion.* New York: Connected Editions, 2016.

Lewis, C. S. *The Abolition of Man.* New York: Harper Collins, 1944.

Limerick, Patricia Nelson. "Imagined Frontiers: Westward Expansion and the Future of the Space Program." In *Space Policy Alternatives*, edited by R. Byerly, 249–61. Boulder, CO: Westview Press, 1992.

Lin, Patrick. "Viewpoint: Look Before Taking Another Leap for Mankind—Ethical and Social Considerations in Rebuilding Society in Space." *Astropolitics* 4, no. 3 (2006): 281–94.

Liu, Cixin. *The Dark Forest.* New York: Tor, 2015.

Liu, Cixin. *Death's End.* New York: Tor, 2016.

Liu, Cixin. *The Three-Body Problem.* New York: Tor, 2014.

Livingston, David. "A Code of Ethics for Conducting Business in Outer Space." *Space Policy* 19, no. 2 (May 2003): 93–94.

Lloyd, William Foster. "Two Lectures on the Checks to Population." 1832. Republished as "W. F. Lloyd on the Checks to Population." *Population and Development Review* 6, no. 3 (September 1980): 473–96. https://www.jstor.org/stable/1972412.

Lombrozo, Tania, Deborah Kelemen, and Deborah Zaitchik. "Inferring Design: Evidence of a Preference for Teleological Explanations in Patients With Alzheimer's Disease." *Psychological Science* 18 (2007): 999–1006.

Longo, Giuseppe. "Chapter 18: The Tunguska Event." In *Comet/Asteroid Impacts and Human Society, An Interdisciplinary Approach*, edited by Peter T. Bobrowsky and Hans Rickman, 303–30. Berlin: Springer-Verlag, 2007.

Loomis, Elias. "The Great Auroral Exhibition of August 28 to September 4, 1859—2nd Article." *The American Journal of Science* 29 (January 1860): 92–97. https://babel.hathitrust.org/cgi/pt?id=uva.x001679511&view=1up&seq=112.

Lucas, Paul. "Cruising the Infinite: Strategies for Human Interstellar Travel." *Strange Horizons*, June 21, 2004. http://strangehorizons.com/non-fiction/articles/cruising-the-infinite-strategies-for-human-interstellar-travel/.

Lupisella, Mark. Conversation at the Social and Conceptual Issues in Astrobiology Conference, Clemson University, Clemson, South Carolina, September 24, 2016.

Lupisella, Mark. "Cosmocultural Evolution: Cosmic Motivation for Interstellar Travel?" *Journal of the British Interplanetary Society* 67 (2014): 213–17.

Lupisella, Mark. "Cosmological Theories of Value: Relationalism and Connectedness as Foundations for Cosmic Creativity." In *The Ethics of Space Exploration*, edited by James S. J. Schwartz and Tony Milligan, 75–92. Switzerland: Springer International, 2016.

Lupisella, Mark. *Cosmological Theories of Value: Science, Philosophy, and Meaning in Cosmic Evolution.* Cham, Switzerland: Springer, 2020.

Lupisella, Mark. "The Rights of Martians." *Space Policy* 13 (1997): 89–94.

Lupisella, Mark, and John Logsdon. "Do We Need a Cosmocentric Ethic?" Paper IAA-97-IAA.9.2.09, presented at the International Astronautical Federation Congress, American Institute of Aeronautics and Astronautics, Turin, Italy, 1997.

MacAskill, William, Krister Bykvist, and Toby Ord. *Moral Uncertainty.* Oxford: Oxford University Press, 2020.

Mace, Mikayla. "Comet NEOWISE Sizzles as It Slides by the Sun, Providing a Treat for Observers." Near Earth Object Wide-field Infrared Survey Explorer (NEOWISE)/Infrared Processing and Analysis Center, California Institute of Technology, July 8, 2020. https://neowise.ipac.caltech.edu/news/neowise20200708/.

MacIntyre, Alasdair. *After Virtue: A Study in Moral Theory.* Second edition. Notre Dame, IN: University of Notre Dame Press, 1984.

Mahoney, John. *The Making of Moral Theology: A Study of the Roman Catholic Tradition.* Oxford: Clarendon, 1987.

Mallove, Eugene, and Gregory Matloff. *The Starflight Handbook: A Pioneer's Guide to Interstellar Travel.* New York: John Wiley and Sons, Inc., 1989.

Marino, Lori. "Humanity Is Not Prepared to Colonize Mars." *Futures* 110 (2019): 15–18.

Marks, Paul. "Satellite Collision 'More Powerful than China's ASAT Test.'" *Space*, February 13, 2009. https://www.newscientist.com/article/dn16604-satellite-collision-more-powerful-than-chinas-asat-test/#ixzz6iEG1BLab.

Marshall, Alan. "Ethics and the Extraterrestrial Environment." *Journal of Applied Philosophy* 10 (1993): 227–36.

Martin, Mike W., and Roland Schinzinger. *Ethics in Engineering.* Fourth edition. New York: McGraw Hill, 2005.

Martin, Mike W., and Roland Schinzinger. *Introduction to Engineering Ethics.* Second edition. New York: McGraw-Hill, 2010.

Martin, Osmel, Rolando Cardenas, Mayrene Guimarais, et al. "Effects of Gamma Ray Bursts in Earth's Biosphere." *Astrophysics and Space Science* 326 (2010): 61–67. https://doi.org/10.1007/s10509-009-0211-7.

Masters, Dexter, and Katharine Way, eds. *One World or None: A Report to the Public on the Full Meaning of the Atomic Bomb.* New York: The New Press (on behalf of the Federation of American Scientists), 2007 [1946].

Mathison, Melissa. *ET the Extra-Terrestrial.* Universal City, CA: Universal Studios Amblin Entertainment, 1982.

Mautner, Michael N. "Life-Centered Ethics, and the Human Future in Space." *Bioethics* 23 (2009): 433–40.

Mautner, Thomas. "How Rights Became 'Subjective.'" *Ratio Juris* 26, no. 1 (March 2013): 111–33.

Mazzini, A., H. Svensen, G. G. Akhmanov, G. Aloisi, S. Planke, A. Malthe-Sørenssen, and B. Istadi. "Triggering and Dynamic Evolution of the LUSI Mud Volcano, Indonesia." *Earth and Planetary Science Letters* 261 (2007): 375–88.

McGee, Robert W. "Thomas Aquinas: A Pioneer in the Field of Law & Economics." *Western State University Law Review* 18, no. 1 (Fall 1990): 471–83.

McGrew, W. C. "Is Primate Tool Use Special? Chimpanzee and New Caledonian Crow Compared." *Philosophical Transactions of the Royal Society B* 368 (2013): 20120422.

McKay, Christopher P. "Biologically Reversible Exploration." *Science* 323, no. 5915 (February 6, 2009): 718.

McKay, Christopher P. "Does Mars Have Rights? An Approach to the Environmental Ethics of Planetary Engineering." In *Moral Expertise*, edited by D. MacNiven, 184–97. New York: Routledge, 1990.

McKay, Christopher P. "Planetary Ecosynthesis on Mars: Restoration Ecology and Environmental Ethics." In *Exploring the Origin, Extent, and Future of Life: Philosophical, Ethical, and Theological Perspectives*, edited by Constance M. Bertka, 245–60. Cambridge: Cambridge University Press, 2009.

McKay, Christopher P. "The Search for a Second Genesis of Life in Our Solar System." In *First Steps in the Origin of Life in the Universe*, edited by J. Chela-Flores, T. Owen, and F. Raulin, 269–77. Dordrecht: Springer, 2001.

McKay, Christopher P., and Margarita M. Marinova. "The Physics, Biology, and Environmental Ethics of Making Mars Habitable." *Astrobiology* 1, no. 1 (2001): 89–109.

McLean, Margaret R. "Reaching Out from Earth to the Stars." In *Geoethics: Status and Future Perspectives*, edited by G. Di Capua, P. T. Bobrowsky, S. W. Kieffer, and C. Palinkas, 508. London: Geological Society of London Special Publications, November 12, 2020.

McQuaid, Kim. "Earthly Environmentalism and the Space Exploration Movement, 1960–1990: A Study in Irresolution." *Space Policy* 26 (2010): 163–73.

Mearns, Robin, and Andrew Norton. "Equity and Vulnerability in a Warming World: Introduction and Overview." In *Social Dimensions of Climate Change: Equity and Vulnerability in a Warming World*, edited by Robin Mearns and Andrew Norton. Washington, DC: World Bank, 2010.

Menosky, Joe, and Phillip LaZebnik. "Darmok." *Star Trek: The Next Generation*. Paramount Domestic Television, 1991.

Midgley, Mary. "The Withdrawal of Moral Philosophy." In *The Essential Mary Midgley*, edited by David Midgley. London and New York: Routledge, 2005.

Mill, John Stuart. Utilitarianism *and* On Liberty*: Including Mill's "Essay on Bentham" and Selections from the Writings of Jeremy Bentham and John Austin*. Edited by Mary Warnock. Malden, MA: Blackwell Publishing, 2003.

Millar, Jason. "Technology as Moral Proxy: Autonomy and Paternalism by Design." IEEE International Symposium on Ethics in Engineering, Science, and Technology, Symposium Record. *IEEE Xplore*, May 2014.

Milligan, Tony. *Nobody Owns the Moon: The Ethics of Space Exploration*. Jefferson, NC: McFarland and Company, Inc., 2015.

Milligan, Tony. "Scratching the Surface: The Ethics of Helium-3 Extraction." Paper presented to the 8th IAA Symposium on the Future of Space Exploration: Towards the Stars, Torino, Italy, July 3–5, 2013. http://uhra.herts.ac.uk/bitstream/handle/2299/12133/Scratching_the_Surface.pdf?sequence=2.

Mitcham, Carl. "A Philosophical Inadequacy of Engineering." *The Monist* 92, no. 3 (2009): 349.

Mitchell, Audra. "Can International Relations Confront the Cosmos?" In *The Routledge Handbook of Critical International Relations*, edited by Jenny Edkins. Abingdon, Oxon: Routledge, 2019.

Mizrahi, Moti. "How to Play the 'Playing God' Card." *Science and Engineering Ethics* 26 (2020): 1445–61.

Mogul, R., P. D. Stabekis, M. S. Race, and C. A. Conley. "Planetary Protection Considerations for Human and Robotic Missions to Mars." In *Concepts and Approaches for Mars Exploration*, abstract #4331. Houston, TX: Lunar and Planetary Institute, 2012. http://www.lpi.usra.edu/meetings/marsconcepts2012/pdf/4331.pdf.

Monod, Jacques. *Chance and Necessity: An Essay on the Natural Philosophy of Modern Biology*. Translated by Austryn Wainhouse. New York: Vintage Books, 1972.

Moore, G. E. *Principia Ethica*. Cambridge: Cambridge University Press, 1959.

Moorman, Thomas S., Jr. "Space: A New Strategic Frontier." *Airpower Journal* (Spring 1992).

More, Max. "The Proactionary Principle, Version 1.0." Extropy.org, 2004. Accessed April 10, 2020. http://www.extropy.org/proactionaryprinciple.htm.

Morton, Adam. *Should We Colonize Other Planets?* New York: Wiley, 2018.

Mullane, Mike. *Riding Rockets: The Outrageous Tales of a Space Shuttle Astronaut*. New York: Simon and Schuster, 2006.

Musk, Elon. @elonmusk, Twitter, 5:56 p.m., January 16, 2020. https://twitter.com/elonmusk/status/1217989066181898240.

Musk, Elon, @elonmusk, Twitter, 6:01 p.m., January 16, 2020. https://twitter.com/elonmusk/status/1217990326867988480.

Musk, Elon. "Making Humans a Multi-Planetary Species." *New Space* 5, no. 2 (June 1, 2017). https://www.liebertpub.com/doi/abs/10.1089/space.2017.29009.emu?journalCode=space.

NASA. "Life Detection Ladder." Astrobiology at NASA. Accessed January 1, 2020. https://astrobiology.nasa.gov/research/life-detection/ladder/.

NASA. "New Map Shows Frequency of Small Asteroid Impacts, Provides Clues on Larger Asteroid Population: Bolide Events 1994–2013 (Small Asteroids That Disintegrated in Earth's Atmosphere)." *NASA Jet Propulsion Laboratory News*, November 14, 2014. https://www.jpl.nasa.gov/news/news.php?release=2014-397.

NASA. "Oral History 2 Transcript: Robert E. Stevenson Interviewed by Carol Butler." Houston, Texas, May 13, 1999, 13–35. http://www.jsc.nasa.gov/history/oral_histories/StevensonRE/RES_5-13-99.pdf.

NASA, Jet Propulsion Laboratory, Center for Near Earth Object Studies. "NEO Basics." Accessed March 14, 2021. https://cneos.jpl.nasa.gov/about/neo_groups.html.

National Commission for the Protection of Human Subjects of Biomedical and Behavioral Research. *The Belmont Report*. Department of Health, Education and Welfare (DHEW). Washington, DC: U.S. Government Printing Office, April 18, 1979. https://www.hhs.gov/ohrp/regulations-and-policy/belmont-report/read-the-belmont-report/index.html.

National Research Council. "Assessment of Planetary Protection Requirements for Mars Sample Return." Washington, DC: National Academies Press, 2009. http://www.nap.edu.

National Research Council. "Division on Engineering and Physical Sciences, Commission on Engineering and Technical Systems, Committee on Space Debris." *Orbital Debris: A Technical Assessment.* National Academies Press, July 7, 1995.

National Research Council. *Evaluating the Biological Potential in Samples Returned from Planetary Satellites and Small Solar System Bodies: Framework for Decision Making.* Washington, DC: The National Academies Press, 1998.

National Research Council. "Preventing the Forward Contamination of Mars." Washington, DC: National Academy Press, 2006. http://www.nap.edu.

Neveu, Marc, Lindsay E. Hays, Mary A. Voytek, Michael H. New, and Mitchell D. Schulte. "The Ladder of Life Detection," *Astrobiology* 18, no. 11 (November 13, 2018),https://www.liebertpub.com/doi/full/10.1089/ast.2017.1773.

Newman, Christopher. "Establishing an Ecological Ethical Paradigm for Space Activity." *Room, The Space Journal* 2, no. 4 (2015): 55–61.

Nicholson, Wayne L., Andrew C. Schuerger, and Margaret S. Race. "Migrating Microbes and Planetary Protection." *Trends in Microbiology* 17, no. 9 (September 2009): 389–92.

Nozick, Robert. *Anarchy, State, and Utopia.* New York: Basic Books, 1974.

Nozick, Robert. "Moral Complications and Moral Structures." *Natural Law Forum* 13, no. 1 (1968): 34–35.

"The Nuremberg Code." Trials of War Criminals before the Nuremberg Military Tribunals under Control Council Law No. 10, Nuremberg, October 1946–April 1949. Washington, DC: U.S. Government Printing Office, 1949–1953. Accessed December 30, 2020. https://www.ushmm.org/information/exhibitions/online-exhibitions/special-focus/doctors-trial/nuremberg-code.

Nussbaum, Martha C. "Non-Relative Virtues: An Aristotelian Approach." *Midwest Studies in Philosophy* 13 (1988): 32–53.

Nussbaum, Martha C. *Women and Human Development: The Capabilities Approach.* Cambridge: Cambridge University Press, 2000.

O'Callaghan, Jonathan. "Life on Venus? Scientists Hunt for the Truth." *Nature* 586 (October 2, 2020): 182–83. https://www.nature.com/articles/d41586-020-02785-5.

Ord, Toby. *The Precipice: Existential Risk and the Future of Humanity.* New York: Hachette, 2020.

Ott, C. M., R. J. Bruce, and D. L. Pierson. "Microbial Characterization of Free Floating Condensate Aboard the Mir Space Station." *Microbial Ecology* 47 (2004): 133–36. http://science.nasa.gov/media/medialibrary/2007/05/11/11may_locad3_resources/Ott%202004.pdf.

Overbye, Dennis. "Reaching for the Stars, Across 4.37 Light-Years." *New York Times*, April 12, 2016. https://www.nytimes.com/2016/04/13/science/alpha-centauri-breakthrough-starshot-yuri-milner-stephen-hawking.html.

Parfit, Derek. "Overpopulation and the Quality of Life." In *The Repugnant Conclusion: Essays on Population Ethics,* edited by J. Ryberg and T. Tännsjö, 7–22. Dordrecht: Kluwer Academic Publishers, 2004.

Parfit, Derek. *Reasons and Persons.* Oxford: Oxford University Press, 1986.

Partridge, Christopher. "Alien Demonology: The Christian Roots of the Malevolent Extraterrestrial in UFO Religions and Abduction Spiritualities." *Religion* 34 (2004): 163–89.

Paton, H. J. *The Categorical Imperative: A Study in Kant's Moral Philosophy.* Philadelphia: University of Pennsylvania Press, 1947.

Patterson, Francine G. "The Gestures of a Gorilla: Language Acquisition in Another Pongid." *Brain and Language* 5, no. 1 (January 1978): 72–97.

Peeters, Walter. "From Suborbital Space Tourism to Commercial Personal Spaceflight." *Acta Astronautica* 66, nos. 11–12 (June–July 2010): 1625–32.

Pellegrino, Charles. *Flying to Valhalla.* New York: William Morrow and Company, 1993.

Pellegrino, Charles, and George Zebrowski. *The Killing Star.* New York: William Morrow and Company, 1995.

Peltzman, Sam. "The Effects of Automobile Safety Regulation." *Journal of Political Economy* 83, no. 4 (August 1975): 677–726.

Perminov, V. G. *The Difficult Road to Mars—A Brief History of Mars Exploration in the Soviet Union.* Washington, DC: NASA History Division, July 1999.

Perrow, Charles. *Normal Accidents: Living with High Risk Technologies.* New York: Basic Books, 1984.

Persson, Erik. "The Moral Status of Extraterrestrial Life." *Astrobiology* 12 (2012): 976–84.

Peters, Ted. "Are We Playing God with Nanoenhancement?" In *Nanoethics: The Ethical and Social Implications of Nanotechnology*, edited by F. Allhoff, P. Lin, J. Moor, and J. Weckert, 173–84. Hoboken, NJ: Wiley, 2007.

Peters, Ted. "ET: Alien Enemy or Celestial Savior [*sic*]." *Theology and Science* 8 (2010): 245–55.

Peters, Ted. "The Implications of the Discovery of Extra-Terrestrial Life for Religion." *Philosophical Transactions of the Royal Society A* 369 (2011): 644–55.

Peters, Ted, and Julie Froehlig. "The Peters ETI Religious Crisis Survey." Counterbalance. Accessed September 11, 2020. https://counterbalance.org/etsurv/fullr-frame.html.

Peters, Ted, with Martinez Hewlett, Joshua M. Moritz, and Robert John Russell. *Astrotheology: Science and Theology Meet Extraterrestrial Life.* Eugene, OR: Cascade Books, 2018.

Peterson, Martin. *The Ethics of Technology: A Geometric Analysis of Five Moral Principles.* Oxford: Oxford University Press, 2017.

Phillips, Tony. "The Mysterious Smell of Moondust." *NASA Science: Science News*, January 30, 2006. http://science.nasa.gov/science-news/science-at-nasa/2006/30jan_smellofmoondust/.

Piaggio, Antoinette J., Gernot Segelbacher, Philip J. Seddon, Luke Alphey, Elizabeth L. Bennett, Robert H. Carlson, Robert M. Friedman, Dona Kanavy, Ryan Phelan, Kent H. Redford, Marina Rosales, Lydia Slobodian, and Keith Wheeler. "Is It Time for Synthetic Biodiversity Conservation?" *Trends in Ecology & Evolution* 32, no. 2 (February 2017): 97–107.

Pimentel, David, et al. "Economic and Environmental Threats of Alien Plant, Animal, and Microbe Invasions." *Agriculture, Ecosystems and Environment* 84 (2001): 1–20.

Pirtle, Zachary, and Zoe Szajnfarber. "On Ideals for Engineering in Democratic Societies." In *Philosophy and Engineering: Exploring Boundaries, Expanding Connections*, edited by Diane P. Michelfelder, Byron Newberry, and Qin Zhu, 99–112. Philosophy of Engineering and Technology series, vol. 26. Cham, Switzerland: Springer, 2017.

Planetary Resources. "About." 2014. www.planetaryresources.com.

Pompidou, Alain. *The Ethics of Space Policy*. Working Group on the "Ethics of Outer Space," UNESCO World Commission on the Ethics of Scientific Knowledge and Technology, UN, 2000.

Pop, Vigiliu. "Lunar Exploration and the Social Dimension." *Proceedings of the ESLAB 36 Symposium "Earth-like Planets and Moons,"* ESTEC, Noordwijk, June 3–6, 2002.

Popova, Olga P., et al. "Chelyabinsk Airburst, Damage Assessment, Meteorite Recovery and Characterization." *Science* 342 (2013).

Powers, Thomas. "Seeing the Light of Armageddon." *Rolling Stone*, April 29, 1982.

"Precautionary Principle." *Glossary of Summaries, EUR-Lex: Access to European Union Law*. Accessed July 6, 2016. http://eur-lex.europa.eu/summary/glossary/precautionary_principle.html.

Presidential Commission for the Study of Bioethical Issues. *New Directions: Ethics of Synthetic Biology and Emerging Technologies*. Washington, DC, December 2010. http://bioethics.gov/sites/default/files/PCSBI-Synthetic-Biology-Report-12.16.10_0.pdf.

Prisco, Giulio. "Uploaded E-Crews for Interstellar Missions." *Kurzweil AI (Accelerating Intelligence) Daily Blog*, December 12, 2012. https://www.kurzweilai.net/uploaded-e-crews-for-interstellar-missions.

"Protocols for an ETI Signal Detection." SETI Institute, April 23, 2018. https://www.seti.org/protocols-eti-signal-detection.

Pulliam, Christine, and Laura Betz. "Astronomers Propose a Novel Method of Finding Atmospheres on Rocky Worlds." NASA, December 2, 2019. https://www.nasa.gov/feature/goddard/2019/astronomers-propose-a-novel-method-of-finding-atmospheres-on-rocky-worlds.

Quirk, Joe, with Patri Friedman. *Seasteading: How Floating Nations Will Restore the Environment, Enrich the Poor, Cure the Sick, and Liberate Humanity from Politicians.* New York: Free Press, 2017.

Race, Margaret S. "Policies for Scientific Exploration and Environmental Protection: Comparison of the Antarctic and Outer Space Treaties." In *Science Diplomacy: Antarctica, Science, and the Governance of International Spaces*, edited by Paul Arthur Berkman, Michael A. Lang, David W. H. Walton, and Oran R. Young, 143–52. Washington, DC: Smithsonian Institution Scholarly Press, 2011.

Race, Margaret S. "Preserving History on the Moon." *Astronomy Beat*, no. 91 (March 13, 2012). www.astrosociety.org.

Race, Margaret S., G. Kminek, and J. D. Rummel. "Planetary Protection and Humans on Mars: NASA/ESA Workshop Results." *Advances in Space Research* 42, no. 6 (2008): 1128–38.

Race, Margaret S., C. McKay, and A. Steele. "Session 26. Mars Sample Return Planning Issues." *Astrobiology* 8, no. 2 (April 2008): 420–21.

Race, Margaret S., J. Moses, C. McKay, and K. J. Venkateswaran. "Synthetic Biology in Space: Considering the Broad Societal and Ethical Implications." *International Journal of Astrobiology* 11 (2012): 133–39.

Randolph, Richard O., and Christopher P. McKay. "Protecting and Expanding the Richness and Diversity of Life, an Ethic for Astrobiology Research and Space Exploration." *International Journal of Astrobiology* 13, no. 1 (2014): 28–34.

Randolph, Richard O., Margaret S. Race, and Christopher P. McKay. "Reconsidering the Theological and Ethical Implications of Extraterrestrial Life." *CTNS Bulletin* 17, no. 3 (Summer 1997): 1–8.

Rappaport, Roy. *Ecology, Meaning, and Religion.* Berkeley: North Atlantic Books, 1979.

Rappaport, Roy. *Ritual and Religion in the Making of Humanity.* Cambridge: Cambridge University Press, 1999)

Rawls, John. *A Theory of Justice.* Cambridge, MA: Harvard University Press, 1971.

Reardon, Sara. "NIH Finds Forgotten Smallpox Store." *Nature*, July 9, 2014. https://www.nature.com/news/nih-finds-forgotten-smallpox-store-1.15526.

Reiman, Saara. "Sustainability in Space Exploration: An Ethical Perspective." *MarsPapers*, Mars Society, 2011. Accessed August 29, 2014. http://www.marspapers.org/papers/Reiman_2011_paper.pdf.

*Report of the Presidential Commission on the Space Shuttle Challenger Accident* (Rogers Report). "Chapter IX: Other Safety Considerations." Washington, DC: U.S. Government, June 6, 1986. https://history.nasa.gov/rogersrep/v1ch9.htm.

Richardson, James. "The U.S. Space Force Logo and Motto." U.S. Space Force Public Affairs, United States Space Force, July 22, 2020. https://www.spaceforce.mil/News/Article/2282948/the-us-space-force-logo-and-motto/.

Rice, Donald B. "The Air Force and U.S. National Security: Global Reach—Global Power." Washington, DC: U.S. Air Force Department, 1990.

Ricoeur, Paul. *Freud and Philosophy: An Essay on Interpretation.* Translated by Denis Savage New Haven, CT: Yale University Press, 1970.

Robinson, Paul, ed. *Just War in Comparative Perspective.* London: Routledge, 2003.

Robinson, Kim Stanley. *Blue Mars.* New York: Bantam Books, 1996.

Robinson, Kim Stanley. *Green Mars.* New York: Bantam Books, 1994.

Robinson, Kim Stanley. *The Ministry for the Future.* New York: Orbit Books, 2020.

Robinson, Kim Stanley. *Red Mars.* New York: Bantam Books, 1993.

Rogers, Thomas F. "Safeguarding Tranquility Base: Why the Earth's Moon Base Should Become a World Heritage Site." *Space Policy* 20, no. 1 (2004): 5–6.

Rolston, Holmes, III. "The Preservation of Natural Value in the Solar System." In *Beyond Spaceship Earth: Environmental Ethics and the Solar System*, edited by Eugene C. Hargrove, 140–82. San Francisco: Sierra Club Books, 1986.

Rosoff, Matt. "Elon Musk Worries That His Kids Are Too Soft to Be Entrepreneurs." *Business Insider*, September 16, 2011. https://www.businessinsider.com/elon-musk-worries-that-his-kids-are-too-soft-to-be-entrepreneurs-2011-9.

Ross, W. D. *The Right and the Good.* Oxford: Clarendon, 1930.

Rummel, J. D., and Linda Billings. "Issues in Planetary Protection: Policy, Protocol and Implementation." *Space Policy* 20 (2004): 49–54.

Rummel, J. D., M. S. Race, C. A. Conley, and D. R. Liskowsky. "The Integration of Planetary Protection Requirements and Medical Support on a Mission to Mars." In *The Human Mission to Mars: Colonizing the Red Planet*, edited by J. S. Levine and R. E. Schild. Cambridge, MA: Cosmology Science Publishers, 2010. http://journalofcosmology.com.

Rummel, J., M. Race, and G. Horneck, eds. "COSPAR Workshop on Ethical Considerations for Planetary Protection in Space Exploration." COSPAR, Paris, 2012. http://cosparhq.cnes.fr/Scistr/PPP Reports/PPP_Workshop Report_Ethical Considerations.pdf.

Rummel, J. D., M. S. Race, G. Horneck, and the Princeton Workshop Participants. "Ethical Considerations for Planetary Protection in Space Exploration: A Workshop." *Astrobiology* 12, no. 11 (2012): 1017–23.

Rumsfeld, Donald H. "Department of Defense News Briefing—Secretary Rumsfeld and Gen. Myers." United States Department of Defense, February 12, 2002. https://archive.defense.gov/Transcripts/Transcript.aspx?TranscriptID=2636.

Sagan, Carl. "Comments on O'Neill's Space Colonies." In *Space Colonies*, edited by Stewart Brand, 42. San Francisco, CA: Waller Press, 1977.

Sagan, Carl. *Contact.* New York: Simon and Schuster, 1985.

Sagan, Carl. *The Cosmic Connection: An Extraterrestrial Perspective.* Cambridge: Cambridge University Press, 2000 [1973].

Sagan, Carl, and Ann Druyan. *Contact*. Novato, CA: South Side Amusement Company, 1997.

Sagan, Carl, and Ann Druyan. *Pale Blue Dot: A Vision of the Human Future in Space.* New York: Ballantine Books, 1997.

Sagan, Carl, Ann Druyan, and Steven Soter. "Episode 5: Blues for a Red Planet." *Cosmos: Collector's Edition.* Studio City, CA: Cosmos Studios, Inc., 2000 [1980].

Salk, Jonas. "Are We Being Good Ancestors?" *World Affairs: The Journal of International Issues* 1, no. 2 (December 1992): 16–18.

Sandler, Ronald L. "Environmental Virtue Ethics." In *International Encyclopedia of Ethics*, edited by Hugh LaFollette, 1665–74. Blackwell Publishing Ltd., February 1, 2013.

Savage, Ian. "Comparing the Fatality Risks in United States Transportation Across Modes and Over Time." *Research in Transportation Economics* 43 (July 2013): 9–22.

Savage, Marshall T. *The Millennial Project: Colonizing the Galaxy in Eight Easy Steps.* Boston: Little, Brown and Company, 1992.

Schelling, Thomas C. "Game Theory and the Study of Ethical Systems." *Journal of Conflict Resolution* 12, no. 1 (1968): 34–44.

Scholz, Carter, and Alan Brennert. "A Small Talent for War." *The Twilight Zone*, January 24, 1986. https://www.youtube.com/watch?v=fbT1fCHOjfI.

Schwartz, James S. J. "The Accessible Universe: On the Choice to Require Bodily Modification for Space Exploration." In *Human Enhancements for Space Missions: Lunar, Martian, and Future Missions to the Outer Planets*, edited by Konrad Szocik. New York: Springer, 2020.

Schwartz, James S. J. "Myth-Free Space Advocacy Part I—The Myth of Innate Exploratory and Migratory Urges." *Acta Astronautica* 137 (August 2017): 450–60.

Schwartz, James S. J. *The Value of Science in Space Exploration.* New York: Oxford University Press, 2020.

Schwartz, James S. J., and Tony Milligan, eds. *The Ethics of Space Exploration.* Switzerland: Springer International, 2016.

Scientific Expert Group on Climate Change. *Confronting Climate Change: Avoiding the Unmanageable and Managing the Unavoidable*. Report for the United Nations Commission on Sustainable Development. Research Triangle, NC, and Washington, DC: Sigma Xi and the United Nations Foundation, 2007.

Scott, James C. *Seeing Like a State: How Certain Schemes to Improve the Human Condition Have Failed.* New Haven, CT: Yale University Press, 1998.

Seebens, Hanno, Tim M. Blackburn, Ellie E. Dyer, et al. "No Saturation in the Accumulation of Alien Species Worldwide." *Nature Communications* 8, no. 14435 (2017).

Seedhouse, Erik. *SpaceX: Making Commercial Spaceflight a Reality.* Springer Science and Business Media, 2013.

Sen, Amartya. "Equality of What?" In *Tanner Lectures on Human Values*, edited by Sterling M. McMurrin, 197–220. Cambridge: Cambridge University Press, 1979.

Sen, Amartya. "Rights and Capabilities." In *Morality and Objectivity: A Tribute to J.L. Mackie*, 130–47. London: Routledge and Kegan Paul, 1985.

Shostak, Seth. "A Signal from Proxima Centauri?" SETI Institute, December 19, 2020. https://www.seti.org/signal-proxima-centauri.

Shue, Henry. *Basic Rights: Subsistence, Affluence, and U.S. Foreign Policy*. Second edition. Princeton, NJ: Princeton University Press, 1980.

Singer, Peter. *Animal Liberation.* New revised edition. New York: Avon Books, 1990 [1975].

Singer, Peter. *Practical Ethics.* Second edition. Cambridge: Cambridge University Press, 1993.

Siddiqi, Asif A. *Beyond Earth: A Chronicle of Deep Space Exploration, 1958–2016.* Washington, DC: NASA History Division, 2018.

Simmons, Beth. "Treaty Compliance and Violation." *Annual Review of Political Science* 13 (2010): 273–96.

Smith, Kelly C. "A(nother) Cosmic Wager: Pascal, METI, and the Barn Door Argument." *Theology and Science* 17, no. 1 (February 2019): 29–35.

Smith, Kelly C. "*Homo Reductio*: Eco-Nihilism and Human Colonization of Other Worlds." *Futures* 110 (2019): 31–34.

Smith, Kelly C. "Manifest Complexity: A Foundational Ethic for Astrobiology?" *Space Policy* 30, no. 4 (November 2014): 209–14.

Smith, Kelly C. "METI or REGRETTI: Ethics, Risk, and Alien Contact." In *Social and Conceptual issues in Astrobiology*, edited by Kelly C. Smith and Carlos Mariscal, 209–35. Oxford: Oxford University Press, 2020.

Smith, Kelly C. "The Trouble with Intrinsic Value: An Ethical Primer for Astrobiology." In *Exploring the Origin, Extent, and Future of Life: Philosophical, Ethical, and Theological Perspectives*, edited by Constance M. Bertka, 261–80. Cambridge: Cambridge University Press, 2009.

Smith, Kelly C., Keith Abney, Gregory Anderson, Linda Billings, Carl L. Devito, Brian Patrick Green, Alan R. Johnson, Lori Marino, Gonzalo Munevar, Michael P. Oman-Reagan, Adam Potthast, James S. J. Schwartz, Koji Tachibana, John W. Traphagan, and Sheri Wells-Jensen. "The Great Colonization Debate." *Futures* 110 (2019).

Smith, Kelly C., and Carlos Mariscal, eds. *Social and Conceptual Issues in Astrobiology.* New York: Oxford University Press, 2020.

Snow, C. P. "The Two Cultures." In *The Two Cultures*. Cambridge: Cambridge University Press, 1998.

Space Adventures. Accessed January 3, 2020. https://spaceadventures.com/.

SpaceX. "Mars & Beyond: The Road Map to Making Humanity Multiplanetary." 2020. Accessed November 23, 2020. https://www.spacex.com/human-spaceflight/mars/.

Sparrow, Robert. "The Ethics of Terraforming." *Environmental Ethics* 21 (1999): 227–45.

Spennemann, Dirk H. R. "The Ethics of Treading on Neil Armstrong's Footprints." *Space Policy* 20, no. 4 (2004): 279–90.

Spielberg, Steven. *Close Encounters of the Third Kind*. Culver City, CA: Columbia Pictures, 1977.

Stannard, David. *American Holocaust: The Conquest of the New World.* Oxford: Oxford University Press, 1992.

Stapledon, Olaf. *Last and First Men.* London: Methuen, 1930.

*Star Trek: The Original Series*. "Title Sequence." 1966.

Stern, Robert. "Does 'Ought' Imply 'Can'? And Did Kant Think It Does?" *Utilitas* 16, no. 1 (2004): 42–61.

Straczynski, J. Michael. *Amazing Spider-Man* 2, no. 38 (February 2002).

Stuhlinger, Ernst. "Letter to Sister Mary Jucunda/Why Explore Space? A 1970 Letter to a Nun in Africa." May 6, 1970. *Roger Launius's Blog*, February 8, 2012. https://launiusr.wordpress.com/2012/02/08/why-explore-space-a-1970-letter-to-a-nun-in-africa/.

Sugiyama, Yukimaru, and Jeremy Koman. "Tool-Using and -Making Behavior in Wild Chimpanzees at Bossou, Guinea." *Primates* 20 (1979): 513–24.

Szocik, Konrad, ed. *Human Enhancements for Space Missions: Lunar, Martian, and Future Missions to the Outer Planets.* New York: Springer, 2020.

Szocik, Konrad, and Tomasz Wójtowicz. "Human Enhancement in Space Missions: From Moral Controversy to Technological Duty." *Technology in Society* 59 (November 2019).

Taleb, Nassim Nicholas. *Antifragile: Things that Gain from Disorder.* New York: Random House, 2012.
Taleb, Nassim Nicholas. *The Black Swan: The Impact of the Highly Improbable.* New York: Random House, 2007.
Taleb, Nassim Nicholas. *Skin in the Game: Hidden Asymmetries in Daily Life.* New York: Random House, 2020.
Tallberg, Jonas. "Paths to Compliance: Enforcement, Management, and the European Union." *International Organization* 56, no. 3 (Summer 2002): 609–43.
Tancredi, Laurence R., and Arthur J. Barsky. "Technology and Health Care Decision Making: Conceptualizing the Process for Societal Informed Consent." *Medical Care* 12, no. 10 (October 1974): 845–59.
Taneja, Hemant. "The Era of 'Move Fast and Break Things' Is Over." *Harvard Business Review*, January 22, 2019. https://hbr.org/2019/01/the-era-of-move-fast-and-break-things-is-over.
Taylor, Richard L. S. "Paraterraforming—The Worldhouse Concept." *Journal of the British Interplanetary Society* 45, no. 8 (August 1992): 341–52.
Tedesco, Edward F., and François-Xavier Desert. "The Infrared Space Observatory Deep Asteroid Search." *The Astronomical Journal* 123, no. 4 (April 2002): 2070–82.
Tempier, Stephen. *Condemnations of 1277*. In "Selections from the Condemnations of 1277," *Medieval Philosophy: Essential Readings with Commentary*, edited by Gyula Klima with Fritz Allhoff and Anand Jayprakash Vaidya, 180–89. Malden, MA: Blackwell, 2007.
Thomas, Brian C. "Gamma-Ray Bursts as a Threat to Life on Earth." *International Journal of Astrobiology* 8, no. 3 (July 2009): 183–86.
Thomas, Jim, and John Thomas. *Predator*. Century City, Los Angeles, CA: 20th Century Fox, 1987.
Thompson, Amy. "SpaceX's Elon Musk and His Plans to Send 1 Million People to Mars." Teslarati, January 26, 2020. https://www.teslarati.com/spacex-ceo-elon-musk-plan-colonize-mars-1-million-people/.
Thomson, Judith Jarvis. "The Trolley Problem." *The Yale Law Journal* 94, no. 6 (May 1985): 1395–1415.
Tierney, Brian. *The Idea of Natural Rights: Studies on Natural Rights, Natural Law, and Church Law 1150–1625.* Grand Rapids, MI: Eerdmans, 1997.
Tingay, Mark, Oliver Heidbach, Richard Davies, and Richard Swarbrick. "Triggering of the Lusi Mud Eruption: Earthquake versus Drilling Initiation." *Geology* 36, no. 8 (August 2008): 639–42.
Tito, Dennis A., et al. "Feasibility Analysis for a Manned Mars Free-Return Mission in 2018." Inspiration Mars. Accessed January 3, 2021. https://web.archive.org/web/20130319140952/http://www.inspirationmars.org/Inspiration%20Mars_Feasibility%20Analysis_IEEE.pdf.
Tolkien, J. R. R. *The Fellowship of the Ring.* New York: Houghton Mifflin, 2004 [1954].
Tough, Allen, ed. *When SETI Succeeds: The Impact of High-Information Contact*. Bellevue, WA: Foundation for the Future, 2000.
Traphagan, John W. "Active SETI and the Problem of Research Ethics." *Theology and Science* 17, no. 1 (February 2019): 69–78.
Traphagan, John W. "Religion, Science, and Space Exploration from a Non-Western Perspective." *Religions* 11, no. 8 (2020): 397.
Turchin, Alexey. "Processes with Positive Feedback and Perspectives on Global Catastrophes." *Social Sciences and Modernity*, no. 6 (2009) (in Russian).
Turchin, Alexey, and Brian Patrick Green. "Aquatic Refuges for Surviving a Global Catastrophe." *Futures* 89 (May 2017): 33.
United Nations Office for Outer Space Affairs. "Convention on International Liability for Damage Caused by Space Objects." General Assembly Resolution 2777 (XXVI) in 1971, entered into force 1972.
United Nations Office for Outer Space Affairs. "Treaty on Principles Governing the Activities of States in the Exploration and Use of Outer Space, including the Moon and Other Celestial Bodies." (The Outer Space Treaty), United Nations General Assembly Resolution 2222 (XXI), agreed upon in 1966, signed and entered into force 1967. https://www.unoosa.org/oosa/en/ourwork/spacelaw/treaties/outerspacetreaty.html.
United Nations. *The Universal Declaration of Human Rights*. 1948.
United States Federal Bureau of Investigation. "Amerithrax or Anthrax Investigation." Accessed August 17, 2020. https://www.fbi.gov/history/famous-cases/amerithrax-or-anthrax-investigation.
United States Space Force. Accessed December 31, 2020. https://www.spaceforce.mil/.
United States Space Force. "About the Space Force." Accessed December 31, 2020. https://www.spaceforce.mil/About-Us/About-Space-Force/.
United States Space Force. "Space Power: Doctrine for Space Forces." *Space Capstone Publication*, June 2020. https://www.spaceforce.mil/Portals/1/Space%20Capstone%20Publication_10%20Aug%202020.pdf.
Vakoch, Douglas A. "Correspondence: In Defence of METI." *Nature Physics* 12 (October 2016): 890.
Vakoch, Douglas A., ed. *Extraterrestrial Altruism: Evolution and Ethics in the Cosmos*. Heidelberg, Germany: Springer, 2014.
Vallero, Daniel A. "The New Bioethics: Reintegration of Environmental and Biomedical Sciences Ethics in Biology." *Engineering & Medicine—An International Journal* 1, no. 4 (2010): 269–71.
Vallor, Shannon. "Moral Deskilling and Upskilling in a New Machine Age: Reflections on the Ambiguous Future of Character." *Philosophy & Technology* 28 (2015): 107–24.
Vallor, Shannon. *Technology and the Virtues: A Philosophical Guide to a Future Worth Wanting*. Oxford: Oxford University Press, 2016.

Vallor, Shannon, Brian Green, and Irina Raicu. "Ethics in Tech Practice: An Introductory Workshop." Slide deck, Markkula Center for Applied Ethics, June 22, 2018. https://www.scu.edu/media/ethics-center/technology-ethics/Updated-Sample-Slide-Deck-for-Ethics-in-Tech-Practice-materials.pdf.

Vallor, Shannon, Brian Green, and Irina Raicu. "Ethics in Technology Practice." Markkula Center for Applied Ethics, June 19, 2018. https://www.scu.edu/ethics-in-technology-practice/.

Vanem, E. "Ethics and Fundamental Principles of Risk Acceptance Criteria." *Safety Science* 50, no. 4 (2012): 958–67.

Velasquez, Manuel, Dennis Moberg, Michael J. Meyer, Thomas Shanks, Margaret R. McLean, David DeCosse, Claire André, and Kirk O. Hanson. "A Framework for Ethical Decision Making." Markkula Center for Applied Ethics, 2009. https://www.scu.edu/ethics/ethics-resources/ethical-decision-making/a-framework-for-ethical-decision-making/.

Verbeek, Bruno, and Christopher Morris. "Game Theory and Ethics." In *The Stanford Encyclopedia of Philosophy*, edited by Edward N. Zalta, Winter 2020 edition. Accessed March 5, 2020. https://plato.stanford.edu/archives/win2020/entries/game-ethics/.

Vinge, Vernor. "Long Shot." *Analog Science Fiction/Science Fact*. August 1972.

Ward, Peter D., and Donald Brownlee. *Rare Earth: Why Complex Life Is Uncommon in the Universe.* New York: Copernicus, 2000.

Weckert, J. "Playing God: What Is the Problem?" In *The Ethics of Human Enhancement: Understanding the Debate*, edited by S. Clarke, J. Savulescu, C. A. J. Coady, A. Giubilini, and S. Sanyal, 87–99. New York: Oxford University Press, 2016.

Whitaker, Albert C. *History and Criticism of the Labor Theory of Value in English Political Economy.* New York: Columbia University Press, 1904.

Whitehead, Alfred North. *Adventures of Ideas.* New York: Free Press, 1967.

The White House. "Introduction to Outer Space." Washington, DC: U.S. Government Printing Office, March 26, 1958.

Wiblin, Robert. "Toby Ord on Why the Long-Term Future of Humanity Matters More Than Anything Else, and What We Should Do About It." *80,000 Hours Podcast*, September 6, 2017. https://80000hours.org/podcast/episodes/why-the-long-run-future-matters-more-than-anything-else-and-what-we-should-do-about-it/.

Wiblin, Robert, Arden Koehler, and Keiran Harris. "Peter Singer on Being Provocative, EA, How His Moral Views Have Changed, & Rescuing Children Drowning in Ponds." Interview with Peter Singer. *80,000 Hours Podcast*, December 5, 2019. https://80000hours.org/podcast/episodes/peter-singer-advocacy-and-the-life-you-can-save/.

Wilde, Gerald J. S. "Risk Homeostasis Theory: An Overview." *Injury Prevention* 4 (1998): 89–91.

Wilks, Anna Frammartino. "Kantian Foundations for a Cosmocentric Ethic." In *The Ethics of Space Exploration*, edited by James S. J. Schwartz and Tony Milligan, 181–94. Switzerland: Springer International, 2016.

Williams, Bernard. *Ethics and the Limits of Philosophy.* Cambridge, MA: Harvard University Press, 1985.

Williams, David, Andre Kuipers, Chiaki Mukai, and Robert Thirsk. "Acclimation during Space Flight: Effects on Human Physiology." *Canadian Medical Association Journal* 180, no. 13 (June 23, 2009): 1317–23. https://www.cmaj.ca/content/180/13/1317.short.

Williamson, Mark. "Space Ethics and Protection of the Space Environment." *Space Policy* 19 (2003): 47–52.

Wilson, David Sloan. *Darwin's Cathedral: Evolution, Religion, and the Nature of Society.* Chicago: University of Chicago Press, 2002.

Wollstonecraft, Mary. *A Vindication of the Rights of Woman: With Strictures on Political and Moral Subjects.* London: J. Johnson, 1796.

World Commission on the Ethics of Scientific Knowledge and Technology (COMEST). "The Precautionary Principle." Paris: United Nations Educational, Scientific and Cultural Organization (UNESCO), 2005.

World Health Organization. "World Health Organization Model List of Essential Medicines, 21st List, 2019." Geneva: World Health Organization, 2019.

York, Richard. "Toward a Martian Land Ethic." *Human Ecology Review* 12, no. 1 (2005): 72–73.

Young, M. Jane. "'Pity the Indians of Outer Space': Native American Views of the Space Program." *Western Folklore* 46, no. 4 (October 1987): 269–79. https://www.jstor.org/stable/1499889.

Yudkowsky, Eliezer. "Cognitive Biases Potentially Affecting Judgement of Global Risks." In *Global Catastrophic Risks*, edited by Nick Bostrom and Milan M. Circovic, 91–119. Oxford: Oxford University Press, 2008.

Yurtsever, Ulvi, and Steven Wilkinson. "Limits and Signatures of Relativistic Spaceflight." Arxiv.org, April 21, 2015. http://arxiv.org/pdf/1503.05845.pdf.

Zedler, Joy B., and Suzanne Kercher. "Causes and Consequences of Invasive Plants in Wetlands: Opportunities, Opportunists, and Outcomes." *Critical Reviews in Plant Sciences* 23, no. 5 (2004): 442–44.

Zhou, Shiyong, Kai Deng, Cuiping Zhao, and Wanzheng Cheng. "Discussion on 'Was the 2008 Wenchuan Earthquake Triggered by Zipingpu Reservoir?'" *Earthquake Science* 23 (2010): 577–81.

Zubrin, Robert. *Entering Space: Creating a Spacefaring Civilization.* New York: Jeremy P. Tarcher/Putnam, 1999.

Zubrin, Robert, with Richard Wagner. *The Case for Mars: The Plan to Settle the Red Planet and Why We Must.* New York: Touchstone, 2011.

# Credits

Figure 1.1. John Webber, from Brother Bertram Photograph Collection, via Wikimedia Commons, Public Domain: https://commons.wikimedia.org/wiki/Category:John_Webber#/media/File:Hawaiian_War_Canoe,_Warriors_Masked,_(2),_from_Brother_Bertram_Photograph_Collection.jpg.

Figure 2.1. NASA, Saturn V Engines and Werner von Braun, public domain, https://commons.wikimedia.org/wiki/File:S-IC_engines_and_Von_Braun.jpg.

Figure 3.1. NASA, Challenger Explosion, public domain, https://commons.wikimedia.org/wiki/File:Challenger_explosion.jpg.

Figure 4.1. NASA, Mark and Scott Kelly, public domain, https://en.wikipedia.org/wiki/Scott_Kelly_(astronaut)#/media/File:Mark_and_Scott_Kelly_at_the_Johnson_Space_Center,_Houston_Texas.jpg.

Figure 5.1. NASA, Space Debris Impact on Shuttle Window, image, public domain, https://en.wikipedia.org/wiki/Space_debris#/media/File:Space_debris_impact_on_Space_Shuttle_window.jpg.

Figure 6.1. U.S. Air Force, Strategic Defense Initiative, image, public domain, https://en.wikipedia.org/wiki/Strategic_Defense_Initiative#/media/File:Ground-Space_based_hybrid_laser_weapon_con imark="D04.5" cept_art.jpg.

Figure 6.2. Created by the author.

Figure 7.1. Leonid Kulik Expedition, Tunguska Event Fallen Trees, image, public domain, https://commons.wikimedia.org/wiki/File:Tunguska_event_fallen_trees.jpg.

Figure 8.1. NASA, ALH84001 Structures, image, public domain, https://commons.wikimedia.org/wiki/File:ALH84001_structures.jpg.

Figure 9.1. NASA, President Nixon Welcomes Apollo 11 Astronauts Aboard USS *Hornet*, image, public domain, https://en.wikipedia.org/wiki/Mobile_quarantine_facility#/media/File:President_Nixon_welcomes_the_Apollo_11_astronauts_aboard_the_U.S.S._Hornet.jpg.

Figure 10.1. Mariordo, image, CC 4.0 license, no changes made, https://creativecommons.org/licenses/by-sa/4.0/deed.en, https://commons.wikimedia.org/wiki/File:Arecibo_radio_telescope_SJU_06_2019_7497.jpg.

Figure 11.1. Steve Jurvetson, image, CC 2.0 license, no changes made, https://creativecommons.org/licenses/by/2.0/deed.en, https://commons.wikimedia.org/wiki/File:The_SpaceX_Factory.jpg.

Figure 12.1. NASA, Space Launch System (SLS), image, public domain, https://commons.wikimedia.org/wiki/File:Space_Launch_System_(SLS)_Mission_Planner%27s_Guide_-_ESD_30000_Baseline_-_12Apr17_106pp_-_20170005323.pdf.

Figure 13.1. NASA, Mars Design Reference Mission, image, public domain, https://commons.wikimedia.org/wiki/File:Mars_design_reference_mission_3.jpg.

Figure 14.1. NASA, Mars and Earth Size Comparison, image, public domain, https://en.wikipedia.org/wiki/Mars#/media/File:Mars,_Earth_size_comparison.jpg.

Figure 15.1. NASA, Apollo 8, December 24, Earthrise, image, public domain, https://commons.wikimedia.org/wiki/File:NASA-Apollo8-Dec24-Earthrise.jpg.

# Index